D1014488

TERRESTRIAL ENERGY

TERRESTRIAL ENERGY

How Nuclear Power Will Lead the Green Revolution and End America's Energy Odyssey

WILLIAM TUCKER

Bartleby Press
Washington • Baltimore

Published and distributed by:

Bartleby Press
8600 Foundry Street
Savage Mill Box 2043
Savage, Maryland 20763
800-953-9929
www.BartlebythePublisher.com

Library of Congress Cataloging-in-Publication Data

Tucker, William.
Terrestrial energy : how nuclear power will lead the green revolution and end america's energy odyssey / William Tucker.
p. cm.
ISBN 978-0-910155-76-2
1. Nuclear energy--United States. I. Title.

TK9023.T85 2008
333.792'4--dc22

2008028129

Printed in the United States of America

Contents

Part Three Solar and Renewables

Part Four Terrestrial Energy

Preface

For the first time in history mankind will be using energy not derived from the sun.

—Albert Einstein, spoken to Leo Szilard and Eugene Wigner, as he signed a letter to President Franklin Roosevelt, August 2, 1939

The book you see before you has taken a long time to get into print. It was bought and paid for twice by publishers that subsequently decided they could not go ahead with it, for opposite but identical reasons. Each found it could accept only one of the book's two premises—but found the other premise unacceptable.

The two premises are this:

a) Al Gore is right. In addition to worrying about the security of our energy supplies—a looming threat that could disrupt our entire economy and throw the whole world into a series of "resource wars"—we also have to be concerned about putting too much carbon dioxide into the atmosphere. Some of the alarms about global warming are undoubtedly exaggerated—South Florida is not likely to end up underwater—but the problem of climate change demands our attention. As long as we continue burning fossil fuels, there is the possibility we may be modifying the earth's climate in a way that could bring wrenching changes and be almost impossible to reverse.

b) The only way we are ever going to supply ourselves with enough energy while reducing our carbon emission is through a revival of nuclear power.

"Soft," "green," and "renewable" forms of energy are all worth pursuing. They will all go a long way toward alleviating the problem. But each has notable limitations. None of these alternative energy sources can address our core problem, which is producing electricity with coal. Only nuclear can replace coal—as it does in France today. Once we have eliminated carbon from our electrical base, we can begin switching our transport sector to the electrical grid through electrical- or hydrogen-powered hybrid vehicles. This will go a long way toward reducing our dependence on oil as well. America should be leading the world in this technology, as we once did. Instead we are lagging behind because of exaggerated fears and misunderstandings.

Neither of these propositions are particularly controversial among scientists. The majority of those who have been polled—and those I have interviewed—agrees to both premises without much hesitation. In the political arena, however, a dichotomy has taken hold that says one cannot embrace both. If you believe in global warming, then you must be opposed to nuclear power. If you favor nuclear power, then you must think global warming is a hoax. The number of public figures willing to accept both premises is small, with the same names appearing over and over—Steward Brand, founder of *The Whole Earth Catalogue*, Patrick Moore, one of the cofounders of Greenpeace, Christie Whitman, former director of the EPA, and James Lovelock, the venerated British biologist who proposed the "Gaia Hypothesis," that the earth is a single living organism.

At one end of this bifurcated spectrum is someone like Richard Lindzen, a maverick professor of meteorology at MIT who has been one of the most prominent skeptics of global warming. After meeting Lindzen at a conference, I wrote him and asked to use something he said in an article I was writing about global warming and nuclear power. He said I could quote him but warned, "You're damaging the cause of nuclear power."

"Just out of curiosity," I replied, "how does promoting nuclear as a solution to climate change damage nuclear?"

"Because global warming will turn out to be a false alarm," he said. "Nuclear should be sheltered from dependency on environmental hysteria and groupthink."

For the purists, even an association with global warming is discrediting.

Lindzen, by the way, is also one of those pessimists—they are on both sides—who say the forces of nature may be overwhelming us. "If man-made global warming is really happening," he says, "it's probably too late to do anything about it anyway"—which seems like stacking the deck in your favor. If we aren't warming the planet, then obviously we shouldn't try to do anything to stop it. And if we are warming the earth, we can't do anything about it anyway. Heads I win, tails you lose.

At the other end of the spectrum is former Vice President Al Gore, who won the 2007 Nobel Peace Prize for two decades of passionate crusading about global warming. Even though he argues that humanity must stretch itself to the limits in order to avert the crisis, Gore has not bothered to learn even the basics about nuclear power. Testifying before the U.S. Senate in March 2007, Gore was asked if nuclear had a role to play. He responded:

> I think it's likely to be a small part of it. I don't think it will be a big part of the solution, Senator.... I'm assuming that we will somehow find an answer to the problem of long-

> term storage of waste.... I'm assuming that we will find an answer to the problems of errors by the operators of these reactors.... But the main problem I think is economics. The problem is these things [nuclear reactors] are expensive, they take a long time to build, and at present, they only come in one size—extra large.

Saying that nuclear reactors only come in "one size—extra large" is woefully uninformed. Reactors can come in any size. Experimental reactors in laboratories and universities can generate 1 or 2 megawatts. (A megawatt—MW—is the standard unit of commercial electricity, able to power about 1,000 homes.) Submarine reactors in the Nuclear Navy generate between 20 and 50 MW, and battleships run on 70 to 100 MW. When Admiral Hyman Rickover, father of the Nuclear Navy, "beached" one of his submarine reactors at Shippingport, Pennsylvania in 1957 to produce the first commercial nuclear plant, it generated 60 MW—about 1/25th the size of todays.

Utility reactors grew to 300 and 500 MW and beyond, with the largest now reaching 1,500 MW—what Gore calls "extra large." This is because *giant generators are the cheapest way to produce electricity*. Coal plants are built to the same size, but this isn't the only way reactors can be built. The Russians are now powering Siberian villages with 80 MW reactors floated in on barges. China and Japan are building modular reactors of 150 MW to power small communities. There isn't any reason reactors can't be built to the neighborhood level, combined with hydrogen production or water desalinization. If we ever colonize the moon, it will probably be with transportable nuclear reactors.

The real problem is public fear of all things nuclear. In truth, nuclear power still terrifies people. It seems unnatural and diabolic, a bastard technology conjured up by guilt-ridden scientists trying to exonerate themselves for inventing the atomic bomb. For many people—even those most concerned about global warming—nuclear remains the embodiment of evil, the symbol of all that is wrong with the modern world.

That is the reason for the title of this book, *Terrestrial Energy*. What I hope to demonstrate is that nuclear power is a perfectly natural phenomenon, as natural as the warmth in the ground beneath our feet. Nuclear energy is the source of the earth's natural heat, the incredible interior furnace that heats the earth's interior to temperatures hotter than the surface of the sun, spitting out volcanoes and lava flows, floating the planet's continents like giant barges on its molten core. The source of this energy is nuclear power, the greatest scientific discovery of the twentieth century. While we have always looked to the sun for our energy, the unlocking of nuclear power has left us with an alternative—terrestrial energy. There is nothing sinful or reprehensible about using this energy. In fact, it has come just in time to help us deal with what may be our twin crises—climate change and the increasing scarcity of world oil. That is the reason for this book.

"Everything is connected to everything," as the saying goes and so it will take some time to present the whole picture. Chapter 1 will review the potential dangers of increased CO_2 levels in the atmosphere and explain why despite a certain amount of hype, the situation is worth addressing. Chapter 2 will introduce the term "terrestrial energy" and show how it is different from all forms of solar energy and why it will leave such a remarkably smaller environmental footprint. Chapter 3 will review the Three Mile Island and Chernobyl accidents, since they lie at the heart of people's fears about nuclear and represent the nadir from which the technology has just begun to recover.

Part Two will expand the picture by looking at the fossil fuels. For more than three decades we have lived with an "Energy Crisis" that most people experience as an unacceptably high price for gasoline. The real explanation is that American oil production peaked in 1970, throwing us onto a world market where we have lived with the consequences, both economical and political, ever since. Now there is talk that *world* oil production may be peaking within a few years or less. That would make the earlier "Energy Crises" look like a relatively benign event.

Before oil came coal, however, and that is what we will discuss first. Coal was the fuel of the Industrial Revolution, and in many respects, *was* the Industrial Revolution. We still produce half our electricity with coal. What that has meant for the environment, even before global warming became a concern, will be the subject of Chapters 4 and 5.

Chapters 6 and 7 will review the history of oil, which at one point was poised to replace coal, but now fuels transportation—40 percent of our energy budget. Chapter 8 will cover the third fossil fuel, natural gas, which is seen as having the potential to replace both coal and oil but is now running into its own supply problems.

In Part Three, we will turn to the "alternate" strategies—conservation, hydro, solar, and wind—each of which merit a chapter. Chapter 13 will cover the more exotic "renewables"—geothermal, tidal power, and biofuel—which has already achieved a national market. Chapter 14 will cover electric cars and the elusive "hydrogen economy," which sometimes seems to promise carbon-free transportation, although this is widely misunderstood. Chapter 15 will show what happened when one state—California—attempted to eliminate new power plants altogether and rely entirely on the conservation-and-renewables strategy. Chapter 16 will review the work of energy guru Amory Lovins and other environmental utopias, and show how their effort to present solar and nuclear as an either/or issue obscures the real choice.

In Part Four we will return to terrestrial energy and trace its discovery as well as development as nuclear power. Chapters 17 and 18 will tell the story of the heroic generation of physicists, beginning with Albert Einstein and ending, perhaps, with Alvin Weinberg, who unlocked the vast stores of energy at the heart of the atom. Chapter 19 will trace the early development of commercial nuclear power through Three Mile Island. Chapter

20 will review the accident and its aftermath. Chapter 21 will tell of the vast overhaul of the industry since the electrical deregulation of the 1990s.

In Chapter 22 we will look at the main source of fear about nuclear power—radiation—showing that public understanding has been confused by claims that have no scientific basis. In fact, radiation is a relatively weak cause of cancer and may actually have beneficial aspects. Chapter 23 will deal with everyone's ultimate question—"What are we going to do with the waste?"—and show how America's development of nuclear electricity has had nothing to do with the proliferation of nuclear weapons.

Finally, in Chapter 24, we will take a tour of France, the country where a transition to an all-nuclear, electrical system has been made.

When I was in college I took a course in the great political philosophers—Hobbes, Locke, John Stuart Mill, and the others. Halfway through the semester I happened to glance at a time chart and realized that each of these great theorists had lived almost exactly 100 years before the era in which they became influential—Hobbes with eighteenth century monarchies, Locke with the American Revolution, Rousseau with the Romanticism of the nineteenth century, Kant with the imperial states of nineteenth century Europe, Marx and twentieth century Communism. The lesson seemed plain—it takes about 100 years for ideas to filter into history.

The Nuclear Era began in 1905, when Einstein posited his famous equation, $E=mc^2$. Everyone now knows the formula but few understand its true implication—that matter and energy are interchangeable, and that the energy obtainable from the conversion of only the smallest quantities of matter can be vast beyond any previous comprehension.

One hundred years later, all this has not entirely sunk into the public consciousness. Everyone understands the destructive power of the A-bomb, but few have looked beyond this to realize that energy from only the smallest amounts of matter will be sufficient to power a world of growing prosperity without environmental degradation. At the dawn of the twenty-first century, as we face the possible peaking of world oil production plus the uncertainties of climate change, it is time for nuclear power to enter history.

Acknowledgments

This book is dedicated to the late Professor Arnold Arons, my freshman physics teacher at Amherst College in 1960–61. College was a different experience then. We arrived expecting to work much harder than we had in high school and we did. But nothing quite prepared us for Physics 1–2.

Twice a week, half the freshman class would file into the huge lecture hall in the old Physics Building where we took our seats, believe it or not, in alphabetical order while monitors took attendance. At the door, Professor Arons, small and wiry with a tuft of gray hair, would stand like a smiling undertaker, greeting everyone as they arrived. Then as the hour struck, he would lock the door against latecomers. A hush would fall on the room. Standing behind a 40-foot-wide lecture table, he would begin.

"Definitions of experience. This is a frictionless puck—a small canister with dry ice leaking gas out the bottom so it sits on a cushion of air. What can we say about it?" he would ask. "How can we describe its motion?" Then he would give it a tap and the puck would glide gently but steadily all the way to the other end of the table. "How do you explain?" he would say, a crooked smile matching the natural irony in his voice. And then the inevitable, "The rest is homework."

Our task was to derive classical Newtonian mechanics. We would observe a phenomenon—like a canister that seemed to float on indefinitely—frame a hypotheses, give it a mathematical relationship, and test it through experiment. I had learned some of the material in high school. I knew $F=ma$ and $E=\frac{1}{2}mv^2$ but it had all been rote memorization. Now the very process of discovery unfolded itself. How did Newton arrive at these ideas in the first place? How did we know they were true? The rest is homework.

It was the days of the "core curriculum" and everyone in the freshmen class took the same four courses. It was the greatest intellectual boot camp of all time. You could strike up a conversation with any other freshman and end up talking about Physics 1–2. Half the class was totally at sea. People regularly took the course two or three times without passing. There were students who left school entirely because they couldn't handle another round of Arons' course.

I was one of the lucky ones. I grasped most of it. It was in the other end of the curriculum—English 1–2—that mystified me. That was an exercise in epistemology devised by Professor Theodore Baird, where I achieved all D's and F's. I was, quite simply, the worst student in the class. And so, in my usual contrary fashion, I decided to become an English major. It seemed more challenging. Years of hard labor got me to the point where I could function fairly well as a newspaper and magazine writer.

Then I decided to write a book about nuclear power. Right away I was back in the classroom, as if I had never left. In the last lecture of the year, Professor Arons had stood in the well of the lecture hall and showed us Louis de Broglie's famous 1924 experiment. We had learned all about the behavior of objects in Newtonian mechanics. We had learned all about waves. Now he showed us an electron can behave both as a particle and a wave! The rest was homework. Explain.

And so I did my homework. Everything we had learned in Professor Arons' course predated Einstein's announcement of the Theory of Relativity in 1905. Now I had to go back to the lecture hall and, taking the same step-by-step approach, learn everything that had happened since. But I was *prepared* for the adventure. For this I thank Professor Arons and dedicate this book to him.

Of course there are many other people I would like to thank as well. First, there is Kim Dennis of the American Enterprise Institute, who had the foresight to initiate this project. Though her successors later passed on it, I was perfectly willing to accept their verdict—that it deserved a more popular audience. Geri Thoma was outstanding in threading the needle in finding a second publisher. That house also offered helpful advice, although ultimately it did not work out either. Finally, there is Jeremy Kay of Bartleby Press, a brilliant publisher and knight-errant who not only rescued the book when all seemed lost but made extensive contributions and added excellent editorial advice. Morgan Young, Irene Oladeinde, and Greg Giroux also worked long and hard to make it happen.

I would like to thank all the editors at various magazines who have handled pieces of this book over the years, going back to Lewis Lapham, who gave me my first break at *Harper's* thirty years ago. The list would include Jon Larsen at *New Times*, Peter McCabe at *Harper's*, Richard Starr at *Insight* and *The Weekly Standard*, Howard Dickman and Erich Eichman at *The Wall Street Journal*, and last but hardly least, Wlad Pleszczynski at *The American Spectator*, who has been a mentor and good friend for many years.

I would like to thank Chris Greene, Ted Rockwell, Mark

Mills and Charles Alexander for reading the entire manuscript and offering cogent comments and criticisms. I would like to thank Denis Bellar and Bernard Cohen for their expert advice. I would like to thank John Baden at the Foundation for Research in Economics and the Environment (FREE), John Entine and Lauren Campbell at the American Enterprise Institute, Michael Burdi at the Applied Finance Group, and Timothy Caspar at Hillsdale College for allowing me to test the arguments in this book before live audiences.

For their hospitality in helping me gather information, I would like to thank the staffs at Los Alamos National Laboratory, Oak Ridge National Laboratory, Idaho National Laboratory, Yucca Mountain Nuclear Waste Repository, the National Renewable Energy Laboratory, the Rocky Mountain Institute, Greenpeace, the Sierra Club, the Natural Resources Defense Council, Environmental Defense, the American Nuclear Society, the Nuclear Energy Institute, the Zimmer Generating Station in Ohio, the Cooper Nuclear Plant in Nebraska, Vermont Yankee, Gary Taylor at Entergy, and the entire organization of Areva in Paris.

I would like to thank my three sons, Kevan, Fritz and Dylan, who have grown up to be not only good friends but sage advisors. I would like to thank Sarah for sticking with me almost the whole way. I would like to thank Kathy and Jim for giving me what amounted to a second home and for tolerating my eccentricities. I would like to thank Irene for her razor-sharp insights and hospitality, and finally Stephanie for all things present and future.

All of these have contributed to making this book possible. The mistakes—and there must be many—are all my own.

Prologue

Golden, Colorado and Oak Ridge, Tennessee

The National Renewable Energy Laboratory is tucked inconspicuously into the foothills on the eastern slope of the Rocky Mountains in Golden, Colorado. A collection of futuristic-looking buildings with an occasional windmill popping up here and there, it has the look of a dude ranch designed by Frank Lloyd Wright.

It's the first day after New Year's, 2006. I arrive early for my 9 AM interview with Larry "Kaz" Kazmerski, director of the National Center for Photovoltaics. With a few minutes to spare, I sit in the car listening to "The World is Flat" by *New York Times* columnist Thomas Friedman on the dashboard CD player.

The chapter ends and as I'm locking up (it's a habit I learned in Brooklyn), I spot a TV crew unloading gear from a van across the parking lot. "Hey, you guys doing a news show?" I ask in journalistic camaraderie. "No, we're with *The New York Times,*" says the bushy-haired crew chief. "We're doing a feature on energy for the Discovery Channel." Suddenly I have an intuition. I scan the group and sure enough, there is Thomas Friedman, bracing himself against the cold wind, nattily attired in a tan leather overcoat and matching gloves.

"Hey, I was just listening to your book in the car," I tell him.

"*The World is Flat? Yes,* that was a lot of reporting," he says. "Three billion new customers just walked onto the playing field in China, India, Russia and Brazil, and if we don't find a way to meet their demands, we're going to burn up this planet in a very short time."

I tell him I am also writing a book on energy and we chat about all the important energy centers in the Rockies. "I was up in Snowmass interviewing Amory Lovins at the Rocky Mountain Institute yesterday," he says. "We're here to talk with people about solar energy this morning and then we go to the National Atmospheric Laboratory in Boulder this afternoon.

"This isn't your grandmother's energy crisis," he adds. "People aren't waiting in gas lines. But it's much more pervasive."

"I'm writing about global warming, too," I say. "I'm putting a big emphasis on nuclear power. Are you doing anything on nuclear?" He gives me a blank look, as if I've just asked if he'll be doing any ice fishing.

"No," he says. He pulls out his cell phone and heads inside with the crew.

Left alone, I pause at the entrance for a minute to look at the displays. By the walkway stand, an eight-foot windmill and a small solar panel. Beneath the panel is a concrete pool just big enough to give a dog a bath. At the edge of the pool sits a ceramic turtle with a small black hose sticking out its mouth. A sign by the pool reads, "The electricity that runs the fountain to your left comes from a 50-watt photovoltaic..." At that point, a typewritten note has been taped over the glass. It says: "The pump is currently not operative due to winterization. Please visit us again in the spring to see it become active when energized by the sun."

"Hmmm…energy that flies south for the winter and doesn't return until the flowers bloom. Somehow this doesn't seem very promising," I think to myself. I head inside.

Soon I find Kazmerski in his office. He is a genial engineer who started with NREL in 1978 when it had just been founded by Jimmy Carter as the Solar Energy Research Institute. He soon became obsessed with photovoltaics and is now one of the world's leading authorities.

Before we almost begin, Kazmerski jumps from behind his desk and unrolls a 15-foot strip of heavy plastic decorated with rows of large silicon wafers. "This is our latest," he enthuses. "It's roofing material with the photovoltaic cells built right in. The Japanese have been experimenting with this for some time. It's cheap—you can stamp it out by the mile—and you don't have to worry about wind damage. We used to talk about people putting solar panels on their roofs. Now the panels are going to *be* the roof."

Settling back behind his desk, Kazmerski talks at length about new technologies and falling costs. "We're just about at the point where we're going to be competitive for peaking power," he says. "It's ideal because the demand for electricity peaks when the sun is brightest—on hot summer days. That's when everybody turns on their air conditioners. With the price of natural gas going up, it's going to be cheaper to lay rooftop panels than to build any more of these gas turbines."

Friedman is about to arrive (TV cameras have enormous precedence in such situations,) and so I pose one last question. "Looking at all these different factors," I say, "what do you think our best strategy should be for dealing with global warming?"

He eyes me cautiously. "You know, you're probably not going to believe this," he replies. "I think we should be doing solar *and* nuclear. We're big enthusiasts of nuclear around here. Dan Arvizu, our director, is very high on it. Nuclear can provide our base load electricity while photovoltaics provide the peaking power. It's an ideal match. If we use them together, we can retire the fossil fuels and reverse global warming."

I am a bit flabbergasted. "You know, that's exactly what I have concluded in writing this book," I say. "We need a solar-nuclear alliance to pass a carbon tax and start phasing out the fossil fuels."

His eyes grow wide. "You're going to write *that book?*" he says. "Good gosh, I've been hoping someone would write that book for ten years! I've spent the last decade trying to get the nuclear and solar people in the Department of Energy to talk to each other. They just won't do it. The nuclear people think the solar people are a bunch of hippies and the solar people think the nuclear people are a bunch of Nazis. There's just no communication." Then Friedman arrives.

Arvizu is gone for the day, so with nothing else to do, I decide to follow Friedman and the camera crew outside. They set up in a large fenced enclosure that looks like a botanical garden—except that the exotic plants are silicon panels etched with aluminum circuitry. One of NREL's responsibilities is to test new products for the manufacturers. On the horizon is the Denver skyline. "It's too bad it's so windy today," whispers George Douglas, NREL's director of communications, who has been ushering me around. "Otherwise they could shoot that brown cloud that always hangs over the city."

After a few takes of "strolling together" through the garden, Friedman and Kazmerski settle in front of a particularly photogenic solar collector. "So the reason Germany and Japan are making so much progress with solar energy and we're not is because their governments are supporting it," Friedman begins.

"Let him say it," the director interrupts.

It takes a couple more takes, but finally they get it right. "Yes, the reason there's so much progress on solar energy in Germany and Japan and not here is because the governments in those countries are supporting it," says Kazmerski. Nothing is said about nuclear power.

Over the next six months, Friedman went on to write a

series of columns proclaiming, "Green is the new red, white and blue" and arguing that the environmental approach is the way ahead for America's economy. He paid particular tribute to Amory Lovins, the sage of Snowmass, who has argued for twenty-five years that conservation and solar—the "soft energy path"—make nuclear unnecessary.

In June 2006, the Discovery Channel broadcast Friedman's show, "Addicted to Oil." The hour-long show featured the Kazmerski interview and urged America to turn to "alternate energy." Among other strategies receiving his blessing were: biodiesel, biomass, "cow power" (methane from manure), ethanol, geothermal, hemp oil, landfill gas, solar, sugar power, veggie oil, and wind. Concluding the show with a plug for windmills, Friedman dismissed their unfortunate killing of migrating birds by saying, "What's a few bird species next to solving our addiction to oil?" Still no mention of nuclear power.

Over the following year, Friedman did begin including nuclear as an alternative, telling a National Public Radio audience, at one point, that he is "in favor of nuclear energy" but that reactors take a long time to build. Then, in April 2007, the Discovery Channel ran another special, "Green is the New Red White and Blue," in which Friedman visited Three Mile Island. Nuclear probably isn't a bad idea, he said, but reactors take at least fifteen years to build and that's too far out to help with global warming. In *Hot, Flat, and Crowded*, published in 2008, Friedman dismissed nuclear altogether. "What other industry do you know whose last major advance was in 1955," he wrote—apparently unaware of the vast overhaul of the industry since Three Mile Island.

To this day, I doubt Friedman realizes that Larry Kazmerski and Dan Arvizu, two of the country's leading experts on renewable energy, are among the biggest supporters of nuclear energy. Why? Because as intelligent scientists, they do not approach the subject with the either/or attitude—either we can develop solar *or* nuclear but not both. Instead, they temper their enthusiasm for solar with an understanding of its limitations and are willing to acknowledge the enormous untapped potential of nuclear power. To both of these leading scientists, retiring the fossil fuels means breaking out of the either/or move and forming a nuclear-solar alliance.

Three months after visiting NREL, I am just finishing up a tour of the Oak Ridge National Laboratory in Tennessee when a thought suddenly hits me.

"Is Alvin Weinberg still alive?" I ask my guide.

"Yes, he lives right here in Oak Ridge."

"Can I interview him?"

"I don't see why not. He's 91 years old, but he's still got his wits about him. I'll get you his phone number."

Weinberg was the last of the Los Alamos generation, the cohort of brilliant physicists who decoded the structure of the atom in the first half of the twentieth century and opened the door to the Manhattan Project. After working under Enrico Fermi to create the world's first nuclear chain reaction at the University of Chicago in 1942, Weinberg migrated to the wartime frontier town of Oak Ridge, eventually becoming director in 1955. Fired in 1972 for raising too many concerns about the safety of nuclear energy, he was quickly appointed by President Nixon as the first head of the U.S. Energy Research and Development Administration, forerunner to the Department of Energy. One of his proudest accomplishments was establishing the Solar Energy Research Institute, which soon became NREL. In 1977, he chaired a federal panel on climate change that urged the government to investigate whether the buildup of carbon dioxide in the atmosphere might lead to global warming.

Weinberg had a knack for coining a phrase. When government began spending huge amounts on scientific research in the postwar era, he called it "Big Science"—a term that eventually became the title of a Laurie Anderson album. Referring to the *ad hoc* nature of progress, Weinberg coined the phrase "technological fix," which Amory Lovins repeated so many times (turning it into a pejorative) that people came to think he had coined it himself.

Then, in a famous 1972 article in *Science*, Weinberg called nuclear technology a "Faustian bargain" and predicted it might require a "priesthood" of technicians to oversee its development.[1] Nuclear opponents loved these frank admissions and repeated them so often that Weinberg used to joke he was their favorite nuclear scientist. Yet he never wavered in his faith in the technology.

On the phone, a nurse tells me Weinberg was at a weekly luncheon with some former colleagues. She gives me the name of the restaurant. Despite unfamiliar streets and a confusing set of directions, I manage to find it before the gathering is disbanding.

The hostess leads me to a small side room where I find three graying nuclear engineers, all aging gracefully into a comfortable retirement. At the head of the table sits Weinberg in a wheelchair. He is an elfin figure, so shrunken that his chin barely reaches the table. It is like meeting Stephen Hawking. Yet he is the center of the conversation, his mind sharp as a tack.

"I like to say we've lived through the First Nuclear Era," Weinberg begins when I ask him about the future of the technology. "We made a lot of mistakes the first time around. All of us were disappointed when the country turned away from nuclear power. Yet to deny its rebirth would be to deny human aspiration.

In 1994 I said we could enter a Second Nuclear Era if we met four requirements.

First, we had to confine reactors to relatively few sites, building four or five on each one. We've already got 60 sites in the U.S., so we probably need only a few more.

Second, we had to improve the security at all these sites so the terrorist threat would be eliminated. I think we've done that pretty well, especially since September 11th. Our plants are now very secure and protected.

Third, I said we needed to professionalize our plant operators. At the beginning, we were running reactors with high school graduates. I thought we could do much better. I like to draw an analogy with airline pilots. We give them extensive training to prepare them for the same kinds of safety responsibilities, yet we pay airline pilots much more money. Now all this has changed as well. Over the last two decades we've become very professional in our plant operations.

Finally, I said we had to separate the business of generating electricity from the business of selling it. That would create a whole cadre of people who worked exclusively with nuclear energy. At that point we just had a bunch of utilities that happened to own nuclear reactors. With electrical deregulation, we've now created an industry that deals exclusively with nuclear power. That's another big step forward.

Altogether then, I think we've done pretty much what we needed in order to begin a Second Nuclear Era. I'm confident it will happen. The only problem now is public opinion. People haven't yet accepted the potential of nuclear energy."

The group hangs on his every word, as if listening to an oracle.

I asked him what he thought could be done to improve public understanding.

"You've got to change the terms of the debate," he said. "All you hear is the same arguments over and over. This isn't going to get us anywhere. You've got to change people's understanding of what nuclear energy is all about."

I tell him about the idea of a nuclear-solar alliance. "I'm thinking of calling my book *Terrestrial Energy*, because nuclear power comes from the earth rather than the sun."

"Yes, that's what Edward Teller had in mind when he titled his book *Energy from Heaven and Earth*," he says. "Almost all our energy comes from the sun but nuclear comes from the earth.

I don't think it will take much," he continues. "With all this concern about global warming, the public will recognize the potential of nuclear soon enough. It's been my life's work. I only regret I won't be here to see it."

Six months later, on October 18, 2006, Alvin Weinberg died peacefully in his sleep, still dreaming of a glorious future for nuclear energy.

Part One
The Crisis

Chapter 1

Global Warming—Hype or Crisis?

"We, the human species, are confronting a planetary emergency—a threat to the survival of our civilization that is gathering ominous and destructive potential even as we gather here." So spoke Al Gore in his Nobel Prize acceptance speech. "These are the last few years of decision, but they can be the first years of a bright and hopeful future if we do what we must. No one should believe a solution will be found without effort, without cost, without change."

To the many scientists, this represents the proper verdict. The Fourth Assessment Report of the Intergovernmental Panel on Climate Change (IPCC), issued in February 2007, said it can now be stated with "very high confidence" that human activity is having an affect on climate.

"The combined radiative forcing due to increases in carbon dioxide, methane, and nitrous oxide…is *very likely* to have been unprecedented in more than 10,000 years," said the report of the working group. "Warming of the climate system is unequivocal, as is now evident from observations of global average air and ocean temperatures, widespread melting of snow and ice, and rising global average sea level.... The last time the Polar Regions were significantly warmer than present for an extended period (about 125,000 years ago), reductions in polar ice volume led to 4 to 6 meters of sea level rise.... Most of the observed increase in globally averaged temperatures since the mid-twentieth century is *very likely* due to the observed increase in anthropogenic greenhouse gas concentrations."[1]

The report made several estimates of possible scenarios for the twenty-first century, with a likely increase in temperature of about 4°C and rises in sea level ranging anywhere from seven to twenty-three inches, depending on future energy policies. As is often pointed out, the difference between the present and the last Ice Age is only 5°C. A rise in sea level of one foot would cause considerable havoc in coastal cities around the world and submerge some mid-ocean islands. The urgency of the situation was emphasized only a few months later when reports from the Arctic, after the summer of 2007, said ice was melting at double the rate predicted by the IPCC Report. The entire Arctic may be uncovered within the next decade.

The general public now seems willing to accept the idea that something unusual is happening to the earth's climate. "Going Green" has become fashionable—so much so that even some skeptics are beginning to question whether anything substantial is being accomplished. Movie stars buy carbon credits to plant trees in Burundi. Policy wonks jet-set to conferences to discuss how to make other people stop burning so much carbon. Diligent authors churn out books arguing that we will all have to go back to a world where we bake our own bread, sew our own clothes, and bicycle everywhere. No one takes these proposals too seriously.

Yet there is a group of conservative critics who *do* take it seriously and argue that trying to do something about global warming will do more harm than good. "[I]f we follow Al Gore's path toward an environmentally obsessed society, it will have big consequences for the world, not least its poor," writes Bjorn Lomborg, the Danish author of *The Skeptical Environmentalist*. "In the year 2100, Mr. Gore will have left the average person 30% poorer, and thus less able to handle many of the problems we will face, climate change or no climate change. Clearly we need to ask hard questions. Is Mr. Gore's world a worthwhile sacrifice?"[2]

Lomborg's argument pretty much defines the conservative position. Conservatives tend to worry more about the economy and are not enchanted by visions of a world run on windmills and biofuels. And in truth, neither is the public. Enthusiasm for doing something about global warming usually wanes rapidly if it means paying more for gas or turning off air conditioning. "The American way of life is not up for negotiation," proclaimed President George Bush, Sr. in 1992 as he reluctantly sent delegates to the Global Summit in Rio, the first UN convention to deal with global warming. That stance has never really changed. The U.S. Senate voted 94–2 to reject the Kyoto Protocol *before* it was adopted in 1998, and the Clinton Administration never dared submit it to Congress for approval. In 2008, a bill to impose a tax on carbon emissions was debated for a week but met strong resistance and never came to a vote.

Nor have the signers of the Kyoto Protocol done much in the way of actual accomplishment. Of the 170 countries that signed the treaty, only six will have achieved their goal of reducing emissions to 1990 levels by 2012. The vast majority has gone right on pumping more and more CO_2 into the atmosphere. (France, with its nuclear infrastructure, is one of the few that is in compliance, having trimmed its carbon production 1.8 percent below its 1990 level by 2007.)[3] There has been a blunt candor to the Bush Administration's refusal to play along with this game.

Yet instead of stonewalling on Kyoto, we could have jumped to the head of the parade with a banner reading, "Global warming is the problem, nuclear energy is the solution." After all, it is *our* technology. We discovered it, and for a long time the world looked to us for scientific leadership. Unfortunately, that era is now passing. While we have created the artificial problem of "nuclear waste" by abandoning nuclear recycling, France and Japan have solved the problem and moved ahead with civilian construction. While we have fretted that using nuclear materials in this country would lead to the spread of nuclear weapons abroad, Russia is *selling* reactor fuel to Iran and gaining a huge economic and diplomatic foothold. Just because we find the technology too frightening does not mean the rest of the world will do the same thing.

Oddly, while we might expect conservatives to take the lead in promoting nuclear technology, they instead expend most of their energy taxing to debunk global warming. Some of this is legitimate science. There are many aspects of climate change that have been overplayed, and visions of South Florida ending up underwater or New York City freezing over, are obviously exaggerated.

A recent example occurred with the correction of some NASA figures by Canadian mathematician Steven McIntyre, who had previously gained notoriety by debunking the famous "hockey stick" graph on global warming (which we'll discuss later in the chapter). For several years, NASA's Goddard Institute of Space Studies in New York had been touting that nine of the ten hottest years on record had occurred since 1990, with 1998 as the hottest on record. In 2007, McIntyre examined the numbers and noted that NASA had changed the standard of measure in 2000 without making a complete adjustment. When the proper compensations were made, only four of the hottest years have occurred since 1990, with 1934—the year of the "Dust Bowl"—the hottest year ever. Surprisingly, NASA quickly acquiesced to the corrections—although it only announced it in a short press release. Conservative critics quickly seized on the admission and touted it as further evidence that man-made global warming is a fraud.

This is the kind of debate that science requires. As several critics immediately pointed out, however, these corrections only applied to U.S. temperatures and, even after this correction, world temperatures remained higher than they have been in tens of thousands of years. And while skeptics have been good at pointing out mistakes and exaggerations, their arguments often move beyond the facts into the realm of scientific uncertainty itself.

It is often argued, for example, that man-made global warming is "only a theory" and has "never been proven." Unfortunately, no scientific theory is ever proved beyond all doubt; it can only be disproved. And there is always the possibility that some new facts may emerge which cannot fit. According to current scientific standards, in order to establish convincing proof of global warming, we would need another earth. We would pump carbon dioxide into the atmosphere of one for a hundred years and leave the other in its natural state as a control. Then we would see what happens. If the first earth warmed up, we would have proof that the theory of man-made global warming is correct. Even then, someone would probably argue that the sampling wasn't large enough.

There will always be an element of existential uncertainty in our evaluation. We will never know with absolute certainty whether the theory is correct or what effect our actions may have in preventing it. The *preponderance* of evidence suggests, however, that even though there have been overstatements, something highly unusual is happening to the earth's climate. Moreover, the geological record shows that there is a chance that such changes could quickly get out of hand.

Surprisingly, there is wide agreement on the basic facts of global warming. To begin with, everyone agrees that the carbon dioxide content of the earth's atmosphere has been rising steadily. Throughout geological history it has ranged between 250 ppm and 350 ppm. Since the advent of the Industrial Revolution, atmospheric CO_2 has climbed to 380 ppm in 2008, above its historic range. Since 1960, the increase has averaged 1.4 ppm per year. At this rate, by the middle of the twenty-first century we will reach 450 ppm—a territory that has not been reached in the last 60 million years.

Carbon dioxide makes up only a tiny portion of the air we breathe. Three hundred eighty ppm constitutes 0.038 percent of the atmosphere. Nitrogen is 78 percent, oxygen 21 percent, and argon .97 percent, making up approximately 99.998 percent. The remainder is trace gases, including water vapor and methane.

What distinguishes these three is their ability to prevent heat from reflecting back into space. All are transparent to ultraviolet rays from the sun. After these solar rays have reached the earth and interacted with electrons, however, they emanate back into space as lower-energy infrared heat waves. These waves are partially absorbed by methane, water vapor, and carbon dioxide, trapping heat in a "greenhouse effect." You see this in winter, when cloudy days tend to be warmer than bright sunny days.

All planetary atmospheres act like an envelope, trapping some heat. The moon is the same distance as the earth from the sun, yet its temperatures range from 250°F in sunlight—above the boiling point of water—to -175°F in the dark. This is because the moon has no atmosphere. If there were a layer of gas around the moon, it would retain some heat and moderate its temperatures. Venus' atmosphere is 98 percent CO_2, creating surface temperatures of 860°F, hot enough to melt lead. The earth's earliest atmosphere is believed to have been once similar to that of Venus. Only as photosynthetic organisms started converting carbon dioxide to oxygen did the greenhouse abate and relatively moderate temperatures prevail. (Carl Sagan once suggested we seed Venus with photosynthesizing blue-green algae so that the same transformation could create an earth-like environment.)

A second point on which there is *no* dispute is that the earth's temperature has risen over the past century. In the 1850s, scientists began taking regular readings at the remote Mauna Loa station in Hawaii, creating a continuous record. Temperatures rose steadily from 1910 to 1940, then dipped slightly until 1980, when they shot up even faster—although they seem to have leveled off again since 2000. There is no argument that we are in a warming period. The dispute is over whether recent temperature increases are the result of human activity—particularly carbon emissions—or whether they are part of a natural cycle.

In 1988, during a particularly hot week in July, James Hansen, director of NASA's Goddard Institute for Space Studies, testified before Congress that the buildup of CO_2 in the atmosphere was going to lead to significant temperature change. This was the first time the issue had been broached to the public. At the time there was little evidence to suggest that such a trend might take place. The 1970s had been a particularly cold decade, so much so that in April 1975, *Newsweek* ran an article entitled, "The Cooling World," warning of a new Ice Age. Over the next two decades, however, global temperatures rose steadily, which gave to Hansen's warnings predictive value.

Where proponents and critics part company is in how this data should be interpreted. Can the rise in atmospheric CO_2 and temperature be correlated? Is the one causing the other? Can these changes be projected forward through computer models? "How can we predict the weather a century from now," say the skeptics, "when meteorologists can't get it right two weeks in advance?"

Richard Lindzen, the maverick meteorologist at MIT, has long argued that computer models were exaggerating the effects that water vapor will have in warming the earth. "Warmer air holds more water vapor and so the presumption has been that as the earth heats up from more carbon dioxide, increases in water vapor will amplify the effects," said Lindzen in an interview. "But we don't know this is true. More water vapor may lead to increased cloud cover, which will cool the earth since clouds reflect sunlight. The science is just too uncertain." Lindzen is another skeptic who worries about the economy. "The risk of messing up the economy and condemning billions to poverty is not worth the cost. If the Kyoto Protocol is to be regarded as an insurance policy, then the premiums may cost more than the damage it's supposed to cover."

Fred Singer, a retired professor of environmental science at the University of Virginia, is another well-qualified skeptic. The founding director of the U.S. Weather Satellite Service, who once received a Congressional Gold Medal for his efforts, Singer devised the basic instrument for measuring stratospheric ozone and was the principal investigator on a satellite experiment retrieved by a space shuttle in 1990. He was also the first scientist to predict the buildup of atmospheric methane as another greenhouse gas.

Like Lindzen, Singer concentrates his attack on the computer models of global warming, claiming they do not accurately represent the earth's climate. One long-standing problem, for example, was that the warming was not happening as fast as early models predicted. Global warming supporters tried to explain this by suggesting that sulfur aerosols from coal smoke were blocking sunlight, countering the greenhouse effect. In a 2004 article in *The New Atlantis*, Singer dismissed this attempt with the following argument:

> It turns out that these supposedly-cooling aerosols are produced mainly in the northern hemisphere, where industrial activity is highest. Therefore, if the models are correct, the northern hemisphere would presumably warm more slowly than the southern hemisphere.... But observations show exactly the opposite. The highest rate of warming in the last 25 years occurred at northern mid-latitudes.[4]

There is a certain forest-for-the-trees quality to this logic. What Singer is saying is that because warming in the northern hemisphere is occurring *faster* than the new models would predict, the whole global warming hypothesis should be dismissed. It's not a very reassuring line of argument.

In 2006, Singer teamed up with Dennis Avery, an agricultural scientist at the Hudson Institute, to write *Unstoppable Global Warming...Every 1,500 Years*. The book laid out an alternate explanation to the recent warming trend. Based on the work of several European scientists, Singer and Avery argued that the earth's climate undergoes a roughly 1,500-year temperature cycle driven by variations in the sun's magnetic activity. "Fred and I are willing to concede some warming from the additional CO_2, but it looks like the models estimate it three times too high," said Avery, who has taken out ads in *The New York Times* challenging Al Gore to a debate. "They do this to make it look more frightening."

Singer and Avery's data explains the Medieval warming, a period from 900 AD to 1300 AD, when both historical records and scientific evidence show the earth was much warmer than today. "The Vikings settled Greenland on territory that is now covered with glacier," says Avery. "Chinese naval squadrons reached the Arctic Ocean and found no ice. The same conditions occurred during the Roman Warming, which occurred from 200 BC to 600 AD. Yet sea levels didn't rise during these eras. Civilization didn't collapse. In fact, they were eras of growing prosperity."

Singer and Avery's data of solar intensity can also account for the lull in temperature increase that occurred between 1940 and 1975 as well as produced the extremely cold weather of the 1970s, which made people think we were undergoing global cooling. This interval has always been a sore point for alarmists. "If the slow buildup of industrial exhausts has been driving global warming, why did the temperatures level off for 35 years?" asks Avery. "The extremists have to do a lot of fudging to account for this."

Still, there is a certain amount of fudging in Singer and Avery's model. Neither the Roman nor the Medieval Warm-

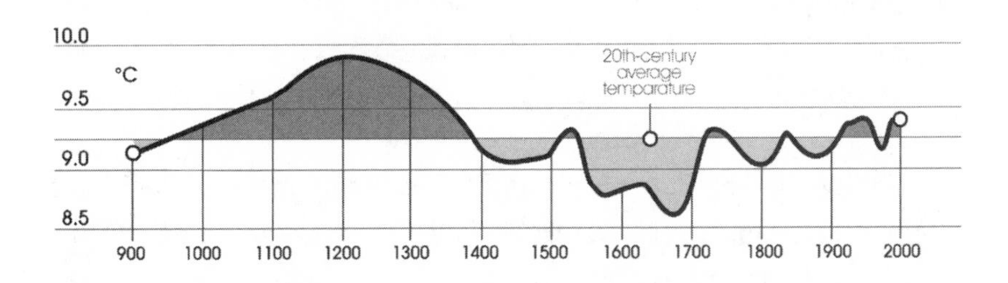

Medieval Warming & The Little Ice Age

Climate changes in Europe over the past thousand years

ing falls clearly into a 1500-year cycle. If they did, the present should be comparable to 500 AD, when the Roman Warming was just starting to cool off. Moreover, even the scientists who provide Singer and Avery with their best data are not entirely in agreement with them.

S.K. Solanki of the Max Planck Institute and M. Fligge of the Institute of Astronomy in Zurich are the two European researchers who determined that over the last century the sun was hotter than during any period in the last 11,400 years. Of particular significance was their discovery that solar activity diminished from 1956 to 1970, which fits well with the temperature lull from 1940 to 1975. Yet even Solanki and Fligge acknowledge that greenhouse gases are having an impact that will gradually overshadow the pattern of solar behavior.

> Since approximately 1975 the situation is clearly different...with solar irradiance showing a comparatively much more modest rise than air temperature.... [T]his most recent temperature increase [probably] reflects the influence of man-made greenhouse gases.[5]

Nir Shaviv, an Israeli astrophysicist often cited by global warming skeptics, also reaches the same conclusion. Shaviv says that while solar radiation probably contributed two-thirds of the temperature increase since 1970, and greenhouse gases the other third, these proportions will soon reverse. "The truth is probably somewhere in between, with *natural causes* probably being more important over the *past* century, whereas *anthropogenic* causes will probably be more dominant over the *next* century," writes Shaviv.[6] Thus, while solar intensity can explain the past, it may not be able to predict the future.

The Medieval Warming has always been the Achilles' heel of the alarmists. David Deming, a climate specialist at the University of Oklahoma, wrote an article in *Science* reconstructing 150 years of North American temperature data through a study of tree boreholes. As he later told Congress:

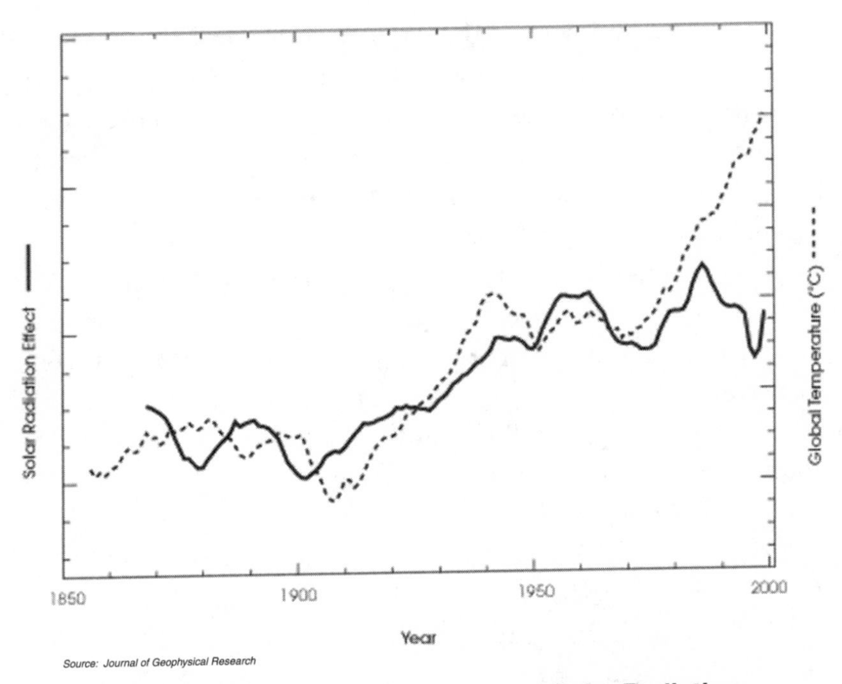

The Divergence of Temperature and Solar Radiation Since 1980, as Discovered by Solanki

With the publication of the article in *Science*, I gained significant credibility in the community of scientists working on climate change. They thought I was one of them, someone who would pervert science in the service of social and political causes. One of them let his guard down. A major person working in the area of climate change and global warming sent me an astonishing email that said: "We have to get rid of the Medieval Warm Period."[7]

In 1998, Michael Mann, an adjunct professor at the University of Massachusetts, and a team of associates did just that. They published a graph that has forever after been known as the "hockey stick." Mann and his group studied tree rings, particularly among ancient redwoods in California's Sierra Nevada Mountains, as a way of extending temperature records back seven hundred years. "To that point, most of the records had shown a very warm period during the early Middle Ages," says Mann. "But most of the data had concentrated around northern Europe. We expanded the database and found that the Medieval Warming was not nearly as widespread as supposed." In Mann's graph, temperatures remained almost completely flat over the last millennium until taking off rapidly over the last few decades.[8]

The graph was immediately seized upon by the Clinton Administration, which featured it prominently in its 2000 National Report on Climate Change. The IPCC made the hockey stick the centerpiece of its 2001 edition of *Climate Change*. Canada distributed a copy of the graph to every household in the country. Meanwhile, Mann was appointed editor of the IPCC's *Journal of Climate*.

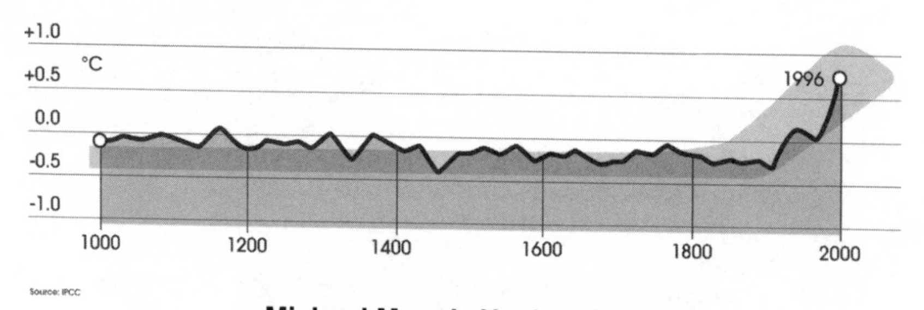

Michael Mann's Hockey Stick
Temperature anomally relative to 1960–1990

The hockey stick itself soon became the subject of even more controversy. Stephen McIntyre, the Canadian mathematician who would later correct NASA's annual temperature figures, teaming with Ross McKitrick, another Canadian mathematician, analyzed Mann's method and found that certain key assumptions he made gave tremendous weight to any tree-ring sequence that showed increased growth during the twentieth century. Feeding random numbers into the model, they found that *any* set of data produced the same hockey stick curve.

In addition, they found that the twentieth century pattern was particularly influenced by a single stand of twenty bristlecone pines at Sheep Mountain, California that had undergone an unusual growth spurt. When these twenty trees were removed from the study, the hockey stick effect disappeared. Reviewing Mann's computer files, McIntyre and McKitrick discovered that Mann himself had run an experiment without the Sheep Meadow data and found the same effect. The graph was stored in a folder titled "Censored."[9]

Reviewing the numbers, McIntyre and McKitrick published a 2003 critique in *Energy & Environment* that concluded, "The major finding is that the [warming] in the early 15th century exceed[s] any [warming] in the 20th century."[10] Several other critics have suggested that Mann's data—particularly at Sheep Meadow—is actually measuring increased growth from carbon fertilization, since the graph closely matches the increase in atmospheric carbon dioxide.

Meanwhile, Mann and his supporters—who called themselves "the hockey team"—set up a website, *www.realclimate.com*, to defend the hockey stick and its data. They point out that several subsequent reports, included in the IPCC's Third Annual Report of 2005, have essentially duplicated Mann's findings.

Stephen Hayward, a non-scientist at the American Enterprise Institute, wrapped up all these arguments in 2006 in "An Inconvenient Truth...Or Convenient Fiction?" a low-budget documentary that was presented as a response to Al Gore's Academy-Award-winning effort. Screened before small audiences in Washington and New York, and broadcasted on the Internet, the film pulls together most of the conservative critiques.

Hayward begins by citing the many scientists who remain skeptical about the crisis, and argues there is no "consensus" on global warming. He points to the many weaknesses of computer models. He argues that meteorologists often have trouble predicting next week's weather. He rehearses the criticisms of the hockey stick and concludes, "By the time of the IPCC's Third Annual Report in 2005, the hockey stick wasn't even mentioned."

"I don't think he's read the whole report," responds Mann, sitting in his office at the Walking Building at Penn State, sur-

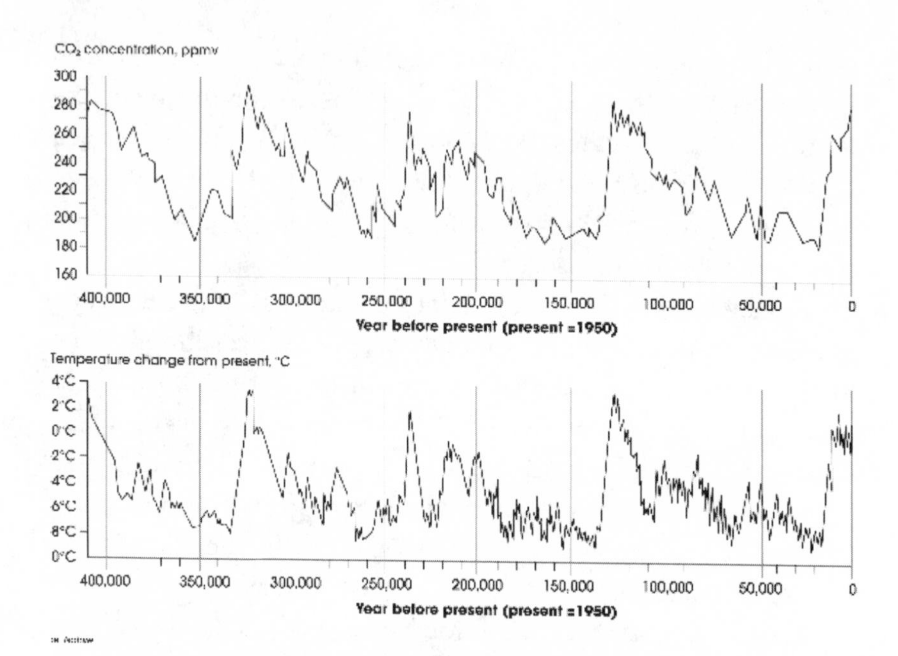

Comparison of CO_2 Content and Global Temperature Over 400,000 Years, as Presented in *An Inconvenient Truth*

rounded by huge slices of Sierra pines. "He only read the executive summary. Figure 6.10 shows the original hockey stick and a dozen later versions."

Finally, Hayward makes a seemingly irrefutable point with relation to Al Gore in *An Inconvenient Truth*. In the early part of his movie, Gore tells the story of a high school classmate who looked at the opposing coasts of North and South America, Africa and Europe, and asked the teacher, "Did they ever go together?" The theory of continental drift, first established in the 1960s, later proved they did.

Now Gore puts up the paired graphs of world temperature and CO_2 content as determined from excavations of the Vostok ice core in Antarctica and asks, "Don't they seem to fit together?" And indeed they do.[11]

There is only one problem. Careful analysis has shown that the rise in CO_2 content *lags* the rise in temperature by anywhere from 400 to 4,000 years. "If carbon dioxide were *causing* the rise in the earth's temperature, wouldn't it be the other way around?" asks Hayward in his film. It is an excellent point. However, it is not the last word on the subject, as we'll see at the end of this chapter.

"I'm not saying global warming isn't happening or that human activity doesn't have something to do with it," said Hayward in an interview. "It's just a question of how serious the effects are going to be. It's a lot like *The Population Bomb* of the 1960s. There was some truth to the idea that population was increasing faster than the earth could handle it, but there were other factors at work and the population bomb didn't go off after all. We didn't need to take draconian measures. I just don't see how we can go off half-cocked and ruin the economy over another theory that has very little basis in fact."

Patrick Michaels, author of *Meltdown: The Predictable Distortions of Global Warming,* is a critic who is willing to give even more ground to the alarmists, but still comes to the conclusion that fears of global warming are exaggerated—or pointless. A fellow at the Cato Institute and a member of the IPCC, Michaels acknowledges that carbon dioxide is producing a greenhouse effect. He also admits that temperatures are rising. But the effects won't be that bad. "It's nothing humanity can't deal with," he argues.

"The increase in carbon dioxide has been exponential to a slight degree, meaning that it is gradually accelerating," said Michaels in an interview. "But the response in temperature is logarithmic, meaning that each increment of CO_2 produces a slightly smaller increase in temperature. When you feed these two effects into a computer, the result is pretty much a straight line. In other words, even if carbon emissions keep increasing, we'll merely have a straight-line increase in temperature, which is what we've had for the past few decades.

"The earth in 2075 will be warmer than it is today," he continues, "but Florida and New York aren't going to be underwater. Since 1900 the surface temperature of the earth has increased 0.8°C. That's the same level increment predicted for

2100 by global warming. What was the result of the warming of the last century? Life expectancy doubled in the industrial world, some crop yields quintupled, and per capita income in the United States grew by a factor of ten. That's some threat! We can probably expect similar results if global warming continues over the next hundred years."

The argument that warming may actually prove beneficial to humanity is often made. "In Europe as a whole, about 200,000 people die from excess heat each year," writes Lomborg in *Cool It*, his 2007 critique of global warming. "However, about 1.5 million Europeans die annually from excess cold.[12] Lomborg notes, "Actually, the direct impact of climate change in 2050 will mean *fewer* dead, and not by a small amount. In total, about 1.4 million people will be saved each year, due to more than 1.7 million fewer deaths from cardiovascular diseases and 365,000 more deaths from respiratory disorders."[13] Lomborg also argues that rising temperatures will reduce the number of people living in water-stressed areas "because a warmer world also means more precipitation"—although he does acknowledge that "some parts will get drier, especially the Mediterranean area, central and southern Africa, and the southern United States."[14]

The health effects of warmer weather is underscored by the constant migration of Americans to warmer climates. Talk to any employee in a Las Vegas or Los Angeles airport and you almost inevitably find a refugee from Minnesota or upstate New York who says, "I never want to see snow again." Older people head to Florida and Arizona for relief from lung conditions, weak hearts, arthritis, and a whole litany of diseases. And they tend to live longer once they get there.

All this can be persuasive. So far, the ups and downs of temperature are not outside historical ranges. Even if the world does get warmer, catastrophic disruption is not inevitable. Humanity will have decades to adjust. Besides—in an argument heard from both sides of the fence—if we are really serious about stopping global warming, it may be too late to do anything about it anyway. The amount of carbon we have pumped into the atmosphere already assures us of a warmer world.

This combination of passivity, fatalism, and looking for the good side of things certainly has its appeal. Yet there is one problem. It is something called "positive feedback loops."

Go back and take another look at those graphs that Al Gore and Steve Hayward have been debating. Notice the pattern in which the lines move up and down—especially up. When temperature and CO_2 content have risen historically, they have climbed *very, very rapidly*. There is no smooth sine wave, but a pattern of abrupt temperature increases followed by a long, jagged descent. What this says is that when the earth warms up, it doesn't occur in a nice, even progression. It happens very, very fast. The reason almost certainly has to do with positive feedback loops.

Feedback mechanisms are part of any complex, adaptive system. Negative feedbacks tend to keep a system in balance and retard change. Positive feedback loops accelerate change and disrupt equilibrium. Richard Lindzen and Fred Singer base part of their skepticism on the existence of two negative feedbacks—the shielding effect from both increased water vapor and particulate matter in coal smoke—that will dampen the impact of increased carbon dioxide. Unfortunately, there are several other *positive* feedback mechanisms that are likely to push us in the opposite direction.

The first is the "albedo" effect of Arctic ice. Glaciers and snow reflect up to 80 percent of the sun's energy into space. As glaciers melt, however, they uncover darker areas of earth or sea. These surfaces then absorb much more solar energy. This will accelerate the warming process.

A second feedback mechanism is the oceans' ability to hold carbon dioxide—something the skeptics now cite as a negative feedback. It is true that more than half of today's CO_2 emissions are apparently being absorbed by the oceans, where they remain dissolved in water or are taken up by marine organisms. Much of the heat from global warming is also being absorbed by the oceans, which have risen about 1°C. But this cannot go on forever. As the oceans warm up, they become *less* capable of dissolving carbon dioxide. (Recall how a bottle of soda loses its "fizz" and can even blow its cap when exposed to heat.) At some point the oceans will become saturated and start *giving back* carbon dioxide. This will increase atmospheric CO_2 and accelerate the process.

Third, Arctic permafrost is a huge carbon sink. Permafrost is essentially a peat bog frozen over tens of thousands of years. As long as its plant material remains locked in ice, carbon is kept out of the atmosphere. Once the permafrost starts to melt, however—as is already happening in Siberia and Alaska—the organic material quickly oxidizes, releasing even more CO_2 into the atmosphere. This creates yet another positive loop.

This is why Michaels' scenario may be too optimistic. The temperature graph may indeed rise in a linear fashion, but the slope of that line could turn almost straight up. That is what has happened in geological history, as the Vostok graphs indicate. It also indicates why Singer and Lindzen's concern that global warming is not happening fast enough may be premature. Once the Arctic ice cover shrinks far enough, once the oceans start to give up carbon dioxide, and once the permafrost begins to melt, the process could become a runaway.

That is also why Steven Hayward's reassurance that CO_2 content had *trailed* temperature changes may be misplaced. Historically, increased solar activity may have *kicked off* world warming trends, but once the process starts, CO_2 content and temperature increases may *amplify* each other, creating a rapid upward spiral. As the National Research Council's 2001 report, "Abrupt Climate Change: Inevitable Surprises," put it:

> Recent scientific evidence shows that major and widespread climate changes have occurred with startling speed. For example, roughly half the north Atlantic warming since the last Ice Age was achieved in only a decade.[15]

The corresponding vertical spikes of temperature and carbon

content in the Vostok graphs describe a world coming out of successive ice ages. During these periods, rising solar activity and CO_2 each seem to reinforce the other, producing a very rapid rise in both. Then the system hits some natural limit—perhaps from declining solar activity, perhaps from reaching a point of diminishing returns in the albedo effect and carbon release from permafrost and the oceans. Temperatures level off and begin their jagged decline.

The problem is that we are now changing the scenario. We are adding *carbon that has been buried for hundreds of millions of years*. The coal laid down in the Carboniferous Age has never been a part of this up-and-down cycle. In fact, the earth was *much hotter* during the Carboniferous, precisely because so much carbon dioxide was still in the atmosphere and had not yet been buried in the ground. The world was indeed a huge greenhouse—which is why plants and insects grew to such monstrous proportions during that early era.

By digging up coal that has been buried in the earth for hundreds of millions of years, we are throwing an entirely new factor into the equation. Al Gore makes the point to illustrate the new situation in *An Inconvenient Truth* by climbing on a ladder. Atmospheric CO_2 is now headed completely outside the normal range of the last 650,000 years. Temperatures could easily follow carbon dioxide up the ladder.

It is highly unlikely that we can throw all that long-buried carbon into the atmosphere without having some impact on climate. Even if all the alarms about drowning Florida are exaggerated, putting three billion tons of carbon dioxide into the atmosphere each year has to force some change somewhere. We are headed into unknown territory. That is why we must take climate change seriously.

Chapter 2

Terrestrial Energy

Except for poets, no one should speak about stars without knowing the properties of matter with which they are constructed.

—George Gamow

The story of civilization has been essentially the story of utilizing solar energy. Each year the sun pours 6 million quads of energy onto the earth's atmosphere, enough to power the world economy 15,000 times over. Most of this, of course, goes to warming the planet. The sun heats the earth's surface to an average of 57°F, with temperatures rising to over 100°F in the hottest regions. Some of this heat is stored for longer periods in the oceans and the atmosphere. But the most accessible reservoir of stored solar energy is in plant material.

Plants store solar energy by using sunlight to split carbon dioxide molecules from the atmosphere, releasing the oxygen and using the carbon to build the long carbon chains of organic molecules. This process—what we call "photosynthesis"—operates at an efficiency of about 1 percent. Once locked in carbon chains, the energy will remain stored until the process is reversed. In the process we call "combustion," the molecules are broken apart and the carbon is recombined with atmospheric oxygen.

As Edward Teller points out in *Energy from Heaven and Earth,* organic molecules are the *only* substances on earth that are not already at their lowest energy level.[1] In other words, they are the only things that will "burn." Other chemicals can be combined with oxygen at high heat, but the breakdown of organic molecules is the only process that *releases* energy. The only place we can find stored solar energy waiting to be tapped is in organic material—or in its fossil remains.

Our ancestors may have used fire as far back as 1.4 million years ago and had domesticated it as recently as 750,000 BC. Charred remains are found in the caves of Peking Man and in all archaeological sites ever after. One recent theory suggests that human brains never could have grown larger without the softening of plant and animal material that comes with cooking food.

In *Energy in World History,* Vaclav Smil, the meticulous Canadian energy historian, calculates that taming fire was the greatest energy input ever made in human history, with the domestication of animals holding second place.[2] The 6,000-year-old Neolithic man discovered frozen in the Alps a decade ago was carrying a lighted coal to help him start a fire. A recent history of the Middle Ages was titled *A World Lit Only By Fire.*[3] Even today, wood is still the principal fuel of the non-Western world, providing 60 percent of the energy.

All this began to change in the sixteenth century when Europeans discovered coal. Coal is organic material boiled down almost completely to its carbon content, so that it burns cleaner and more intensely. During this era, another process was invented of burning wood underground to produce a second pure-carbon product called "charcoal." The solar energy stored in coal has been there for geological ages.

On a planetary scale, the amount of carbon stored in coal is immense—4,000 gigatons (Gt). This is about the equivalent of all the carbon stored in the atmosphere (750 Gt), forests (610 Gt), the soil (1580 Gt), and the surface of the oceans (1,030 Gt), combined. It is also four times the amount stored in the other fossil fuels, oil (500 Gt) and gas (500 Gt). The only larger reservoir of carbon is the deep oceans, which contain 38,000 gigatons.[4] Coal is so common in some parts of the world that no one even bothers to prospect for it. The entire western United States is underlain with thick layers of coal.

Oil and gas are not as abundant. Both are more highly concentrated forms of energy, however—the octane molecule in gasoline being the most concentrated form of stored solar energy ever discovered. As a liquid and a gas, both are more flexible and can be transported more easily than coal. Oil and gas are not as abundant in the earth's surface, however, and there is some question about their long-term sustainability. We will never run out of coal, but oil and gas supplies could eventually start tapering off.

The drawback of fossil fuels, of course, is that burning them means turning the original carbon back into the atmosphere. This is the indisputable source of rising levels of CO_2 and the principal suspect in global warming.

In search of other ways of utilizing solar energy, we have turned to a variety of technologies that tap the sun's rays directly or draw on physical processes driven by the sun's heat. Wind is an example. Wind is created by the sun's varied heating of the earth, coupled with the motion from the earth's rotation. Wind energy has been tapped since ancient times. Medieval gristmills ran on windmills. Their rotary motion is now being used to generate electricity.

Hydropower is another form of solar energy that also blossomed in the Middle Ages. It is created when the sun's heat evaporates water and then it returns to earth as rain. Running water has tremendous kinetic energy and was harnessed to run early factories. Water also has an additional advantage in that it can be stored behind dams. Large reservoirs are the only means ever invented for storing massive quantities of energy.

"Biofuels" are another form of short-term stored solar energy now being touted as a replacement for oil and gas. In fact, the term is basically new name for firewood. Biofuels attempt to accelerate the natural cycle by taking fast-growing crops and distilling them into burnable fuels such as diesel oil or ethanol. Whether they do anything to reduce carbon emissions—or whether they even produce any net energy gain—is a subject of fierce debate.

"Passive solar" involves designing buildings to trap the sun's rays for space heating. Usually this means redesigning houses so they absorb more solar heat in cold climates or are sheltered where the sun is too hot. "Active solar" involves employing rooftop panels to capture solar energy for space heating or hot water—or to run heat exchanges that can provide air conditioning.

The more exciting prospect is to use solar energy to generate electricity—the most flexible, versatile, and easily transmitted energy carrier ever discovered. The "power tower" is a strategy that employs mirrors to concentrate the sun's heat to produce steam. This drives a turbine in the same way that is done in a coal, oil, gas, or nuclear generator.

"Photovoltaics" is a method of turning the sun's rays directly to electricity, according to principles originally identified by Albert Einstein. Fulfilling its promise will mean covering hundreds of square miles of desert or millions of rooftops with solar panels. This process is still in the early stages.

All these forms of solar energy—wind, hydro, biofuels, tidal, and direct use of the sun's rays—are called "renewable," meaning it does not require vast geological ages to recreate them. The term, however, can be somewhat misleading. They can only be renewed at a pace that natural cycles allow. The amount of solar energy that shines down upon the earth is seemingly inexhaustible, but it is highly dilute. In order to match the highly concentrated power of fossil fuels, it must be collected over vast areas of land.

It is the *collection process* that is not inexhaustible and not always renewable. Hydroelectric dams, the most successful form of non-fossilized solar energy, back up reservoirs covering hundreds of square miles in order to generate the same amount of electricity produced by a square mile coal plant (not counting the area required to mine the coal). Wind farms will have to cover almost a hundred square miles to do the same thing. Fueling only a portion of the nation's automobiles would require dedicating almost our entire inventory of agricultural land to growing biofuels. Obviously, being "renewable" is not the only standard by which energy sources can be judged.

People often talk of harnessing the ocean waves and tides. The waves are a form of solar energy, since they are driven by the winds, but the tides are more correctly called "lunar energy," since the pull of the moon's gravity is the biggest factor. Once again, these forms of energy are extremely dilute and would require covering vast tracts of ocean with a collection infrastructure.

There is one other form of energy that is often grouped as solar or renewable, however, that actually has nothing to do with the sun. This is "geothermal" energy. Geothermal energy is created when groundwater comes in contact with the interior heat of the earth. Sometimes this produces "hot springs," which were long believed to have medicinal qualities. Where the heat comes close to the surface, steam may emerge in natural vents called "fumaroles." In rare instances, groundwater superheated deep in the earth explodes periodically to the surface as a "geyser." Long regarded as tourist attractions, fumaroles and geysers are now being tapped as "geothermal" sites.

What is the source of geothermal energy? The interior of

earth, it turns out, is a very hot place. Below the surface, the earth's temperature increases 16 degrees for every 1,000 feet of depth. By the time we reach the world's deepest mine shaft two miles down—the Robinson gold mine in South Africa—the temperature reaches 150 degrees and the tunnels must be air-conditioned for the miners to survive.

That is only the beginning. At 80 miles down, we hit the Mohorovicic Discontinuity, discovered by Yugoslav seismologist Andrija Mohorovicic, in 1909. At this point the temperature reaches 900°C and rock turns to liquid "magma." At 1,500 miles deep the temperature rises to 3,700°C, and another discontinuity—the Gutenberg—marks the place where molten rock becomes pure iron and nickel. Below that, tremendous pressures turn the iron core solid once again and temperatures reach 7,000°C—hotter than the surface of the sun.

Where does all this heat energy come from? Some of it is due to gravitational forces. As the earth is pulled inward, some of this force is translated into heat. Another portion is residual heat from the earth's formation. According to the commonly accepted theory, originally proposed by Immanuel Kant, the solar system precipitated out of a huge swirling dust cloud, where particles kept colliding with each other until they agglomerated into the sun and the planets. In the later stages, this involved huge collisions among very large objects. These impacts generate large amounts of heat, a residue of which still remains in the earth's core. Together, gravitational forces and residual heat may account for as much as 40 percent of the earth's temperature—the exact figure has still not been determined.

But the other portion of the earth's heat, anywhere from 60 to 90 percent by most estimates, comes from a remarkable diminutive source—the slow breakdown of two of the ninety naturally occuring elements found in the earth—uranium and thorium. With 92 protons, uranium is the largest natural atom, while thorium (90) is the third largest. Because of their size, they are "radioactive," meaning they are unstable. The internal "binding energy" that overrides the mutual repulsion among positively charged protons is occasionally overcome itself. This releases large quantities of energy, which sets subatomic particles in motion, creating large amounts of heat. Incredibly, the slow breakdown of these two radioactive elements, uranium and thorium, is enough to raise the earth's internal temperature *beyond the level of the surface of the sun*.

Most geothermal sites are at natural steam vents that are created along geological fault lines, but there is now talk of drilling further down to tap the earth's internal heat. Drill down ten miles almost anywhere on earth and you will encounter enough heat to boil water. (The deepest oil wells now only go down about five miles.) But here's a better idea. Why don't we just take the source of that heat—the uranium or thorium—bring it to the surface, and reproduce or even accelerate the process that produces this heat in a controlled environment?

This is what we do in a "nuclear reactor."

The process of tapping terrestrial energy is really no different than tapping fossilized solar energy. What do we do when we build a coal plant? We find stored solar energy beneath the earth's surface. We mine it, we bring it to the surface, we concentrate it. We ignite it, starting a chain reaction where the energy released from one molecular breakdown triggers a breakdown of the next. We capture the heat to boil water, to produce steam, to drive a turbine, to generate electricity.

What do we do in a nuclear plant? We discover accumulations of *terrestrial* energy in the earth. We mine it, we bring it to the surface, we concentrate it. We ignite it, starting a chain reaction in which the energy released from one atomic breakdown triggers a breakdown of the next. We use this heat to boil water, to produce steam, to drive a turbine, to generate electricity.

A nuclear reactor is nothing more than terrestrial energy brought to the surface, just as a coal plant is stored solar energy brought to the surface. There is nothing sinister or diabolical about it. We are not defying the laws of nature. Rather, we are working with a process that already takes place in nature. Many geothermal plants are now almost indistinguishable from nuclear reactors. They have the same parabolic cooling towers that throw excess heat into the atmosphere. The steam rising from a nuclear reactor is just as harmless and potentially beneficial as the steam rising from a geothermal vent. In fact, they are the same thing. They are both terrestrial energy.

There is one great difference between terrestrial energy and solar energy, however, and that is the energy *density*. Terrestrial energy is far more concentrated—by a factor of about *two million*. This is because terrestrial energy draws its energy from transformations with the *nucleus* of an atom, while fossilized solar energy is stored in the electron *orbit*. This distinction is the key to understanding why terrestrial energy offers such a remarkable possibility for powering the planet without disrupting the environment.

The place to begin understanding energy density is in the most famous equation of the twentieth century—and perhaps of all time—$E=mc^2$. Albert Einstein first posited this equation in 1905 almost as an afterthought to his Special Theory of Relativity, written while he was still a clerk in the Swiss patent office. Everyone is familiar with $E=mc^2$, but how many people understand exactly what it means?

What $E=mc^2$ says is that *matter and energy are interchangeable. They are different aspects of the same thing.*

Before Einstein, the laws of the conservation of mass and the conservation of energy had been separately developed. In the eighteenth century, Antoine Lavoisier did careful experiments showing that the sum of the weights of gas, liquid and solid in a chemical reaction always remains the same before and after it takes place. He posited the Law of Conservation of Matter, which says that matter is never created or destroyed. It can take different forms—solid, liquid, or gas—it can change its chemical composition, but the sum total of mass always remains the same.

In the early nineteenth century, a series of brilliant European scientists determined the same law for energy—the First Law of Thermodynamics—which says energy is never created or destroyed. The big impediment had been the "phlogiston" theory, which described heat as an invisible fluid—a form of matter. Since the amount of heat that could be produced from boring a cannon, for example, seemed unlimited, the amount of phlogiston stored in matter could be nearly infinite.

Count Rumford, an American expatriate, was observing one such cannon being bored in 1798 when he decided that so much heat could not possibly be a fluid. It must be *matter in motion*. After that, everything fell into place. Putting matter in motion requires energy so that energy can be transformed into heat. Soon, a whole series of scientists—Robert Julius Mayer, James Joule, and Hermann von Helmholtz—had performed careful experiments showing that energy was never gained or lost in a closed system. It might transmute from motion to heat to an electrical current, but the amount of energy always remained the same.

By the dawn of the twentieth century, the laws of the conservation of matter and energy were both well established. Then Einstein suggested something completely unexpected. He proposed that *Matter and energy are the same thing*. Matter can become energy and energy can become matter. $E=mc^2$, means that energy equals matter, governed by the coefficient, c^2. As Einstein himself put it: "The law of the conservation of the mass of a system becomes identical with the law of the conservation of energy."[5]

The important thing, however, is the coefficient—the *speed of light squared*. This is a very, very large number. The speed of light is 186,000 miles per second. In scientific notation, the square of this number is 90,000,000,000,000,000 m^2/sec^2—*90 quadrillion.* This means that a *very, very* small amount of matter converts into a *very, very* large amount of energy.

All this meant little as long as no one knew much about what was inside an atom. As scientific investigation began to outline its structure, however—protons and neutrons at the nucleus, the electrons in the outer orbits—it became clear that almost all the mass of an atom was concentrated in the nucleus. "Chemical" reactions involve changes in energy levels among the electrons—small amounts of matter are transformed into energy in the electron orbits. But the electrons only contain .01 percent of the mass of the atom. Meanwhile, the remaining 99.99 percent of the mass is locked in the nucleus. For a long time it was believed this energy could never be tapped. Einstein himself insisted it was nearly impossible. But as the probing continued, it became clear by 1939 that in certain very special circumstances the energy at the core of the atom *could* be released. This was the discovery of "nuclear power."

Because Europe was already entangled in war, the most immediate task became to build a cataclysmic weapon that could release vast quantities of energy in a single instant. This was the "atomic bomb." But the promise that nuclear energy could be released *slowly* under controlled circumstances that would *not* lead to an explosion was always uppermost in the researchers' minds. Although we have developed this potential more deliberately, nuclear power has unfortunately always lived in the shadow of nuclear weaponry.

The dimensions of that bomb, however, do give an indication of the vast stores of *non-chemical* energy that lie at the core of the atom. The bomb dropped at Hiroshima was the equivalent of 15,000 tons of TNT. The largest conventional (chemical) bomb ever exploded to that point had been one ton. This vast release of energy was produced by the transformation of a very small amount of matter into a very large amount of energy. The amount of matter completely transformed into energy at Hiroshima was *one gram.*

Consider a gallon of gasoline. The "orbital" energy stored in this gallon is enough to propel a 3,000-pound automobile about 30 miles—an extraordinary feat when you think about it. A team of human beings would exhaust itself in the effort. Yet if we could tap the binding energy stored in the nucleus of those same molecules, it would be enough to propel the same automobile *60 million miles*, almost all the way to Mars.

What does this mean for the environment? Basically, it is the best news possible. Because so much energy is stored at the nucleus of the atom, tapping this energy leaves an astonishingly small environmental footprint.

Let us look at how this works in real life. The North Omaha Power Plant in Omaha, Nebraska, produces 500 megawatts (MW) of electricity, about one-fifth of the power needed to run the city. Every three days, a 110-car unit train arrives, each car is loaded with 125 tons of coal. One car produces twenty minutes of electricity. The plant occupies more than two square miles—much of it needed to store the mountains of coal.

Each day's consumption of 4,500 tons of coal at North Omaha will combine with atmospheric oxygen to form 15,000 tons of carbon dioxide. (The two oxygen molecules in CO_2 are slightly heavier than the carbon.) Across the country, America has 600 similar coal plants that provide half our electricity and put 3 billion tons of CO_2 into the atmosphere each year—10 percent of the world's total. This is the greatest single source of global greenhouse gases on the planet.

About thirty miles south of Omaha lies the Cooper Nuclear Station on the banks of the Missouri River. The plant occupies two square miles, slightly less than the coal station. Every eighteen months, a single tractor-trailer arrives carrying several dozen bundles of 18-foot nuclear fuel rods. These rods are only mildly radioactive and can be handled safely with gloves. They are loaded into the reactor core, where they will undergo nuclear fission for three years. After the fuel rods are spent, they will be removed from the reactor core looking exactly as they did when they went in, except they will be highly radioactive. They can be stored in a 40-foot-deep, on-site "swimming pool," where their radioactivity dissipates in six feet of water. There, they can remain for decades. After three years, when the radioactivity has dropped by half, they may be moved to nearby outdoor dry casks. There they may remain for almost a century. The Cooper Station produces no sulfur emissions, no mercury, no soot, no particulate matter, no ash, no slag, and no greenhouse gases. And it does produce more electricity than North Omaha—750 MW.

Terrestrial energy is something completely new in human history, qualitatively different from anything we get from the sun. That is why there has been such a lag in public understanding. Solar energy, in its many forms, has accustomed us to the idea that using energy must create huge environmental impacts, either by polluting or by occupying vast tracts of land. Terrestrial energy is so highly concentrated that it can provide us with enormous amounts of energy while barely leaving a trace. Combined with the contributions of solar power, terrestrial energy offers us the opportunity to power the world while eliminating all manners of environmental degradation.

Chapter 3

Three Mile Island and Chernobyl

"They're all so young," says Jane Fonda, looking from the plate-glass-enclosed balcony at the floor of the nuclear control room below.

"There's one old guy down there," says her cameraman, portrayed by Michael Douglas.

"Yes, he's a former submarine commander in the Navy," says the utility executive.

Once again, *The China Syndrome* got it right. Even today, the control rooms of nuclear power plants across the United States are stocked with Navy veterans, just as they were in 1979 when the movie was released in theaters—just two weeks before the nuclear accident at Three Mile Island.

The Three Mile Island accident marked the turning point in America's nuclear development. To that point, our nuclear "fleet" was expanding rapidly—often with the encouragement of environmental groups. The 1960s was an era when both dam construction and air pollution from coal were regarded as major environmental problems. Nuclear power was seen as a solution to both.

Resistance was growing, however, particularly around the issue of safety. A great deal of debate took place during the 1970s over whether a catastrophe at a nuclear plant was possible. *The China Syndrome* was Hollywood's effort to add to this debate—and in truth, it didn't do a bad job. The movie was excellently researched, well crafted, and dramatically compelling without overstating the case.

Fonda plays a big-haired TV news reporter trying to be taken seriously by her supervisors. With Douglas, her cameraman sidekick, she is assigned to visit a nuclear power plant for a routine promotional piece, when suddenly something goes awry. People on the control room floor are suddenly scurrying around in a panic. From the balcony, Douglas surreptitiously films the crew as it frantically tries to control a "scram," meaning an unplanned shutdown. (The origin of that word is a fascinating story to be elaborated on later.)

Jack Lemmon, as the former submarine commander and manager of the control room, finally realizes the problem. A spring-operated gauge showing the level of cooling water has gotten stuck. He gives it a smack and it snaps to a completely different position. Instead of trying to replenish the coolant, the crew has been draining it. Lemmon catches the error just in time and the plant just misses a serious accident. Almost the exact same thing happened at Three Mile Island.

Confident they have a great story, Fonda and Douglas rush back to the studio to prepare the script. The first technological shock comes when Fonda starts her story by rolling a piece of paper into the typewriter. You mean they didn't have computers in those days? No, they didn't have many computers in nuclear reactors, either.

Fonda's supervisors don't like the story, however, and decide not to run it, saying it will result in a lawsuit. A utility executive reaches higher-ups at the newspaper and the whole incident is squelched. The cover-up is significant, since the utility is about to receive a license to operate a second reactor on the other side of Los Angeles. Determined to get the story out, Douglas steals his own footage while Fonda goes out to investigate.

Hanging out at a country-and-western bar near the plant, Fonda runs into Lemmon and starts grilling him. "I love that plant," Lemmon tells her as she pries him for information. "It is a beautiful line. Men do fall in love with the machines they operate." Lemmon is a loyal company man and has no question about his mission, but Fonda's inquiries arouse his curiosity. He goes back and starts checking the welds on the cooling pipes, only to discover that some inspection photos have been faked. He confronts the construction company boss, who tells him to mind his own business. When Lemmon says he's going to the Nuclear Regulatory Commission, the construction boss warns him, "We've got security forces. Don't you understand? We've got security forces."

Lemmon goes back to his own supervisors, but they aren't interested either. "It's costing this company half a million dollars every day this plant is shut down," one tells him. So Lemmon decides to send the faked inspection photos to the Nuclear Regulatory Commission. The "security forces" are closing in, however, and almost murder Douglas's sidekick as he is trying to deliver the videotapes to Lemmon. When Lemmon goes to investigate, they go after him.

After the obligatory car chase, Lemmon makes it back to the plant. Shaken by his narrow escape, he is appalled to realize the company is about to restart the reactor without fixing the welds. He grabs a security guard's gun, hijacks the control room and demands to speak to Fonda on TV.

Fonda arrives to do the interview as Lemmon holds the control room hostage. Once on camera, however, he becomes muddled and inarticulate and ends up sounding like the "lunatic" corporate executives are saying he is. A moment later a

SWAT team breaks into the control room and shoots him dead. Meanwhile, revved up to full speed, the reactor is falling apart downstairs like Dr. Frankenstein's laboratory at the end of a Boris Karloff movie. (It does get a little hokey at this point.)

As the fire engines and TV transmitter trucks gather outside, utility executives once again assure everyone that "the public was never in any danger." "Jack Godell [Lemmon] was drunk and emotionally disturbed," they tell reporters. But Fonda sees things differently. She gets Lemmon's subordinate—another tired old company man—to tell her audience, "He wasn't emotionally disturbed, he was a hero."

"I met Jack Godell only two days ago," Fonda wraps up the broadcast. "He was about to present evidence that he believed would show this plant should be shut down. Let's hope it doesn't end here."

And it didn't.

The China Syndrome anticipated Three Mile Island so accurately that, watching it today, it is surprising to find that the movie *wasn't* set in Pennsylvania. It was Los Angeles. At one point, however, a scientist does tell Fonda a meltdown would make "an area the size of Pennsylvania uninhabitable"—an eerily prescient detail that was repeated over and over in the press.

At Three Mile Island, operators experienced an almost identical scram. A relief valve stuck open, drained cooling water from the core. Despite a hundred blinking lights and wailing sirens—or maybe because of them—the operators couldn't figure out what was going on. Nothing on the control panel told them exactly how much water was in the reactor vessel. Just as in the movie, the operators guessed the core was *getting too much* water when it actually was *not getting enough*. At Three Mile Island, the mistake wasn't caught in time. One-third of the core melted.

For almost a decade before, critics of nuclear power had been arguing that such an accident could happen. Early reactors had operated at relatively low temperatures. As they grew toward 1,000 MW, however, temperatures rose and "loss of coolant" was pinpointed as the main danger. If the core lost its cooling water, it could overheat, causing the uranium fuel rods to melt. When the design flaw was identified, manufacturers added an Emergency Core Cooling System (ECCS) that would flood the reactor vessel if anything sprung a leak. Critics argued that the high-pressured water would just blow past the core and flush right out the hole, especially if the core had developed air pockets. This scenario seemed to be supported by some early testing. Against this evidence, the industry and the Atomic Energy Commission insisted the ECCS would work.

Trying to get an overview, the Atomic Energy Commission published a Reactor Safety Study in 1974 that came to be known as the Rasmussen Report, after Dr. Norman Rasmussen, its principal author. The report used probabilities and complex event-trees to evaluate whether a nuclear accident was possible. Famously, the report concluded that the chances of being killed by a plant "transient" were about equal to that of being killed by a meteorite.

> All non-nuclear accidents examined in this study, including fires, explosions, toxic chemical releases, dam failures, airplane crashes, earthquakes, hurricanes and tornadoes, are much more likely to occur and can have consequences comparable to, or larger than, those of nuclear accidents.[1]

Critics pounced on the report, arguing that it concentrated only on equipment and ignored the possibility of human error. They charged it excluded "common-mode failures," where multiple systems were knocked out simultaneously. Less than a year after the report, exactly such an accident occurred at the Brown's Ferry Nuclear Station in Alabama. A maintenance worker, using a candle to check for air leaks (that's right, a candle!) set fire to the electrical insulation and knocked out *seven* independent safety systems, including the ECCS and its backup. Believing it was an electrical fire, AEC spent eight hours trying to smother the flames with carbon dioxide. Finally, federal officials took the advice of the local fire chief, who said it was an *insulation* fire, and used water. The flames were doused in twenty minutes.

The AEC eventually withdrew the Rasmussen Report in January 1979, three months before Three Mile Island. Meanwhile, another Canadian report, written by Herbert Inhaber, backed up the general conclusions of the Rasmussen Report, arguing that, per kilowatt-hour of electricity, nuclear was actually safer than any other means of generating power, except natural gas.

When Three Mile Island finally hit the headlines, *Harper's Magazine*'s droll cartoonist, David Suter, offered his own fault-tree analysis on the possibilities of such a nuclear accident:

> We think the chance of *that* [i.e., a complete meltdown] is nearly infinitesimal, about one in 104 reactor-years of operation, or roughly the same as, say, the likelihood that a feature-length movie describing a reactor accident should open in New York just as a similar accident actually happens in a neighboring state.[2]

If the possibilities of a nuclear accident proved to be much more likely than anticipated, however, the *consequences* of such an accident proved to be less than everyone believed. The key was "the China syndrome" itself. The concept was originally put forth in 1963 by William K. Ergen, a theoretical physicist at the Oak Ridge. Since the earliest days, commercial nuclear reactors have been surrounded with a "containment structure," a concrete- and reinforced-steel housing that—by the latest standards—can withstand the full impact of a jet plane. Ergen theorized, however, that if the core of uranium rods lost their coolant and overheated, it could reach a temperature that would melt through the steel reactor vessel and then through the concrete floor of the containment—"all the way to China." Anti-nuclear activists picked up this idea and ran with it. By 1979, the script of *The China Syndrome* had a sober physics professor explaining to Fonda and Douglas:

> If the core is exposed for whatever reason, the fuel heats beyond core heat tolerance. [I]n a matter of minutes, nothing can stop it and it melts right down through the bottom of the planet, theoretically to China. But of course as soon as it hits groundwater, it blasts into the atmosphere and sends out clouds of radioactivity. The number of people killed would depend on which way the wind is blowing, rendering an area the size of Pennsylvania uninhabitable, not to mention the cancers that would show up later.

A nuclear engineer who is with him adds, "I may be wrong but I would say you're lucky to be alive. For that matter, I think we could say the same for the rest of Southern California."

Few people realize that by this time the idea of a nuclear accident immediately killing hundreds or thousands of people had become a runaway fantasy. At Three Mile Island, almost one-third of the core *did* melt down and nothing resembling the China syndrome even began to happen.

A runaway reactor, one where the fuel keeps fissioning even though the cooling water has been lost, is impossible with present commercial designs. There's a good reason for this. All nuclear reactors—at least the kind built in this country—have a fail-safe mechanism. If the water used to cool the reactor is lost, the reactor also stops "reacting."

The water circulating through the core does three things:

1) It cools the fuel rods so they don't melt.
2) It carries away the excess heat to create steam to drive the electric turbine.
3) It *moderates* the speed of the neutrons.

We'll go into the details of nuclear fission in much greater detail later, but suffice to say, when a uranium atom splits in two, it releases a pair of neutrons that have the potential to split other uranium atoms. This is the "chain reaction." The neutrons are flying so fast, however, that they cannot be absorbed by other uranium atoms. They must be "moderated" by bouncing off much smaller atoms that can't absorb a neutron. In a stroke of intuitive genius, Enrico Fermi used paraffin—mostly carbon and hydrogen—as the first moderator in 1933. The result was the first "splitting" of the atom. Water—hydrogen and oxygen—is an ideal moderator and is used for this purpose in nearly all commercial reactors.

The point is this. *If the coolant water is lost, the chain reaction stops.* Since the water serves as both coolant and moderator, a loss of coolant means no more nuclear fission. There is another fail-safe mechanism as well. The fuel rods in a nuclear reactor must be arranged in precise geometric alignment so that the flying neutrons have a maximum opportunity of striking other uranium atoms. If the core starts to melt, this alignment is lost and the chain reaction peters out. This is yet another reason why a reactor cannot "run away" into an uncontrolled reaction.

If the coolant is lost, what remains is "decay heat." The by-products in the uranium rods will continue to go through radioactive decay, keeping temperatures at around 400° to 500°F for weeks. These by-products are the so-called "nuclear wastes." They cannot be handled without extreme precautions. It was five years before anyone was able to enter the containment structure at Three Mile Island, and then workers could only stay a few minutes. A nuclear meltdown is truly an industrial disaster. But the decay heat is nowhere near enough to burn through a steel vessel or a concrete floor, let alone begin its route to China.

At Three Mile Island, about one-third of the core did melt to the bottom of the stainless steel reactor vessel—the egg-shaped container that houses the coolant and the nuclear rods. However, the molten rods *didn't even melt through the chromium-and-steel lining of the reactor vessel.* Steel melts at 2,750°F, concrete at 2,900°F. Even after the core had been exposed intermittently over eight hours, temperatures didn't rise above 2,000°F. Three Mile Island proved that *The China Syndrome* can't happen.

Just for the sake of argument, let's assume that the molten core did manage to burn its way through the reactor vessel and onto the concrete floor of the containment structure. And then let's suppose that it kept right on burning through the concrete and on down into the ground. What would happen?

The scenario depicted in *The China Syndrome*—a "steam explosion"—occurs when a super-hot object is dropped into a vessel of water. Instead of gradually boiling, the water "flashes" into steam, creating an explosive effect. This was the explosion that was supposed to contaminate all of Los Angeles.

Would such a scenario occur underground? Water does not lie beneath the earth in large pools, but saturates the pores of permeable rock. As a melting core *slowly* approached groundwater, it would not cause an explosion. Instead, the water would heat gradually, eventually boiling into steam. The resulting steam would slowly rise in a plume out of the ground.[3]

Does it seem as if we are back in familiar territory? Indeed we are. What we are describing is a fumarole or a "geothermal" site. Natural steam vents occur all over the world, where groundwater is heated by radioactive uranium and thorium in the ground. A melted core would accomplish the same thing. Only traces of radioactive material would be dissolved in the steam, just as there are traces of radioactive material in a geyser. In this case, the steam would rise back into the containment structure, where it would condense into water. You could probably attach a turbine to it and produce a little electricity. A "China Syndrome" geothermal site would be an awkward arrangement, not nearly as efficient as a well-designed nuclear reactor, but it would be a useful form of generating steam, not a regional-wide catastrophe.

Three Mile Island was a serious industrial accident. The reactor was ruined, costing General Public Utilities $5 billion to cleanup, almost bankrupting the company. Unlike most other industrial accidents, however, no one died or was even slightly injured. Three decades of monitoring the Harrisburg area have found no trace of health effects on the surrounding population.

What Three Mile Island taught us is that the worst does not have to happen. In fact, the accident—a meltdown with no injuries—was right within the parameters of the Rasmussen Report, which said such an event might occur every thirty-or-so years. Even with the relatively primitive technology of the time, there was still a margin for error. As we shall see later, the handling of nuclear technology has since improved by orders of magnitude. Three Mile Island was a frightening experience for the nation, from which we emerged without any serious consequences for the public.

The same cannot be said of Chernobyl.

The Chernobyl Nuclear Power Station in the Ukraine consisted of four operating units with two more under construction. It was one of the Soviet Union's largest nuclear complexes, supplying the Ukraine with 10 percent of its electricity.

Blatantly confident of Soviet science, the engineers had not even bothered to build *containment structures* around their reactors. Moreover, one of the main purposes of the six reactors was to produce plutonium for nuclear bombs. In order to speed up plutonium production, graphite was substituted for water as the moderator. This is a dangerous practice. A graphite moderator creates a *positive* feedback loop. If the coolant is lost, the reaction actually *speeds up*. On top of all this, graphite is flammable at high temperatures. It is, after all, pure carbon.

On April 26, 1986, two teams of operators were struggling with each other to use the plant for two contradictory purposes. One team was supplying power to the grid while the other was running an experiment to determine whether the momentum remaining in the turbines would be enough to power the cooling system during an accidental shutdown. In the tussle over the reactor, the water in the cooling system stopped circulating momentarily. It quickly overheated, sending a burst of steam through the turbines. This revved up the power, which overheated the core even more. The fuel rods melted, dropping right into the remaining coolant. This caused the dreaded "flash" steam explosion, blowing the lid off the reactor.

The explosion ignited a fire in the graphite moderator. A plume of radioactive smoke rose into the updraft, billowing radioactive debris 3,000 feet into the air. Prevailing winds spread the cloud across the Ukraine, neighboring Byelorussia and eastern Europe as far as Poland. Tight-lipped Soviet officials refused to acknowledge any accident until two days later, when radioactive debris was detected in Sweden. It was only then that the world realized something terrible had happened behind the Iron Curtain.

People living within ten miles of the plant received a dose of radiation up to 200 times normal background. Elsewhere, the fallout was lighter but detectable. (More radioactive material fell on Harrisburg, Pennsylvania as a result of Chernobyl than from Three Mile Island.) Most of this was iodine-131, strontium-90, and cesium-137—all dangerous because they are taken up in the food chain. Iodine-131 is particularly harmful because it migrates to the thyroid. About 4,000 Ukrainian children developed thyroid cancer from the exposure and ten eventually died.

Hundreds of cleanup workers were rushed to the scene. After two days, they managed to extinguish the fire by bombing the reactor with 5,000 tons of lead, boron, sand, and clay. A concrete foundation was quickly constructed under the reactor to prevent groundwater contamination. Then, workers built an enormous concrete and steel sarcophagus over the damaged reactor—the containment structure that should have been there in the first place. Some of the workers were sent in unprotected to throw radioactive debris off the roof. Thirty-one workers died from the fire and acute radiation poisoning. During the three-month clean up, thirteen more died.[4]

Nearby Pripyat (pop. 35,000) was abandoned in thirty-six hours, and within a week, Soviet officials had expanded the evacuation zone to eighteen miles. Eventually, 100,000 people left the region, relocating in other parts of the Ukraine and neighboring Byelorussia. Officials in Byelorussia—then still part of the Soviet Union—began to suspect that Soviet authorities were understating the danger, and eventually 200,000 people were evacuated.

Farmland, surface water, and groundwater were all dusted with radioactive debris. In Byelorussia, 20 percent of the nation's farmland was removed from production. More than a million workers eventually participated in the cleanup. Dislocation, poor nutrition, fear, and anxiety took an enormous toll.

For once, a Communist regime asked for international help. The International Atomic Energy Agency became involved, upgrading operating and safety standards throughout the Soviet Union and Eastern Europe. Some historians argue that, by discrediting the leadership and opening the empire to international contacts, Chernobyl actually hastened the downfall of the Soviet Union.[5]

Chernobyl is the worst nuclear accident that ever happened—and probably the worst accident that *could* happen. The fire and scattering of radioactive debris essentially duplicated the worst-case scenario of *The China syndrome*. So what were the consequences?

In 2005, the UN convened a panel of 100 scientists to do an intensive study of the accident on the eve of its twentieth anniversary. According to the 600-page Chernobyl Report, the impacts were surprisingly mild. The report put the number of deaths at fewer than fifty, mostly among emergency workers. An additional 4,000 cases of thyroid cancer were recorded in the surrounding population, mostly among children. Thyroid cancer is easily treated, however, and only ten died. (The condition can be avoided altogether by taking thyroid pills immediately after exposure, but Soviet health authorities were not prepared.)

In terms of long-lasting effects, the UN panel found no fertility problems or increases in birth defects in neighboring populations. It also found no upsurge in cancer, although it acknowledged that radiation exposures might eventually lead to an additional 4,000 cancers among the 100,000 cancers that would be expected in the population. "There was, however, an increase in psychological problems among the affected population, compounded by insufficient communication about radiation effects and by the social disruption and economic depression that followed the breakup of the Soviet Union."[6]

Critics quickly condemned the report as a whitewash. "The most recent published figures indicate that in Belarus, Russia and the Ukraine alone the accident resulted in an additional 200,000 deaths between 1990 and 2004," claimed Greenpeace in a 138-page counter-report issued in 2006.[7] "The worst effects are still to come," the report predicted, and said tens of thousands more would eventually die from cancer due to radiation exposure from the accident.

It is a strange thing indeed when one world agency can say Chernobyl caused sixty deaths and another organization can claim 200,000. It seems like there should be some neutral body that could resolve such wide differences. One explanation seems to be that Greenpeace relied on the reports of Ukrainian and Belarus doctors, who were trying to discredit the Soviet regime, were looking for international sympathy, and who blamed almost every health condition in the country on Chernobyl for the next several years. In *Chernobyl Heart*, the Academy-Award-winning documentary, Adi Roche, founder of Ireland's Chernobyl Children's Project, visited hospitals, orphanages, and mental asylums in Belarus and was told by doctors that most of the children were suffering from conditions resulting from Chernobyl. One of the most common was cerebral palsy, a disease that occurs everywhere, and which, in America, trial lawyers have blamed on doctors' errors and the failure to deliver by Caesarean sections.

Most critical, however, is Greenpeace's adoption of the theory that there is "no safe dose" of radiation and that *any* exposure to low levels of radiation will inevitably cause cancer. As we shall see late, this single assumption—which has no grounds in scientific evidence—is the core of most exaggerated fears of nuclear power. Two decades have now passed since Chernobyl, and except for the brief outbreak of thyroid cancers, no such pandemic of cancer has emerged anywhere. In fact, as we shall see, radiation is something that has been with us since the dawn of time, and the distant exposures that people in Europe and America received from Chernobyl are indistinguishable from the exposures that come from natural sources that constantly surround us.

Against the recommendations of authorities, many people have moved back to their homes near Chernobyl without suffering any visible consequences. The abandoned farmlands have undergone an explosion of wildlife, so that it is now called the "Land of Wolves." Obviously, it has not become the biological desert that was predicted. As the UN's Chernobyl Report concluded, the victims' greatest suffering has come from the dislocation of losing of their homes and the fear that they have suffered some kind of irreparable damage.

It would be wrong to downplay the horrors of an accident such as Chernobyl. Nonetheless, it is important to put it in perspective. Chernobyl was probably the worst nuclear accident that could happen. Nevertheless, it was not devastating on an unearthly scale. As one columnist put it after the release of the UN report, "A bus accident in India last week killed more people." The Bhopal chemical plant explosion in India in 1984 killed 2,000 people, yet it did not close down the chemical industry.

Three Mile Island and Chernobyl, however, essentially closed down nuclear construction in the United States. At the time we had 62 reactors with 129 more being built. Only 53 of these were ever completed. No new construction license has been issued since. In Europe, voters in Germany and Sweden elected to close down their nuclear reactors—although they have not yet followed through. Only now that global warming and energy strategies have become issues are these countries beginning to recognize that this decision may have been premature.

We'll come back and take a much longer look at the potential dangers and misconceptions about nuclear energy in the final section of this book. For now, however, it seems important to set the stage. How did we arrive where we are today and where can we go from here?

Until the seventeenth century, mankind still lived in "a world lit only by fire." Then, coal was discovered and the Industrial Revolution commenced. The past four centuries have witnessed the amazing advancements in civilization that have come with our ability to tap this form of stored solar energy. In many ways, however, the age of fossil fuels seems to be cresting, both due to supply concerns and because of the inevitable wastes that come with using any form of chemical energy.

All living things draw energy from the environment and discard some of it as "waste." This is not just a characteristic of industrial civilization but of life itself. A corollary to this proposition is that "no living organism can live in an environment composed entirely of its own wastes." Global warming is a confirmation of this law of life. The carbon dioxide buildup in the atmosphere is the inevitable waste product from the growth of our civilization. Yet we cannot stop using energy. All living things require it. Instead, we must learn to use energy more efficiently and develop new sources that do not produce the same volume of waste products.

Let us spend the next two sections, then, taking a longer look at the different forms of solar energy that have served us so far—the fossil fuels and the various forms of "renewable" solar energy that are supposed to take their place. Then, we will return to take a more extensive look at terrestrial energy.

PART TWO
The Fossil Fuels

Interlude

Zimmer Power Station, Moscow, Ohio

The Zimmer Generating Station sits on the banks of the Ohio River about twenty miles southeast of Cincinnati. Driving down U.S. Highway 52, the parabolic cooling tower looms far in the distance.

The plant is easy enough to enter. A sign directs you to the visitors' parking lot. Right on its edge, the water cascading down the inner edge of the 20-story tower generates a constant roar as it gives up its heat to the surrounding atmosphere before being dumped back in the Ohio River. Although it is December, the weather is unseasonably warm—or maybe it isn't so unseasonable anymore. High above, the moisture-laden air hardly traces a wisp of steam as it rises into the atmosphere.

At the reception desk, a single woman guard greets me and summons Ron Frey, the plant supervisor with whom I have been corresponding on the phone for about a month. The place has the feel of a lonely colonial outpost. If there is a town of

Moscow, I haven't yet seen it. Still, this isolated 800-megawatt station pumps out enough electricity to light the homes and power the factories of most of southwestern Ohio.

Frey comes down a long hallway and greets me. He is a stocky man, probably about fifty, with graying hair, wire glasses, and a fringe of beard. "You didn't have any trouble finding the place?" he asks.

"It's the only power plant I saw," I reply.

"Here, let me show you around." He hands me a pair of earplugs. "You'll need these," he says.

We retreat down the long hallway, go outside for a moment, and then reenter a larger building where the faint hum of machinery suddenly becomes a deafening roar. "There are the turbines," says Frey. They are huge glistening structures, almost a block long and several stories high. Inside, their whirling magnets spin around a forest of electric coils, forcing numberless electrons to dance along huge cables—a principle that Michael Faraday, a blacksmith's son, discovered in 1830 when he waved a magnet through a loop of wire and found it induced an electric current. Alongside one of the long walls leans a bicycle. "This place is so huge, some people get tired of walking it," he shouts over the din.

We pass through another passageway and into the generating building. It is a maze of grilled platforms and stairways ten stories high. At the center is a huge dun furnace that stretches almost a hundred yards on each side. The air is hot, although not suffocating, and there is a grim nineteenth century feel to the whole place. As we climb the maze of stairways, I notice my hands are becoming smeared with black dust.

"It's more than 1,000 degrees in there," Frey shouts. "Here, you want to see inside?" He leads me to a small opening in the wall like a ship's porthole and lifts a metal cover, revealing a glowing red window. The glass must be a foot thick. "Here, take a look," he says. I press my face against the glass and see a whirling inferno of tiny, red-hot particles flying about at what seems like supersonic speeds.

Finally, Frey leads me to an outer balcony that rims the top of the building. Standing in the open air, we can see the huge black piles of coal at the end of a series of conveyor belts leading from the river. "That's the sulfur-scrubbing facility down there," he says. As we watch, a few tiny trucks piled with black sludge are wending their way toward the exit gate. "They run twenty-four hours a day," he says. "We're dumping it in a big excavation up the other side of the mountain."

And above it all towers a 25-foot smokestack belching a huge column of thick white smoke. It doesn't have the wispy quality of steam. Even scrubbed of its sulfur gases, the exhaust has a density that suggests carbon dioxide. Only about a thousand feet in the air does the smoke finally thin out and disappear invisibly into the atmosphere.

The Zimmer coal plant stands at a crossroads of American energy history. In 1985 it was 95 percent complete as a nuclear reactor. The Three Mile Island accident had occurred in 1979, however, and a tide of protest was rising around the country. The Nuclear Regulatory Commission was demanding unprecedented levels of safety, often requiring utilities to tear up work that had already been done in order to add new features that had been developed after the construction was completed.

After sinking more than $3 billion into an 800-MW reactor, Cincinnati Gas and Electric found itself facing the arduous task of securing an NRC license to operate the plant. Intervener groups were filing lawsuits and the possibilities for delay seemed endless. So in 1985, CG&E threw in the towel and said it would cancel the reactor and build a coal plant instead. The opposition immediately melted away. After spending more than a decade trying to complete the nuclear facility, a coal plant was built in two years. The decision looked even more brilliant a year later when the Chernobyl accident in the Soviet Union spooked the whole world on nuclear power. The Long Island Lighting Company, which plunged ahead with its $5 billion Shoreham Nuclear Reactor, never saw the facility open, and eventually tumbled into bankruptcy.

The Zimmer Coal Plant was built to the highest standards, then and now. Its fluidized-bed system first pulverizes the coal into dust-like particles, then blows them into the furnace at high speed, giving them maximum exposure to oxygen, so they burn completely. The old soot and black smoke—actually unburned carbon—is almost completely gone. Particulate matter is captured by electrostatic precipitators and bag houses.

Then, the sulfur is eliminated. All coal contains some traces of sulfur, which combines with oxygen to produce sulfur dioxide. Rising into the atmosphere, this combines with water vapor to form a mild sulfuric acid, which damages human lungs, and is the main component of acid rain. SOx had long been an environmental concern, and Zimmer was equipped with sulfur scrubbers, which draw off the sulfur by combining it with calcium. This produces gypsum, which can sometimes be marketed as a building material. More often than not, however, the sludge is of poor quality and must be hauled off to acidic landfills by trucks that roll night and day.

Zimmer represented the best of "clean coal," an energy strategy adopted in the 1970s when oil was suddenly unavailable for electricity, and when nuclear power seemed too frightening. For the most part, it worked as advertised. Then, in 1988, only two years after Zimmer was fired up, James Hansen, the director of NASA's Goddard Space Center, told Congress that human industrial activity was beginning to affect the earth's climate. Specifically, he said, carbon dioxide exhausts from burning coal and oil were building up in the atmosphere, creating a "greenhouse effect" that trapped heat. Hansen predicted that "the 1990s will be hotter than the 1980s and the decade after that will be hotter still."

Almost twenty years later, Hansen's predictions have been

generally borne out. The 1990s were hotter than the 1980s and 1998, while 1999 and 2006 are among the six hottest on record, although overall temperatures have not risen much since 2000. Meanwhile, Hansen was back in the news in 2006 when he charged the administration of President George Bush had been trying to silence him on global warming. In 2008 he proposed that corporations that tried to discredit global warming should be prosecuted.

After returning to Frey's office, we began talking about comparative technologies. "What do you think of nuclear?" I asked.

He shot me a glance and then lowered his voice to a conspiratorial level. "Look," he said quietly, "I'm a nuclear engineer by training. I was here in Moscow when they were building the reactor back in 1984. I was as disappointed as everyone else when they changed plans. We knew the technology. We knew we could have done it. But the public wouldn't allow us.

"After they changed plants, my wife and I moved down South. I worked at the Edwin I. Hatch Nuclear Reactor in south Georgia for ten years because it's what I wanted to do. I loved it. The technology is inspiring. But it was too damned hot down there and the bugs were awful. My wife couldn't take it. She has family in this area, so ten years ago she persuaded me to move back. But I'll tell you, if I had the chance to work at a nuclear plant somewhere else, I'd do it in a heartbeat.

"Look at this stuff," he said, holding up his hands. They were covered with the same black dust. "You can't walk around this plant without getting dirty. And all that stuff is going into your lungs as well. Think about that. And then think about what we're putting into the atmosphere. A nuclear plant is a totally different animal. Everything is kept clean and shipshape. It's like being aboard a ship in the Navy. There's tremendous camaraderie because you know you're doing something that's important and you're working with the best. There's tremendous respect for the technology. Here you're just shoveling dirt. With nuclear, everything is clean, believe me, it's clean. There's nothing better for the environment than a nuclear reactor."

As I headed out the long loop from the parking lot, I realized that the cooling tower and the smokestack were lining up behind me, forming a perfect juxtaposition. The thin wisp of steam from the cooling tower squared precisely with the thick white column of carbon dioxide and water vapor from the smokestack. I took out my camera and snapped a picture. Framed against the ultra-blue sky, they seemed to symbolize the choice this country now faces—coal or nuclear? The invisible haze of greenhouse gases continued to rise into the atmosphere while weather remained unseasonably warm.

Chapter 4

Coal: The Industrial Revolution

Coal is the most important fossil fuel in history. The Industrial Revolution would never have occurred without it. In fact, for all intents and purposes, coal *was* the Industrial Revolution. Only a few nations have ever industrialized without shifting most of its' energy dependence to coal, as the experience of China and India proves again today.

At the same time, coal has also been the most environmentally destructive substance in history. Even before global warming became a concern, coal had caused far more pollution and environmental deaths than any other material ever used. This legacy continues today in the smoke-clogged cities of China and India—and in the United States—where an estimated 30,000 people still die each year from coal pollution.

If we do not find a way to replace coal in our energy economy, the coughing victims in the smoke-shrouded cities of the Industrial Revolution may not be its worst legacy.

Coal is the remains of plant-choked swamps that covered the earth during the Carboniferous period from 360 to 280 million years ago. The two major geological timescale divisions of the period—the Pennsylvanian and the Mississippian—are named after coal formations in the United States. Life on earth was still relatively young. Many of the bacteria and soil organisms that break down organic material into compost had not yet evolved. As a result, most of the earliest plant material that grew on land was buried undigested in geological layers. That is why coal is so common today. In the early nineteenth century there was some talk about coal shortages, but it has proved far more abundant than anyone ever imagined. We will probably never run out of coal.

Not coincidentally, the earth was a much hotter place at the time coal was formed—about five degrees hotter than it is today. This was probably because the carbon dioxide content in the earth's atmosphere was much higher. One effect of burying all that carbon in the ground was to reduce the greenhouse effect and make the earth's climate more temperate.

Coal comes in three varieties—lignite, bituminous, and anthracite. Lignite is loosely packed and lies close to the earth's surface. Bituminous is more solid with traces of other elements, notably sulfur. Anthracite is buried deep in the earth and is essentially a rock. It is so dense that one of its main drawbacks is it is often difficult to ignite.

Oddly, the first coal discoveries were made on beaches. In northern Europe, coal seams run near the ocean and the coal often washes ashore. For that reason, it was originally called "sea coal." It was not regarded of much use.

Then, sixteenth century England found itself facing an energy crisis. The forests, which had provided wood and fuel since time immemorial, were beginning to disappear as the country tried desperately to cope with the "Little Ice Age" that began in the fifteenth century. During the reign of Queen Elizabeth, several commissions were dispatched to the countryside to investigate. They all returned alarmed over the decline of the forests.[1]

The shortage was felt particularly in the cities, where wood was used both for hearths and for manufacturing. Urban culture might have failed altogether, except that—prodded by the ever-rising price of wood—England underwent an energy transformation and began burning sea coal. By the end of Elizabeth's reign in 1620, wood fuel had largely been abandoned.

All this did not happen easily. As Barbara Freese outlines in *Coal: A Human History*, the English aristocracy was extremely reluctant to make the transition:

> The rich in London tried to avoid using coal, still despised for its smoke, as long as they could. It was said in 1630 that thirty years earlier, "the nice dames of London would not come into any house or room when sea coals were burned, nor willingly eat of the meat that was either sod or roasted with sea coal fire." Within a few years, though, the nice dames and the nice gents had succumbed. By the second decade of the 1600s, coal was widely used in the homes of the rich as well as of the poor.[2]

Nor were these forebodings unfounded. Coal quickly became an enormous environmental hazard, turning the city air into a "foul and pestilent canopy of vapors."

> Thanks to coal smoke belching forth from various sources, [John Evelyn, a noted English writer] observed that "the City of London resembles the face rather of Mount Aetna, the Court of Vulcan, Stromboli, or the Suburbs of Hell... [A] traveler could smell it miles from London, long before the city was visible."[3]

Timothy Nourse, a London pamphleteer, complained, "When Men think to take the sweet Air, they Suck into their Lungs this Sulphurous Stinking Powder, strong enough to provoke Sneezing." Evelyn blamed coal for Londoners' incessant "Coughing and Snuffing," and reported that singers who came in from the countryside complained they lost the top three notes of their register. He believed that half the people who died in London suffered from lung disorders.[4]

Yet there was little choice. "Coal pollution may have been killing them slowly, but a lack of heat would have killed them quickly," concludes Freese. "It's been estimated that a poor family in London had to spend at least a tenth of its meager income on coal... To enjoy the same warmth from firewood, they would have had to spend perhaps two to five times as much."[5] In addition, coal had saved the forests of northern Europe. Where it was not available, as in the Mediterranean basin of Italy, Greece, Spain, Lebanon, and North Africa, the landscape was often stripped bare of wood.

As frequently happens, the adoption of one new technology opened the door for others. Without realizing it, by adopting coal the English had launched the Industrial Revolution.

One problem with coal mines is that they tend to fill up with water. At first, horses were used to pump the seepage from the ground. Then, in the 1720s a small-town English ironmonger named Thomas Newcomen invented an "atmospheric engine," which burned coal to boil water. The steam was injected into a cylinder where it expanded to move a piston. Cold water was then squirted into the chamber to condense the steam, creating a vacuum that pulled the piston back again. This movement could be used to pump water out of the ground. Newcomen's engine did the work of fifty horses, so that ever after the strength of such engines has been measured in "horsepower."

Forty years later, James Watt, a Glasgow builder of scientific instruments, made a key improvement in Newcomen's engine. The problem was the steam chamber, which had to be heated and cooled with each cycle. Walking in the park one day, Watt envisioned a separate "condenser" in which the steam would precipitate while the piston chamber remained hot. With the help of Matthew Boulton, a manufacturer of belt buckles, and Joseph Wilkinson, an ironmaster who had just started manufacturing swords, Watt built the first two modern steam engines, just as the Declaration of Independence was being signed in Philadelphia. One was used to pump water from a mine in northern England, while the other operated the bellows in Wilkinson's sword factory.

Manufacturing was not new to Europe. Since the Middle Ages, grain had been ground and cloth woven at "mills" powered by wind or water. Now, the mills were converted to coal. By 1829, Manchester had more than fifty coal-powered cotton mills, and Birmingham and Newcastle had become grimy "mill towns." Health conditions were so appalling that many observers believed humanity was spiraling toward destruction. In *The Condition of the Working Classes in England*, Frederick Engels, whose father owned a textile mill in Manchester, wrote:

> If anyone wishes to see in how little space a human being can move, how little air—and such air!—he can breathe, how little of civilization he may share and yet live, it is only necessary to travel hither.[6]

Life expectancy in the cities was thirty-eight years, as opposed to more than fifty in the countryside. Almost 90 percent of the population suffered from rickets, a disease that softens the bones of children, bowing their legs and curving their spines. The cause of rickets is vitamin D deficiency, brought on by lack of sunlight. Rickets was so common in smoky English factory towns that it was commonly called "the English disease."[7]

Yet manufacturing was enormously productive. Britain was soon turning out finished goods in previously unimaginable quantities. By 1830, the country produced four-fifths of the world's coal, and in 1848 it manufactured more iron than the rest of the world put together. Germany, with comparable coal reserves, was the only other European country that industrialized to the same degree. Where coal was scarce—in France, Spain, and Italy—manufacturing never achieved the same foothold.

As much as coal was to revolutionize industry, it played an

equal role in transportation. Throughout the eighteenth century, horses were the means of travel on land, while heavy freight was almost always hauled by water. The canal-building mania of the early nineteenth century extended water transport inland and created what was essentially the first system of national highways. Around the coal mines, however, a different technique was evolving. Owners began laying wooden rails along well-trodden pathways so that horses could pull heavy loads without getting stuck in the mud. By the mid-1700s, the coal mines of Newcastle were honeycombed with what were called "wagonways."

George Stephenson grew up in a coal cottage where a wagonway ran right past his door. As a young engineer in the 1820s, he was hired by a syndicate of mine owners to come up with a better means of moving coal. Stephenson mounted James Watt's steam engines on a wagon, substituted iron for wooden rails, and in 1825 built the world's first railroad, a 26-mile course connecting the coal town of Darlington to the river town of Stockton. The first train pulled an incredible thirty-four coal cars plus some 600 passengers. It moved so slowly it was preceded by a man walking a horse. Nonetheless, the railroad had been invented. Over the next twenty years, Britain laid 2,000 miles of track. The technology jumped the Atlantic and soon gained a foothold in America as well.

Anthracite coal was discovered in the Lehigh Valley of eastern Pennsylvania in the late eighteenth century and the nation's first railroad tracks were laid to haul coal down a mountain into Mauch Chunk, an unfortunately named river-town that later changed its moniker to Jim Thorp. Soon, anthracite was being barged to heat homes in Philadelphia, New York, and Boston. Then, vast reserves of bituminous coal were discovered in western Pennsylvania. Before Thomas Jefferson was laid in his grave, coal-fired iron foundries were blackening the skies of Pittsburgh, forever clouding the Sage of Monticello's vision of a pastoral America.

Forests were still abundant and the first commercial American railroads actually ran on wood. Anthracite was difficult to ignite in the firebox and wood burned much cleaner, allowing conductors to wear spiffy uniforms. The only problem was the shower of sparks, which regularly set neighboring fields afire. Trains carried buckets of sand in case the embers landed on passengers' clothing. When coal finally supplanted wood in the 1850s, a trail of dense black smoke replaced the pure white wood smoke, and conductors switched to black uniforms instead.

Water transport, challenged by the "iron horse," fought back with the coal-driven steamboat, a purely American invention. After giving up a career as an artist, Robert Fulton went to France, where he invented a submarine, christened the Nautilus, and tried unsuccessfully to convince the Emperor Napolean to build a fleet of steamboats to invade Britain. After failing to modernized warfare, Fulton returned to America where he built the *Clermont*, the world's first commercial steamboat. On August 7, 1807, it sailed 150 miles up the Hudson, from New York to Albany in thirty-two hours, a journey that took several days by horseback. In 1811, Fulton moved on to Pittsburgh, where he built the *New Orleans* and opened traffic on the Ohio and Mississippi. By 1834, 1,200 steamers were docking in New Orleans every year.

While coal was revolutionizing manufacturing and transportation, coal mining was turning out to be one of the most dangerous occupations ever undertaken. Working sometimes more than a mile underground, coal miners were in constant peril from the three "damps"—"choke damp" (carbon dioxide), which could suffocate, "smoke damp" (carbon monoxide), which poisoned, and "fire damp" (methane), which exploded. Miners tried all kinds of strategies—lowering dogs into newly dug pits to check for choke damp, carrying canaries to warn of fire damp, and even befriending mine rats, which always seemed to sense catastrophe. At one point, they even experimented with phosphorescent fish to avoid the candles that set off explosions. Still, in the twentieth century alone, 100,000 men died mining coal.[8]

When the Civil War broke out, both sides laid rail lines to move troops and ordnance, and regularly tore up each other's tracks. Greater access to coal and railroad lines that wed the Northeast to the Midwest helped the North defeat the South. Then, with the war over, the country began industrializing at a frantic pace. Within four short years, the "golden spike" was driven at Promontory, Utah, connecting both ends of the continent.

At the Centennial Exposition of 1876 in Philadelphia, the most popular attraction was Machinery Hall, a 14-acre factory featuring eight miles of shafts and pulleys that performed a myriad of industrial tasks. As David McCullough described it in *The Great Bridge*, "The amount of activity boggled people's minds. The *New York Herald*, the *Sun*, and the *Times* all printed their daily editions in the hall. Machines started sewing, pins got stuck into paper, wallpaper printed, logs were sawed. What really amazed people, though, was the Corliss Engine. The machine had only one attendant, who sat calmly on the platform and read newspapers."[9]

The Corliss Engine was the latest iteration of James Watt's steam engine, now three stories high, but still burning coal. On opening day, President Ulysses S. Grant and Emperor Don Pedro of Brazil climbed to the stage and pulled the levers, setting the gigantic factory in motion. They were cheered by a crowd of 10,000 that had just heard the "Hallelujah Chorus." The entire exhibition can still be seen at the Smithsonian Institute's Arts and Industries Building on the Mall in Washington.

In an obscure corner of the Smithsonian exhibit sits a small instrument in a glass case. It is labeled an "electric dynamo." The text explains that it is founded on the principle of "electromagnetism." Within a few years, Thomas Edison was experimenting with this mysterious force, finding it could perform all kinds of novel tasks. Two decades later, this unassuming device was in the process of replacing the entire network of belts and pulleys that so awed visitors to the 1876 Exhibition. Electricity became the latest method by which coal would energize the world.

Like the belts and pulleys in Machinery Hall, electricity is not a source of energy, but a *carrier* of energy; electricity only distributes energy to a variety of tasks. The power must still come from some primary resource—coal, oil, gas, wind, or water. (This is important to remember when we talk about replacing gasoline-driven vehicles with electric cars.) Electricity had long been employed in the telegraph, but now it proved far more versatile in applying energy to a wide variety of tasks. Lighting, heat, refrigeration, mechanical motion, the transmission of sound, the production of motion pictures—all became possible with the harnessing of electric power.

In 1882, Thomas Edison opened the first electrical generating station on Pearl Street in lower Manhattan, providing electric light to fifty-nine customers in a square mile area. It burned coal. Small utilities soon sprung up everywhere. Factories and institutions often found it easier to generate their own electricity. In 1900, when the Carnegie Institute of Technology was founded a few miles from Pittsburgh, students were required to spend two hours a week shoveling coal into the school boiler because utility lines had not yet extended out from downtown.

As the generation of electricity became centralized, the air did clean up a bit. Instead of smokestacks sprouting from every factory, combustion could be concentrated in boilers at a few large power plants. This was important because the Progressive era was already witnessing an uprising against coal exhausts. The effort was spearheaded by middle-class women in the "municipal housekeeping" movement. Although unable to vote, these activists formed volunteer groups that focused on sewage, air pollution, and other public health problems no one else was willing to address.[10]

For the most part, though, the rising tide of industrialism overwhelmed any effort to clear the air. Clean-burning anthracite from the Lehigh Valley was retained in Boston, New York, and Philadelphia, but everything mined west of the Appalachians, from Pittsburgh to Birmingham, was soft and hideously polluting bituminous coal. A famous 1913 photograph of downtown Pittsburgh shows electric streetlamps straining to dispel a midnight darkness. The picture was taken at 3 o'clock in the afternoon. When a 1902 coal strike temporarily halted the flow of anthracite to New York City, Fifth Avenue resident Andrew Carnegie—who had made his fortune selling bituminous in Pittsburgh—had the nerve to write the newspapers: "If New York allows bituminous coal to get a foothold, the city will lose one of her most important claims to preeminence among the world's great cities, her pure atmosphere."[11]

World War I pushed these reforms into the background, but they soon returned with postwar prosperity. Engineers noted that most of the smoke and soot that enveloped industrial cities was unburned carbon, and recommended larger, more efficient boilers. Utility companies began building generating stations to the size of 500 megawatts (MW), and moving them to the outskirts of town and beyond. Transmission lines soon marched across the countryside and large generating stations began serving whole regions.

In the early years, hydroelectricity rivaled coal in powering the new industrial empire. The first dams were built in the 1880s, and in 1894 George Westinghouse harnessed Niagara Falls to provide electricity to neighboring Buffalo. In 1900, 40 percent of the nation's electricity was being produced by water power, and as late as 1940, dams still provided 33 percent nationally and 75 percent in the far West. This figure slowly eroded, though, as good dam sites were used up and power requirements continued to grow.[12]

The real difficulty was that, although electricity might resolve some of the worst industrial pollution, coal remained the primary source of residential heat. Except for rural areas where wood was plentiful, almost every home and apartment house in

America had a "coal cellar" to feed its furnace. From the roof of almost every building, a plume of black smoke blossomed. Coal trucks made regular deliveries to every neighborhood, while children stole stray pieces to warm their families. (My high school chemistry teacher recalled jumping aboard railroad tender cars mounting a long hill near his home in rural Pennsylvania during the Depression and throwing chunks of coal to friends running alongside.) Although it was powering the entire nation, coal was a subject of universal contempt. A lump of coal was the worst thing a child could find in his or her Christmas stocking.

By the 1940s, coal was producing more than half of America's energy and King Coal was a nationwide political giant, not so much through the companies as their labor unions. John L. Lewis, the bushy-eyebrowed head of the United Mine Workers (UMW), was often called the second most powerful man in America, negotiating on equal terms with President Franklin Roosevelt. He finally overplayed his hand, however, by calling the UMW out on national strike in the midst of the Korean War. In Britain, the dominance of coal unions was even more pronounced, with coal miners regularly bringing the country to its knees by depriving the British of their primary fuel.

Yet the reign of King Coal was about to be challenged. As the nation became more affluent, the pollution from coal burning was becoming intolerable. As mid-century arrived, the Municipal Housekeeping League of the Progressive era was about to experience rebirth as perhaps the most powerful citizen-based campaign ever to appear in American history—the great environmental movement.

Chapter 5

Coal: The Environmental Era

On December 5, 1952, a blanket of unseasonably warm air settled over London. Since the cold air beneath could not rise, the fumes that made up the famous "London Fog"—actually, coal smoke—became trapped underneath.

Within three days, visibility was reduced to less than a foot. Coal smoke seeped into the Royal Opera House and performances had to be cancelled because the singers could not see the conductor. By the time the "Black Fog" lifted five days later, 4,000 people had died of respiratory conditions and heart failure. Fifty people were found dead in one small city park. Hospitals were overwhelmed. Parliament responded by outlawing the burning of soft coal.

One year later, a similar inversion settled over New York City, leading to 200 deaths. This time, auto exhausts also played a part. The 1955 National Air Pollution Control Act was the result. Then, over the Thanksgiving weekend in 1966, another four-day inversion over New York killed eighty people. The ordeal received national television coverage and caused widespread alarm. Air pollution was becoming a powerful political concern.

The long neglected plight of coal miners also began receiving attention. On November 20, 1968, an explosion killed seventy-eight miners in West Virginia. The incident was old hat for mining communities but for the first time the television networks showed up. The image of this routine carnage in Appalachia touched the heart of the nation. Union officials seized the moment and began a campaign that led to the passage of the Federal Coal Mine Health and Safety Act of 1969.

The new concern for coal miners finally focused attention on black lung disease, the scourge of a lifetime spent breathing coal dust. Although more than half the coal miners in the country were affected by this terrible condition, it wasn't even covered by worker's compensation. Finally in 1977, a Black Lung Disability Trust Fund was added to the Mining and Safety Act. Even today, 1,500 miners still die annually. "That's the equivalent of a Titanic going down in the nation's coalfields each year," says the United Mine Workers Union Web site. "However, unlike the sensational news coverage of such disasters, black lung victims die an agonizing death in isolated rural communities, away from the spotlight of publicity."[1]

Traditionally, air pollution had been the concern of the states. Unfortunately, state and local governments often solved the problem by placing their worst facilities right next to their borders and letting their neighbors suffer the consequences. President Lyndon Johnson took the first step in 1967 by proposing regional "air sheds," and giving state health agencies power to establish and enforce air standards.

Johnson's effort stalled in Congress, however, when Eastern coal interests realized that substituting low-sulfur western lignite coal for high-sulfur eastern bituminous might be the solution. This stalemate lasted three years until the first Earth Day in

1970 gave the environmental movement huge public momentum. President Richard Nixon climbed on board, signing the Clean Air Act of 1970. The bill set national air quality standards and gave the states the power of enforcement.

Research had identified three main pollutants—sulfur oxides (SO_x), nitrous oxides (NO_x), and particulate matter. Sulfur is a contaminate of coal and is much higher in Eastern bituminous. Nitrous oxides are formed when the heat of combustion fuses nitrogen and oxygen in the air. Both attack the cilia, the tiny hairs in the nose and throat that beat back pollutants, clearing the way for more dangerous particulate matter to enter and damage the lungs.

Particulates were first thought to be ordinary dust and soot, which are fairly easily removed by "baghouses" that were installed in the 1970s. Further research however, proved that the most dangerous particles are "PM-2.5," particulates smaller than 2.5 microns (the width of a human hair). These slip by the baghouses and enter people's lungs, causing a variety of respiratory diseases and lung cancer. Even today, the Environmental Protections Administration estimates that 30,000 people a year die as a result of coal exhausts.

To protect Eastern coal interests, the 1970 Clean Air Act mandated that all new coal plants must be equipped with sulfur scrubbers, whether they used Western low-sulfur coal or not. However, the bill created one loophole that was to have enormous consequences. Any coal plant licensed before 1973 would be exempt from the requirement. Utilities rushed to secure licenses, then sat on them over subsequent decades. By the 1990s, coal plants were still being built without sulfur scrubbers, even though it was 1970s technology. Today, more than three decades after sulfur scrubbers were introduced, 60 percent of coal plants still don't have them.[2]

Meanwhile, in the late 1960s, grassroot environmental groups had begun coming up with their own solution to coal's sulfur emissions—substituting oil. The effort began in Redondo Beach,

California in 1968, when a group of residents found their cars and homes being ruined by soot from a coal plant operated by Southern California Edison. The citizens demanded the utility switch to low-sulfur oil—only to be told that none was available. Researching the subject, they soon found that huge new supplies of low-sulfur crude had just been discovered in Libya and Indonesia.[3]

To their chagrin, the activists discovered this oil could not be imported. The problem was a ten-year-old program called "oil import quotas," which limited oil imports to fifteen percent of the nation's consumption on grounds of national security. Offended that their health and welfare were being endangered by what was essentially a scheme to protect the domestic oil industry, the activists appealed to Congress and an exception was made.[4] Soon, other cities up and down the East and West Coasts were making the same appeal, and by 1972, utilities were rapidly converting from coal to low-sulfur oil to solve the air-pollution problem.

From 1965 to 1971, coal consumption dropped for the first time in the nation's history. Meanwhile, oil consumption surged forward. The nation seemed well on its way to alleviating air pollution problems by importing more oil. Unfortunately, the Arab Oil Boycott of 1974 put an end to this strategy. Far from saving the environment with oil, the nation found itself battling a much larger Energy Crisis.

By the time the Carter Administration took office in 1977, it had long been clear that oil imports no longer offered a cheap and easy solution to air pollution. The other candidate for replacing coal was nuclear power. The technology, although still in its early years, had already won the approval of groups such as the Sierra Club, which had proposed reactors as an alternative to new Western dams. Some environmentalists had their doubts, however, and David Brower, executive director of the Sierra Club, led a rump faction out of the organization to found Friends of the Earth, which opposed nuclear power.

The anti-nuclear faction soon got the President's ear. In his famous fireside address of April 1977, Carter celebrated the nation's huge coal reserves and promised to double America's consumption from half a billion tons in 1976 to a billion tons by 1996. We were, after all, the "Saudi Arabia of coal." The President also proposed a "synfuels" initiative that would use coal to make gasoline. Coal was the solution to the Energy Crisis. There would be environmental problems, of course, but these would be resolved by developing "clean coal."

Simultaneously, Carter undercut the development of nuclear power. Within weeks of his energy speech, the President cancelled the Clinch River Breeder Reactor, ending the recycling of nuclear fuel rods, thereby creating the everlasting problem of "nuclear waste."

Thirty years later, we still live with the legacy of these decisions. President Carter's promise for increasing coal consumption has been more than fulfilled. We now burn more than a billion tons of coal a year, twice what we consumed in 1976. What was not anticipated at the time was that cleaning coal would prove much more difficult than expected and that the primary component of coal exhaust, carbon dioxide, although not technically a "pollutant," would become the principal player in a new environmental crisis, global warming.

Enforcement of the Clean Air Act lagged significantly during the Reagan Administration. At one point the President famously argued that air pollution was "caused by trees"—one of those tidbits of gossip that ricochet around the echo chamber of conservative think tanks until it becomes gospel. Trees do give off nitrous oxides that can form the smog of the Great Smoky Mountains, but this hardly covers the broad range of coal's various forms of pollutions. SO_x was the principal cause of concern at the time and where it was addressed, solutions proved difficult. When scrubbers were finally introduced on some plants, they produced gigantic mountains of calcium sulfate sludge that itself became a disposal problem.[5]

For the most part, however, utilities were encouraged to follow the Environmental Protection Agency's early dictum that "the solution to pollution is dilution." Huge smokestacks wafted coal exhausts high into the atmosphere. Only later was it realized the sulfur dioxide was returning to earth as acid rain.

When President George H.W. Bush arrived in office in 1988, he announced his intentions to return to the Nixonian model and become an "environmental President." Heavy-handed environmental regulations were still out of favor with Republicans, however, so Bush negotiated with the Democratic Congress for a system that would attempt to reduce SO_x emissions through economic incentives.

The idea had been discussed in economic circles for more than a decade. The problem of pollution, economists said, was one of "externalities." People were allowed to dump garbage into the atmosphere or the nation's rivers without paying any cost. If a price were attached to polluting, it would "internalize the externalities" and force polluters to come up with economic solutions. A tax on pollution was one approach, but that would give large revenues to the government. Another strategy would be for the government to sell "rights to pollute" and then allow polluters to trade among themselves. Those that find it cheaper would apply new technologies, while those who found it more expensive would buy the rights.

These "marketable rights" would avoid bureaucratic logjams and put a premium on inventing new technologies. They would encourage polluters to cleanup even beyond the regulatory requirements, since the financial incentives for more clean-up would always remain in place. Finally, they would offer municipalities, environmental groups and other organizations the opportunity to enter the fray by buying up rights and "retiring" them, forcing industries to cleanup even further.

After months of negotiation, a "cap-and-trade" system for cleaning up sulfur emissions was included in the 1990 Clean Air Act Amendments. The results proved spectacular beyond anyone's anticipation. Original estimates had placed the cost of meeting the new sulfur standards at between $3 billion and $25 billion per year. Actual costs turned out to be $0.8 billion. By 2003, the cleanup had gone 22 percent *beyond* the goals set by the 1990 act. Sulfur permits were traded on Wall Street and the price had sunk from an original $1,000 per ton to $150—making it even more possible for municipalities and citizens' groups to buy them up. Environmentalists were so exhilarated, they have become the biggest enthusiasts of cap-and-trade, promoting it worldwide and including them in the Kyoto Protocals.[6]

There is one caveat to the sulfur success story, however. Emissions were not generally reduced by employing new technologies or finding new ways to cleanup coal. Instead, the success came almost entirely from converting utility boilers to natural gas, which contains no sulfur. *Ninety-five percent of the new electrical generating capacity built in America between 1990 and 2008 employed natural gas.* This progress was bought at the price of a new problem—diminishing gas supplies. When natural gas prices quadrupled after 2000, many gas plants were idled, and coal once again became the first option for utilities.

With coal in revival once again, the effort to mine it has reached gargantuan proportions. About half the nation's coal now comes from low-sulfur lignite deposits in the Powder River Basin of Wyoming and Montana. Huge square mile open pits dot the landscape. Mile-long, 110-car "unit trains" deliver coal to utility plants as far away as Indiana and Arkansas. In 1999, a unit train left Cheyenne every twenty-five minutes. It is now every six minutes. Almost 40 percent of the nation's rail freight is now coal and the railroads are being stretched to capacity. When a series of derailments and floods disrupted service in 2006, utilities from Arkansas to North Dakota were told to prepare for rolling blackouts. "We're just one derailment away from cutting output or shutting down entirely," said Floyd Robb, a spokesman for Bismarck-based Basin Electric Power Cooperative.[7] There are now plans to rebuild major portions of the nations' rail system.

On the East Coast, the bituminous industry has responded to the competition from Wyoming by inventing "mountaintop mining," a system where whole mountains are decapitated to get at the underlying coal. Huge quantities of sludge are bulldozed into neighboring valleys, where they clog the streams and disrupt drainage. Mudslides are common. *Ten percent of the mountains in West Virginia have now been decapitated through this process.* Although there is a huge outcry at the national level, coal interests essentially run the state. When there was talk of federal legislation in the late 1990s, the coal companies mounted a massive campaign in support of George W. Bush, swinging the state into the Republican Presidential column for one of the few times in history. There has not been much talk in Washington about curbing mountaintop mining since.

Even so, the vision that we can power the nation on "clean coal" has not yet faded. The latest strategy is "carbon sequestration." Sulfur dioxides, particulate matter, and mercury (now a source of grave EPA concern) are all contaminants of coal that can be removed at great expense. Carbon dioxide, however, is the inevitable by-product of coal combustion. There is no way of eliminating it. So, the solution has become to *bury it permanently underground.*

Robert Socolow, head of the Carbon Mitigation Initiative at Princeton University and one of the world's leading experts on global warming, described the procedure in *Scientific American*:

> A new, large (1000-megawatt-generating) coal-fired power plant produces six million tons of [carbon dioxide] annually (equivalent to the emissions of two million cars). The world's total output (roughly equivalent to the production of 1,000 large plants) could double during the next few decades.... Each year at [such a plant], modified for carbon capture, about 50 million barrels of supercritical [i.e., partially fluidized] carbon dioxide would be secured—about 100,000 barrels a day. After 60 years of operation, about three billion barrels (half a cubic kilometer) would be sequestered below the surface.

> An oil field with a capacity to produce three billion barrels is six times the size of the smallest of what the industry calls "giant" fields, of which some 500 exist. This means that each large modified coal plant would need to be associated with a "giant" CO_2 storage reservoir.[8]

The technology will have its risks. Lake Nyos, in the Cameroon, sits in an old volcanic crater, where carbon dioxide constantly seeps into the bottom of the lake and is held in the depths by water pressure. On August 26, 1986, the lake "flipped," sending a 200,000-ton bubble of CO_2 to the surface. The gas shot 250 feet into the air, then, because it is heavier than air, settled back to earth and slid down the mountain into a valley below. Thousands of cattle and 1,700 people in nearby villages were asphyxiated. Carbon dioxide, after all, is the "choke damp" that once terrorized coal miners.

The Department of Energy is currently funding seven sequestration experimental facilities, the first, a two-mile deep hole next to a coal plant in West Virginia. Socolow estimates that carbon sequestration will add at least 20 percent to the cost of coal-fired electricity. Any accident similar to Lake Nyos would create untold havoc—not to mention undoing years of progress on global warming. But that is a problem for future generations. Such facilities will have to be monitored for thousands of years.

One perennial conceit is that new technologies will somehow produce clean coal. The most commonly suggested is the Integrated Gasification Combined Cycle (IGCC), a system that turns coal into methane and then is run through a combined cycle electrical generator, where both the steam and gas fumes run turbines, achieving energy conversions of 60 percent. This would supposedly solve coal's problems and give us access to our huge coal reserves.

Such a system would certainly make use of our coal reserves, but how would it solve any problems? Coal gasification produces huge amounts of sludge, even greater than ordinary coal burning. The same amount of carbon dioxide would be released, unless there is carbon storage. Gasifying coal means bathing it in huge quantities of hydrogen, which must be produced from natural gas—with carbon dioxide another waste product. In addition, the whole process would consume vast amounts of energy, which means multiple sources of carbon emissions instead of just one. Coal is coal and the carbon has to go somewhere. Basically, there is no such thing as "clean coal."

Despite all the efforts to reinvent coal, it remains essentially a nineteenth century technology. From the time it was introduced in the sixteenth century to save Elizabethan forests, coal's environmental impact has always nipped right at the heels of its advantages. None of coal's pollution problems have ever been resolved—as we see in China today. Now, the greenhouse effect has to be added to the roster. As Robert Socolow says, "As long as we rely on coal, it's going to be very difficult to solve the problem of global warming."[9]

Since the Energy Crisis of the 1970s, environmental enthusiasts, such as Amory Lovins, have promoted coal as a "bridge fuel" to some far-off solar future. Yet more and more, that bridge is starting to feel like solid ground. The nation now operates 600 coal plants with 150 more on the drawing boards. Utilities prefer coal to nuclear because coal is less expensive and does not arouse much public opposition. What makes coal cheap, however, is that it does not have to pay any of the costs of its environmental impact.

If we were to construct a bridge to the solar future, it would be much better built on terrestrial energy. Retiring coal as the prime source of industrial energy, however, will not be easy. Next to farming, coal is probably the most politically embedded industry in America. Perhaps the only political force that could possibly make a dent in coal's hegemony would be a nuclear-solar alliance.

Chapter 6

Oil: Crossing Hubbert's Peak

In 1954, M. King Hubbert, one of the premier geologists at the Shell Oil Corporation, began asking questions about the way oil companies were estimating their reserves.

After World War II, the Securities Exchange Commission had become concerned that oil companies were overestimating their readily available oil, using inflated numbers to attract investors. In order to establish uniform standards, the SEC requested that the companies divide their holdings into two categories: 1) "resources," referring to all the oil that might be available at some point in the future, and 2) "economic reserves," meaning what could be pumped out of the ground economically today.

The oil companies dutifully complied. When new discoveries were made, they now tallied the whole field as "resources" and switched them over to "reserves" as they became economically available.

Hubbert noted that the new practice was having a deceptive effect. As resources were regularly upgraded to reserves, it created the impression of a continuously smooth pattern of discovery. In reality, Hubbert noted, oil was getting harder and harder to find all the time. Collecting the figures from existing reservoirs and backdating them to the year in which the drill bit first hit, Hubbert found that the discovery of new oil in the United States had actually peaked in 1935.

Carefully calibrating the pace of development, Hubbert found a 34-year cycle between the day in which wildcatters hit paydirt and the year in which the oil field finally began to decline. Projecting forward, Hubbert calculated that probably more than half of America's oil had already been discovered. At this rate, American oil production would peak in 1969.

When Hubbert presented his analysis to the American Association of Petroleum Geologists in 1954, he was met with universal skepticism. There was good reason. Pessimists had been predicting the end of oil almost since the day Colonel Drake drilled the first well near Pittsburgh in 1859. The rapid decline of the early Pennsylvania oil fields seemed to fulfill this prophecy. By the 1890s, production was leveling off and Tsarist Russia was surpassing the United States in production. Then in 1897, wildcatters discovered the Nellie Johnstone well in Bartlesville, Oklahoma, and the whole cycle began again.

In 1901, Patillo Higgins, a one-armed geologist, and Captain Anthony Lucas, an Austrian engineer, drilled 1,024 feet into a salt dome named Spindletop outside Beaumont, Texas. There they hit a 150-foot gusher that wasn't capped for nine days. The biggest wells in the world at that moment were yielding twenty-five barrels a day. Spindletop spewed 100,000 barrels a day, more than all the wells in the United States combined! The idea that oil might one day run out quickly faded into the background.

Still, the U.S. Geological Survey remained pessimistic. In the 1920 edition of its annals, USGS executive David White stated, "[W]ithin 5 years—perhaps 3 years only—our domestic production will begin to fall off with increasing rapidity, due to the exhaustion of our reserves."[1] Only after World War II did the USGS become converted to the idea of unlimited resources. By the time Hubbert presented his paper in 1954, the Survey was proclaiming we had enough oil to last for centuries. A chorus of skeptics joined in proclaiming Hubbert wrong.

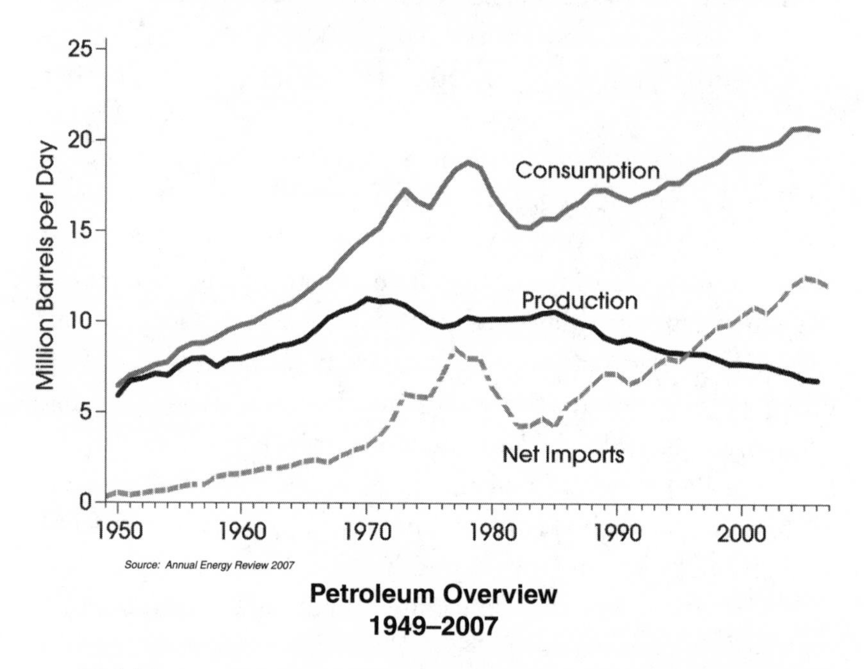

Petroleum Overview
1949–2007

Hubbert was wrong, but not by much. American domestic oil production did not peak in 1969. It topped out a few months later in 1970. *Crossing Hubbert's Peak* introduced Americans to an experience unlike anything in our history—the nation's first "Energy Crisis."

The Industrial Revolution was already more than 100 years old when Colonel Drake drilled sixty-nine feet into the soil at Titusville and began producing fifteen barrels per day. At first, this new product had no particular use. Coal had powered steam engines since the eighteenth century and continued to do so long after 1860. Instead, "petroleum" (from "rock oil" in Greek) was first used as a substitute for whale oil, which had illuminated the darkness since the invention of the wick lamp in 1783.

By the 1840s, Atlantic whales were growing scarce, forcing New England vessels to sail around Cape Horn and into the Pacific. As prices rose, whale oil became vulnerable to competition from kerosene, the first product refined from Titusville oil. The outcome was an economic transition that would "save the whales" for another century.

It wasn't long, however, before a glaring weakness of the new oil industry became apparent. There was too much oil and no way of limiting its production. It was the "tragedy of the commons," as ultimately described in a famous 1965 essay by historian Garret Hardin at the dawn of the environmental era.[2] Commonly owned resources, Hardin wrote, were always overexploited because everyone felt free to take them while no one had any interest in preserving them. Oil reservoirs, for example, could stretch for hundreds of miles underground, but anyone who owned a few square feet of earth could drill into them. As a result, once discovered, new oil fields were quickly exhausted. In Pennsylvania and Ohio, wildcatters regularly dumped oil into the rivers when new discoveries ruined the market.

Resources are conserved only by people who own them. They do this by holding resources off the market, "speculating" on a higher future price. In the absence of private ownership, it was necessary for government to assume the role, rationing out supplies with some account for the future. Governments were not very active in the nineteenth century, however, and so the role fell instead to a private individual named John D. Rockefeller.

The nation's first captain of industry did not try to buy up all the oil in Pennsylvania. Instead, he realized oil wasn't much good unless it was refined. After building his own refineries, Rockefeller ruthlessly cut prices, driving all his competitors out of business. Then, he began "rationalizing" the market by slowing down production in order to optimize the price. With output under control, he also began expanding his markets beyond the nation's borders—"oil for the lamps of China."

Rockefeller's efforts left him hated as a monopolist, but it tempered production and saved oil for future use. The public eventually turned against the "trusts" and Theodore Roosevelt broke Standard Oil into seven pieces. Yet by that time, Standard's dominance had been upset by Spindletop and domestic markets were expanding rapidly. Oil began to substitute for coal in transportation and industrial boilers. The Santa Fe Railroad went from a one diesel locomotive in 1901 to 227 in 1910. In 1905, Roosevelt ordered the U.S. Navy to convert from coal to oil. Every wildcatter in Texas now had an opportunity to market oil. Then came the automobile.

Gasoline was originally only one of three fuels competing to power the new vehicles. The Stanley Steamer, which burned crude oil in a modified steam engine, set a speed record of 127 miles per hour in 1906. It had a long warm-up time, however—forty-five minutes—and soon faded. The electric car outsold all competitors in 1900 and 1901. It could make 30 mph and required no gear changes. It needed lengthy recharging after 200 miles, however, and was impractical for long trips. Most electrics were sold to women, who used them to buzz around town.[3]

The internal combustion engine, meanwhile, had several drawbacks. It was noisy, smelly, and produced a nasty exhaust. The starter required hand cranking and the driver constantly had to change gears. But Spindletop decided the issue. After the Texas gusher, gasoline flooded the market. When Charles Kettering invented the electric starter in 1912, the contest was over. The internal combustion engine had won.

Without a John D. Rockefeller to rationalize the market, the job of conserving oil quickly fell to the Texas Railroad Commission. Established in 1891, the commission was originally designed to regulate rail freight. As oil began to dominate commerce, however, the Texas State Legislature passed several laws trying to prevent price collapses, but they were always ignored. Then, in 1917 open warfare broke out when the large pipeline companies refused to carry wildcatters' oil. The legislature intervened, converting the pipelines to common carriers and turning the whole show over to the Railroad Commission.

A year after Spindletop, 200 property owners had sunk wells within a quarter-mile of the original strike and the small hill looked like a pincushion. The field was exhausted in three years. Now, as the East Texas Field was discovered in the early 1930s, the Railroad Commission moved to establish control. Its authority was not always accepted. At one point, state militia had to be called out to hold off angry wildcatters, but the government won a series of court decisions and the Railroad Commission took command.

Populists saw this as a conspiracy between government and the oil industry. Advocates of the free market will argue such government-created monopolies restrict trade. But with natural resources it is important to remember that *somebody has to establish ownership* and exercise property rights if resources are going to be conserved. Someone has to speculate, holding the commodity off the market in hope of a better price tomorrow. Otherwise the resource will be squandered. In the future, an organization called "OPEC" would play the same role on a world level.

With oil production desperately needed during World War I, Congress added "oil depletion allowances" to the tax code to stimulate production. As commonly happens, the temporary wartime measure became permanent. Over the next five decades, oil depletion allowances would frustrate efforts to conserve oil and draw the wrath of environmentalists for subsidizing domestic overproduction.

In the meantime, the "Seven Sisters"—Esso, Mobil, Gulf, Shell, Texaco, British Petroleum, and Standard Oil of California (Socal)—began scouring the world looking for new supplies. Scorned as intruders by the poorer, oil-producing countries, and distrusted as mercenaries in the consuming countries, the oil companies honed their role as "multinationals," calling themselves "the oil that smoothes the friction" between producers and consumers.[4] In 1933, Socal (today's Chevron) signed an agreement with Saudi Arabia to explore its forbidding deserts. After a series of dry holes, the company sold half the concession to the Texas Oil Company (Texaco) in 1936. Then in 1938, roughnecks finally brought home a well called Damman Number 7. Although the geologists did not realize it, they had penetrated the world's largest reservoir, the 2000-square-mile Ghawar Field, 75 billion barrels, encompassing one-eighth of the world's known resources. "Old Number 7" is still pumping today.

America's superior oil resources proved an overwhelming advantage against Germany and Japan during World War II. The Germans made an early dash for Romania's reserves, then invaded North Africa as a route to the Middle East—only to run short of petrol on the way. The Japanese attacked Pearl Harbor after we cut off their access to Indonesian oil. Then, in 1945, after conferring with Winston Churchill and Joseph Stalin at Yalta, President Franklin Roosevelt made a stop at the Great Bitter Lake along the Suez Canal. There, on board an Arabian yacht, the most powerful man in the world paid tribute to King Ibn Saud, the new owner of the world's largest oil reserves. Although details of the conversation have never been made public, Roosevelt apparently promised to protect the Saudis from both German and Russian incursion in exchange for unlimited access to the Saudi's oil. Neither Churchill nor Stalin was informed of the intended visit. Ever since, America and the Saudis have had a "special relationship."

All this did not sit well with the independent oil industry back home. To producers in Texas, Oklahoma, Louisiana, and California, the prospect of oil flowing out of the Arabian Desert at pennies a barrel only promised the nightmare of another glut. In 1952, the industry approached newly elected President Dwight Eisenhower asking for protection against foreign competitors. Oil was crucial for defense, they argued. It would be against the national interest to become dependent on foreign sources.

No great fan of government intervention, Eisenhower responded cautiously. As a compromise, he proposed *voluntary* quotas that would keep imports below 15 percent. Even as M. King Hubbert was making his prediction to the Society of Petroleum Engineers, the domestic industry's main preoccupation was avoiding a price collapse. As with all such voluntary restraints, the quotas had little impact and imports kept climbing. When Egyptian President Gammal Abdul-Nasser seized the Suez Canal in 1956 and temporarily cut off supplies from Saudi Arabia, opinion in the Eisenhower Administration began to turn. As imports crept up to 20 percent in 1959, the government cracked down and made import quotas mandatory on the basis of national security.

None of this sat well with the producing countries, most of which were in what was then called the "Third World." Although politically liberated, these newly independent states essentially remained economic colonies of the West. Natural resources were their only wealth and Europe and America their only customers. If the United States would not buy more than 20 percent of its oil abroad, there were few other places to sell it. When Vice President Richard Nixon was stoned by mobs in Caracas, Venezuela in 1958, it was partly in reaction to voluntary import quotas.

In fact, the producing countries could hardly even be said to own their own oil. The Seven Sisters drove hard bargains and paid only nominal royalties. If the host country got any ideas about nationalizing the industry, the West would crack down, as happened in 1952 when Harvard-educated Dr. Mohammad Mossadeq was elected prime minister of Iran. Mossadeq immediately nationalized the Anglo-Iranian Oil Company, a co-concession with British Petroleum. The British blockaded the Persian Gulf in retaliation and within a year, the CIA had helped depose Mossadeq. Power was handed to 33-year-old Reza Pahlavi, who claimed the title of the Shah of Iran.

One practice that particularly galled the producing countries was "discounts." Whenever prices dropped, the majors would force the producers to accept a discount in their royalties, even though they had long-term contracts. Tired of being bullied, King Saud put out a call in 1961 for the producing nations to send representatives to Baghdad, where a group of army officers had recently overthrown King Faisal II. Five countries—Saudi Arabia, Iran, Iraq, Kuwait, and Venezuela—representing 80 percent of the world's oil exports, attended. They decided to call themselves the Organization of Petroleum Exporting Countries—OPEC.

In the late 1960s, gasoline cost twenty-five cents a gallon. You could fill your tank for $3. Families were expanding their garages to hold two and three cars. Residents of Redondo Beach were campaigning to have oil replace coal in their local utility plant. Then, almost unnoticed, the United States passed a landmark. In the first few months of 1970, American oil production reached an all-time high of 10 million bbd ("blue barrels per day," after the color of Standard Oil's original barrels in Pennsylvania). Then it tapered off.

Like a train of Conestoga wagons crossing the Continental Divide on its journey to the Pacific, the American oil industry had finally crossed Hubbert's peak. Oilmen recognized it right away. In the spring of 1971, the *San Francisco Chronicle* carried a one-sentence report: "The Texas Railroad Commission announced a 100 percent allowable for next month."[5] For the first time in history, the Commission was lifting its production quotas and telling the companies to pump all they could.

It was already too late. Since landing on the North American continent in the fifteenth century, European explorers and their successors had encountered nothing but an endless abundance of natural resources. In 1954, even as M. King Hubbert was addressing the Society of Petroleum Geologists, historian David Potter wrote a well-received book entitled *People of Plenty*, in which he argued that access to abundant resources was the major factor in shaping the American character.[6] Now, for the first time in our history we were going to have to deal with an entirely new phenomenon—resource scarcity.

Chapter 7

Oil: The Energy Crisis and the World Peak

On October 6, 1973, as Israel was observing Yom Kippur, Egyptian and Syrian troops launched a coordinated attack across the Suez Canal and into the Golan Heights. Backed by Russian technical support, the Arabs made good progress for four days, recapturing much territory lost during the humiliating Six-Day War of 1967.

Then, the Israelis counterattacked at the hinge between two divisions along the Suez Canal. An entire Egyptian army was cut off and left stranded in the Sinai Desert. Within two weeks, the Israelis had also beaten back the Syrians from the Golan Heights. At this point, both sides accepted a UN cease-fire offer.

After being elated by their early victories, the Arabs were dismayed at Israel's quick recovery and enflamed at the United States for providing support. On October 17th, the Organization of Arab Petroleum Exporting Countries (OAPEC), an offshoot of OPEC, declared it would reduce oil production by 5 percent per month in retaliation. The next day, President Nixon defied the Arabs by asking Congress to ship 2.2 billion dollars in weapons to Israel. Two days later, on October 20th, Saudi Arabia announced a complete oil embargo against the United States and the Netherlands, the only European country that was allowing us to use its airfields for the resupply operation.

At first the threat seemed idle. We had never been dependent on foreign countries for anything. Yet the three short years since crossing Hubbert's Peak had brought radical changes. With oil import quotas scrapped, imports had rocketed from 20 percent to over 30 percent almost overnight. By December, long lines were forming at service stations, tempers frayed, motorists scheduled their whole day around finding gas. An alternate-day rationing system based on odd-even license plates was devised and public officials urged people not to "top off their tanks." Rumors of an oil company conspiracy abounded.

For Americans raised on the stories of Colonel Drake, John D. Rockefeller, Spindletop, and *Giant*, it was almost incomprehensible. Suddenly, we were importing one-third of our oil.

All this was not without context. Only a year before, the Club of Rome Report had published a startling report, *The Limits to Growth*, in which a huge MIT computer had predicted that the world was headed for a civilization-wide crash triggered by vanishing resources. We would run out of gold by 1981, mercury by 1985, tin by 1987, zinc by 1990, and oil by 1992. The Arab Oil Boycott only seemed to move things ahead a few decades.

What the Arab Oil Boycott revealed, in fact, was a jarring shift in the terms of trade. For decades the major oil companies had twisted the arms of the producing countries, making them accept price reductions for the benefit of Western consumers. Those days were now over. Oil prices quadrupled over the course of a few months from $3 to $12 a barrel. Because the price of crude is only a small portion of the final product, this translated into a rise from 32 cents to 57 cents a gallon at the pump. Still, the jump reverberated throughout the economy, kicking off a two-year recession and setting off a chain of inflation that would swell until the prime rate reached 21 percent in 1980.

President Nixon quickly announced Project Independence, the first of many proclamations that we could achieve "energy independence." When his advisers came back and told him the effort would require 1,000 new coal plants or nuclear reactors, the project was quietly shelved.

It was not as if people had never given the matter any thought. In 1952, President Harry Truman, concerned about America's growing dependence on foreign supplies for raw materials, had appointed a Resource Policy Commission, chaired by William S. Paley, president of CBS.

The Paley Commission took a long inventory of America's situation and published a book-length report in 1954 entitled *Resources for Freedom*. The 400-page volume concluded that, except for a few rare minerals, America would remain self-sufficient in resources for some time. Then circumstances would gradually change. By the 1970s, the balance of trade would shift and America would start importing a larger percentage of its raw materials. It wasn't a bad bit of prognosticating. The commissioners were so pleased with their work that they settled into Washington and founded a permanent organization, Resources for the Future (RFF).

In 1963, RFF published a second book entitled *Scarcity and Growth*, still regarded as a classic in the field.[1] The authors, Harold Barnett and Morse Chandler, came to the seemingly paradoxical conclusion that, even though we are always using more of the earth's bounty, natural resources are constantly growing *less* scarce. Although this might seem illogical, the authors insisted it was true.

One reason, they wrote, was ever improving technology. Technological advances continually allow us to access lower and lower grades of raw materials. We may exhaust the easy iron deposits building railroads, for example, but those railroads allow us to mine even more remote iron resources. Improved technology also makes consumption more efficient. Cars once weighed 4,000 pounds and got 12 miles per gallon. Now they are made with lightweight plastics and get 30 miles per gallon. Tape recorders used to come in suitcases and weighed 20 pounds. Today you can hide one in the palm of your hand. All consume fewer resources.

Another factor is substitutability. As one resource becomes scarce, others become more economical. When whales became hard to find, kerosene was substituted. When firewood became scarce, it was replaced by coal. In each instance, consumption was shifted to a more abundant resource base. As Barnett and Morse concluded, "There may be relative scarcities, but there is no reason to think there is a *general* scarcity."[2]

Business Professor Julian Simon, elaborated on this argument in *The Ultimate Resource* (1981), written in response to *The Limits of Growth*, Paul Ehrlich's *The Population Bomb*, and other doom-mongering of the 1970s.[3] There are actually two infinities, Simon wrote, the infinity that the universe is without end and the infinity that any substance can be divided into an infinite number of parts. Our consumption can move in either direction. Richard Feynman, the great explorer of the quantum world, expressed the possibilities of the latter in his famous dictum, "There's plenty of room at the bottom." We have just begun to enter the seeming infinity of the sub-universe of quantum physics in our development of microelectronics and nanotechnology.

Simon proved his point in the famous 1980 bet with Paul Ehrlich over commodity prices. Challenging Ehrlich's warnings that humanity was exhausting the world's resources, Simon proffered Ehrlich a futures' contract in which he would sell Ehrlich a basket of raw materials at 1980 prices and then buy them back in 1990 at market prices. Confident that the price of raw materials would be rising, Ehrlich accepted the offer. To his immense chagrin, he found that by 1990 the inflation-adjusted prices of nearly all the raw materials had *declined,* proving Simon's and RFF's point. Raw materials are steadily growing more abundant.

One final key to resource abundance is recycling. As Simon pointed out, we never really "use up" minerals such as copper or aluminum. We just turn them into different forms. For decades, the telephone companies laid cables of copper wire. Now we are replacing them with fiber-optic cables. But the copper does not disappear. The companies tear it all out and recycle it. Once again, resources become less scarce all the time.

All this rests on a distinction first made by Aristotle, that the world is made up of both *form* and *matter*. Matter can change form, but the amount of matter always remains the same. Elaborated into more precise scientific form, of course, this became the eighteenth century's Law of the Conservation of Matter. As mentioned previously, the nineteenth century extended this principle to the Law of Conservation of Energy and then Einstein combined the two into something for which we don't yet have a name but which might be called the "matter-energy continuum," or simply "the Creation."

All this might start making us believe we can live comfortably in a world of endless recycling, except for one thing—a consideration that neither the Paley Commission nor Resources for the Future nor Julian Simon ever completely addressed—the Second Law of Thermodynamics.

We're going to meet the Second Law several times in this book, so let's get acquainted right now. Developed over the course of the nineteenth century by a series of brilliant scientists—Robert Mayer, Sadi Carnot, Rudolf Clausius, James Clerk Maxwell, and Ludwig Boltzman—the Second Law is said to be the only part of classical Newtonian physics that survived Einstein's Theory of Relativity. Therefore, it is worthy of our attention and respect.

The essence of the Second Law can be seen in a bouncing rubber ball. Hold the ball at the height of your head and it has potential energy. When you drop it, that energy translates into downward motion until it hits the ground and bounces up again. Since the energy within the system always remains the same, a perfectly elastic ball would bounce up and down to the same height forever. This would be perpetual motion. But it doesn't. Instead, its bounces to a lesser and lesser height each time, lower and lower, until there is a little drum roll at the bottom and it stops altogether. What has happened to all that energy?

A careful measurement would show that the temperature of the ball and the ground have been raised ever so slightly by the collision between them. This heat is very low-grade and cannot be used to do work. It is essentially irrecoverable and therefore "waste." The Second Law says that no energy transformation—from chemical to kinetic to mechanical to high-grade heat or back again—ever takes place with 100 percent efficiency. There is always some friction in the system, which dissipates the energy as waste heat.

In terms of everyday life, the Second Law comes down to this: Energy cannot be recycled. There is no such thing as a perpetual motion machine, even though geniuses as diverse as Nikola Tesla, James Clerk Maxwell, and perhaps even Amory Lovins have tried to invent one. The concept of infinite resources or a "bottomless well" applies only to material resources. If we use copper to make telephone wires, we can always rip it out again and use it for circuit boards. Once we have burned a lump of coal, however, the energy is gone forever. That is why it is somewhat misleading to talk about "renewable energy." Energy from a source such as the sun may appear inexhaustible but it is not renewable. We will always be searching for more energy supplies.

The Second Law is usually expressed in terms of "entropy," which means "disorder." A state of low entropy is a state of high order, which makes the whole concept a bit confusing. If matter is in a state of low entropy or high order, it has *energy potential*. If it is in a state of high entropy or great *disorder*, its energy potential has been exhausted. The classic example is mixing hot and cold air or water. Carnot's great observation was that a steam engine draws its power from the temperature *difference* between the steam chamber and the surrounding atmosphere. Any transfer of heat from one body to another can be exploited to do useful work. But if hot and cold are mixed, they will soon randomize their distribution to a uniform level. When this "lukewarm" air or water reaches the temperature of the surrounding environment, there is no remaining energy potential.

Edward Teller begins his book, *Energy From Heaven and Earth*, by noting that most chemical substances on earth are already at their lowest energy state, so no energy *potential* remains. We cannot extract chemical energy from rocks or water (despite fantasies like David Mamet's play, "The Water Engine.") Minerals, such as magnesium, can be "burned"—oxidized by high heat—but more energy must be added than is needed. Only where solar energy is stored in highly organized organic molecules do we find the potential that the energy released will exceed the energy originally added through a chain reaction we call "fire."

The Second Law offers a clearer way of viewing global warming. The carbon dioxide that emerges from fossil fuel burning is waste, in the sense that its chemical energy is exhausted. But the real problem is that it is also *scattered randomly* into the atmosphere. Randomness is high entropy. *It would take almost unimaginable amounts of energy to recover this carbon dioxide into an orderly form where it could be safely sequestered or stored.*

Virgin Airways founder Richard Branson has offered a $25 million prize to anyone who can discover a way to take carbon out of the atmosphere. We can be sure no one will ever claim it. *Whatever method is devised, it will take stupendous amounts of energy to accomplish.* This energy will produce its own "exhausts" and we will be worse off than when we started. The only option would be to find a stupendous source of energy that does not produce carbon exhausts—like terrestrial energy.

Although he may not have realized it, all these issues lay before President Jimmy Carter when he took office on the frigid morning of January 20, 1977. (The country was undergoing the coldest winter in decades. *Time*'s cover story the week of the inauguration was "The Big Freeze.") Three months later, on April 18th, Carter, wearing a cardigan sweater, sat in front of a fireplace and informed Americans that we faced the "moral equivalent of war." "Tonight I want to have an unpleasant talk with you about a problem that is unprecedented in our history," he began. "I know that many of you have suspected that some supplies of oil and gas are being withheld from the market. You may be right, but suspicions about the oil companies can't change the fact that we are running out of petroleum." Then he laid out his solutions.

As *Harper's* editor Lewis Lapham would point out six months later in an article titled, "The Energy Debacle," Carter's Energy Plan was largely shaped by S. David Freeman, who had headed up a 1973 Ford Foundation study called *A Time to Choose*.[4] Like Amory Lovins, who was about to emerge as the apostle of the solar future, Freeman's work was an effort to describe a world without nuclear power. Instead, our energy strategy would be a three-pronged effort—coal, conservation, and "alternate" energies.

Freeman's prognostications for the possibilities of energy conservation had already attracted huge controversy. While in 1973 we had consumed 73 "quads" of energy (quadrillion BTUs, the standard measure) and seemed headed for 100 quads by 1985, *A Time to Choose* made the startling pronouncement that consumption could be held to 80 quads in 1985 by cutting energy growth from 3 to 1 percent per year. (In fact, conservation was so successful that we only hit 74 quads by 1985, and didn't reach 100 until 2003.)

To implement this coal-conservation-solar strategy, Carter recommended that:

- Coal consumption be revived and doubled by 1976 in order to provide the nation's electricity
- Conservation efforts be implemented on a broad scale
- The nation undertake a "synfuels" program to make gasoline out of coal, and
- Solar and renewable energy—from windmills and solar panels to an effort to derive ethanol from corn—be subsidized

Before it even began, the Carter effort was badly undercut by the 1975 decision of President Ford and Congress to extend oil price controls. This was a continuation of President Nixon's across-the-board wage and price controls imposed in 1971. When that law expired in 1973, oil price controls had been retained because oil had become "too important to be left to the free market." The result was the same as with all price controls—producers sat on their hands while consumers went merrily along consuming as if there were no resource problem. The result was a "shortage" of domestic oil. Unfortunately, we now had another source. We could *import more oil.* From 1974 to 1979, under the influence of price controls, our oil imports accelerated from 33 percent to an incredible *50 percent* of consumption. This ticking time bomb would eventually bring down the Carter Administration. For the time, however, the President temporized. Oil price controls would remain in place but the Energy Plan would impose a "windfall profits tax" that would raise the price of American oil to world levels, thus discouraging consumption.

Although the Democrats had a large majority in Congress, the Carter Energy Plan met formidable resistance from an alliance of Republicans and Texas and Louisiana Democrats who felt the government was punishing the oil industry. This coalition fought Carter to a standstill for three years. Thus the tax never got imposed and consumers continued to gobble up foreign oil. When the Ayatollah Khomeini overthrew the Shah of Iran in 1979, world oil production dropped 3 percent, but this was enough to precipitate another "gas shortage." With public frustration at a height after three years of seemingly endless dickering, Arkansas Senator Dale Bumpers made his famous pronouncement: "The public wants us to do something, even if it is wrong."[5] Congress quickly adopted the Carter Energy Plan. It was too late for the Administration, however. The second gas shortage, combined with the Iranian Hostage Crisis, torpedoed the Carter presidency.

Conservatives and economists had argued all along that price controls were the problem, and removing them would promote conservation as well as bring forth new supplies. Carter eventually saw the light and began a slow phaseout in 1979 but once again was too cautious. One week after taking office, President Reagan swept aside the phaseout and immediately abolished price controls. It was a remarkable act of courage. Oil was at $45 a barrel ($95 in 2008 dollars), a record that was only equaled again in 2007. Not one person in a thousand would agree that abolishing price controls was the way to bring prices *down*. Yet Reagan knew his economics. The immediate result was a stampede back into the domestic oil business. Anyone with a few thousand dollars in their pocket was trying to find a drilling rig. The number of rentals spiked to 4,000 in 1981, almost twice as high as it has ever been before or since.

The result was a huge upsurge in production; the only time domestic output has risen since we passed Hubbert's Peak. Added to the first returns from Alaska's North Slope, which came online in 1977, this spurt almost pushed American production back to the 1970 levels. At the same time, consumers, finally disabused of the notion that price controls were solving the problem, began changing their habits and buying smaller cars. Saving energy turned out to be much easier than anyone expected and demand slackened. While some of this was the result of the brutal 1979–1982 recession, permanent efficiency improvements were also achieved.

With price controls gone, our demand for foreign oil fell off a cliff. Between 1979 and 1986, imports fell from 7 million bbd to 3.5 million, and purchases from Saudi Arabia plummeted from

1.4 million bbd to 200,000. The 1974 run-up in world prices had sent geologists scrambling to the remotest corners of the planet. Mexico revealed it had discovered huge deposits in the Gulf of Mexico in the 1960s, but kept quiet because it did not want them squandered at the low world price—another classic example of a resource owner practicing conservation. By 1984, Mexico had become our largest foreign supplier. Within five short years, the oil industry was back on familiar territory—a world glut. Even "Dallas" changed its plotline to reflect the oil slump. All human progress is built on investors losing their shirts.

Saudi Arabia, finding itself undercut by other OPEC countries, flooded the market, afraid of losing market share. Oil prices fell from $24 to $14 a barrel in 1986, not much above the $12 at the end of the Arab Oil Boycott.

And so, after more than a decade of doing little more than ratifying price increases, OPEC finally had to start acting like a real cartel. At the behest of the Saudis, the producing nations began the task of assigning production quotas. Faced with competing claims, the ministers decided to parcel out production quotas on the basis of oil reserves. When the countries submitted their updated figures, OPEC's reserves had jumped a remarkable 40 percent!

The Gulf War sent prices soaring again, but this time the Saudis opened the spigots and made up the difference, quickly bringing down prices. Trying to keep OPEC together, the Saudis had emerged in the old role of the old Texas Railroad Commission, holding back supplies and modulating production in order to stabilize the market. It was not an easy job. By 1997, world prices had fallen again to $16 a barrel, almost all the way back to pre-1973 prices. Conservation slackened and the Energy Crisis was almost forgotten.

The remarkable 40 percent jump in OPEC reserves in 1986, however, had raised the old question—just how much do the numbers reflect reality? How much oil is really out there? M. King Hubbert had used the hidden rate of discovery to predict the peaking of American oil production. Was it possible that world production might eventually peak as well?

In the 1990s, even as the world experienced oil glut, Colin Campbell, a veteran Scottish geologist, began applying Hubbert's analysis to world production. He found that discovery had peaked in the early 1960s. Using Hubbert's production cycle of thirty-eight years, he predicted the world would cross its own Hubbert's Peak around the turn of the century. Campbell popularized these views in several forums, including a 1998 *Scientific American* article entitled, "The End of Cheap Oil."[6]

When the millennium passed without a crisis, Campbell updated his prediction to 2006. Kenneth Deffeyes, a veteran geologist at Princeton University, chose the same date in his book, *Hubbert's Peak*. A slew of Web sites have popped up—monitoring world oil production almost to the minute—*drydipstick.com, oilempire.us, lapostcarbon.org,* etc. Doom was predicted by an endless parade of books—*The End of Oil, The End of Suburbia, The Final Energy Crisis, Blood and Oil, It's the Crude Dude, The Party's Over*—most recommending a return to the simple life. In 2007, the GAO issued an 82-page report, saying that world oil would probably peak by 2040 and that, at best, sufficient substitutes would not be in place until 2015.[7] Congressman Tom Udall (D-N.M.), co-chairman of the

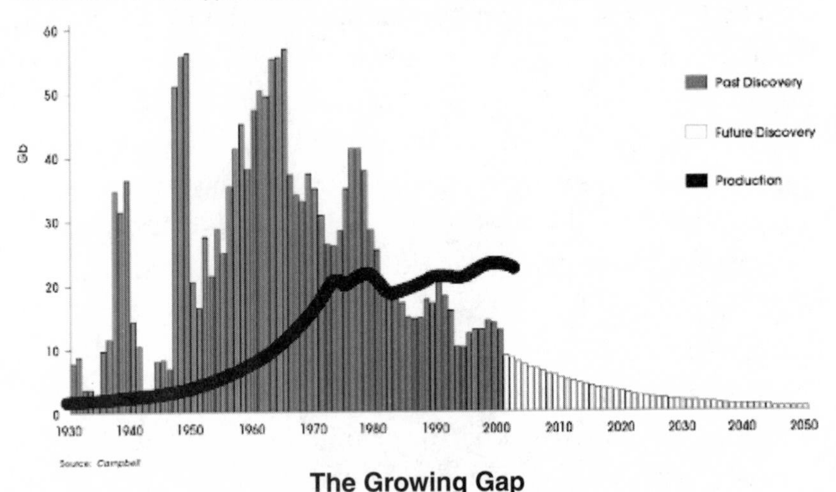

The Growing Gap

World oil discovery and production

Congressional Peak Oil Caucus, called for "conservation, higher fuel efficiency standards…[a] revival of passenger and freight rail, [and] expanding research into and use of biofuels."[8]

Then in 2005, Matthew Simmons, a Texas banker who has followed the oil industry for decades, committed the ultimate heresy and questioned whether the Saudis themselves have as much oil as they claim. Simmons noted that the Saudis refuse to divulge production figures from individual wells and even keep secrets from themselves. The most widely accepted production figures for OPEC, he found, come from a small consulting firm named Petrologistics that operates over a grocery store in Geneva and claims to have spies in every port. "History has shown," Simmons concluded, "that, when such levels of secrecy are maintained, it is often the keepers of secrets themselves who are most surprised when the true state of things emerges."[9]

A consensus has now emerged among oilmen that world production will probably peak within the next twenty years. The best estimate is that there are probably 2 trillion barrels of oil in the world. Of this, *90 percent* has been discovered. Output has exceeded discoveries every year since 1973. The last million-barrel-a-day field was discovered in the Gulf of Mexico in 1976. (Chevron's 2006 find in the Gulf, the best in decades, was only 800,000.) No new discoveries were made in 2004, the first time since exploration began in the nineteenth century. There are few corners of the world left to explore.

As Barnett and Moore predicted long ago, we can keep developing lower grades of resources. After decades of speculation, Canada has opened the Athabasca Tar Sands in Alberta, which contains 174 billion barrels of thick hydrocarbon, enough to vault Canada from twenty-second to third place in world reserves. Ahead of it still are Saudi Arabia and Venezuela, whose Orinoco Belt (270 billion barrels of tar sands) puts it in first place. Canada is now producing a million barrels a day with room for expansion. Venezuela could produce more.

Yet the environmental cost is agonizing. In Alberta, whole

forests are being leveled so that steam can be injected to boil the oil out of the ground. Huge piles of acidic sludge are shoved into artificial lakes, where cannons must be fired every five minutes to keep migrating birds from landing on their acidic waters. The whole process is carbon-intensive, and by 2015 the tar sands region will emit more greenhouse gases than Denmark.[10]

Although no one wants to talk about it, the world's largest reservoir of hard-to-get hydrocarbons is the Green River Formation on the western slope of the Rockies. A thousand feet below ground lies the world's largest deposit of fossil fuel—2 trillion barrels of shale oil. Developing it will mean ruining hundreds of square miles of western landscape. Without any alternative for powering our transportation, however, its development became an issue in Colorado's 2008 election.

What the world faces now seems eerily similar to what the United States confronted in the late 1960s. Demand for oil is rising at a gallop. Both China and India, the world's two largest countries, are rapidly industrializing. China is now Saudi Arabia's largest customer and is signing long-term contracts all over the world. Indonesia, a long-time member of OPEC and the source of all that low-sulfur crude in 1970, is now an importer and has dropped out of OPEC. The world price of oil rose from $11 per barrel in 1998 to $25 in 2003 and $130 in 2008 before falling back. Some calculations show world production leveled off at 85 million bbd in December 2005, although there is dispute over whether Canadian shale should be added to the equation.

"We won't know when we've crossed the world peak until it's already behind us," says Matthew Simmons, sitting in his spacious office overlooking downtown Houston. "The only way you'll ever see it is in the rearview mirror.

"But I'll tell you this," he adds. "If we were crossing Hubbert's world peak, I can't see how it would feel much different than it does right now."

Chapter 8

Natural Gas

As coal's pollution and oil's supply problems have mounted, the hope has arisen that natural gas can solve both our environmental and energy problems.

Natural gas is indeed the most environmentally benign fossil fuel. It is easy to transport through pipelines. It has no sulfur, mercury, or other impurities that mar coal and oil. It does not create the particulate matter of coal or diesel fumes. Because it burns so cleanly, it puts out only two-thirds as much carbon dioxide as oil and half as much as coal. Instead, the problem with natural gas is likely to lie in its supplies—not necessarily in the world at large but definitely in North America.

Through a set of peculiar circumstances, natural gas ended up under federal price controls for almost thirty years. This created huge market distortions and an artificial scarcity that culminated in the "natural gas shortage" of 1977. When the price was finally deregulated in the 1980s, the country suddenly found itself awash in gas. This created the impression that natural gas could be used to solve every environmental and energy problem. Gas was the principal tool for reducing sulfur emissions under the cap-and-trade system implemented in 1990. For the last twenty years, 90 percent of our new electrical generating capacity had been built to burn natural gas.

This broad effort hit a wall in 2000 when production suddenly leveled off. Prices have since sextupled—roughly the same run-up that has occurred with oil over the same period. (The prices of both tend to move in tandem.) Major portions of the fertilizer and plastics industries have left the country because gas prices were too high. We now import 15 percent of our supplies from Canada. And although gas has many advantages, it is different to transport across the ocean. There is still plenty of natural gas in Russia and Iran, but bringing it to America will require the construction of large liquid natural gas terminals. Even then, the U.S. will have a difficult time competing with China and Europe because those countries can import Asian gas much more cheaply by pipeline. The premium paid by Americans for ocean transport may make it prohibitively expensive in the coming decades.

In short, the hope that natural gas would be the solution to all our fossil fuel problems is now fading. There are clear-cut limitations. Many gas-fired electrical generating stations built during the 1990s are now standing idle because they can no longer afford the fuel.

Methane ("natural gas") is usually found in association with oil and coal deposits. It is the "fire damp" that terrified coal miners. (The rotten-egg smell in household gas is a sulfur compound added to warn of leaks.) In early oil drilling gas was an unwelcome by-product and any well that hit only gas was considered a dry hole. Gas was uselessly "flared off" at the wells and refineries until pipeline construction finally made it saleable to other parts of the country.

The Greeks and Persians noticed gas vents coming out of the ground and often built shrines around these "burning springs." In China, gas was being piped to the ocean to boil salt from seawater in 500 BC. Although it was not recognized, all these vents were commonly coming from coal or oil deposits. When Europe finally got around to burning gas in the nineteenth century, it was not the gas found in the ground. Because natural supplies were so difficult to transport, commercial gas was first synthesized from coal. Only when pipelines were built in the twentieth century did we revert to using the real stuff from the ground, which is why we still call it "natural" gas.

In 1792, a young British inventor named William Murdoch boiled coal in his mother's teapot and captured methane. He used it to light his small cottage in Soho, and in 1797, illuminated the Manchester Police Station. In 1812, the London and Westminster Gas Light and Coke Company started making coal gas and transporting it through wooden pipes beneath the streets. The company festooned Westminster Bridge with gas lamps on New Year's Eve. In 1816, painter Rembrandt Peale helped found the Gas Light Company of Baltimore, lighting the city streets with coal gas. Then in 1821, William Hart of Fredonia, N.Y. dug America's first well that tapped "natural" gas from the ground. Yet the Fredonia Gas Light Company did not open its doors until 1858. Until gas could be transported from the field, it was considered unusable.[1]

Soon, cities across America and Europe were turning coal into gas at the "gas house," a place so rough-and-ready that it came to symbolize society's lower depths. (Baseball's St. Louis Cardinals of the 1920s were known as the "Gas House Gang.") With it came the inevitable scourge of coal dust. Touring the Horseferry Gas Works in London in the 1830s, French visitor Flora Tristan wrote:

> The foreman told me that stokers were selected from among the strongest, but that nevertheless they all became consumptive after seven or eight years of toil and died of pulmonary

> consumption. That explained the sadness and apathy in the faces and every movement of the hapless men."[2]

It would be more than a century before exposure to coal dust was finally restricted.

After the Civil War, coal gas almost completely replaced kerosene for indoor lighting, inaugurating the "Gaslight Era." By the turn of the century, however, gaslight was giving way to electricity. Then, in 1909 German chemists Fritz Haber and Carl Bosch invented a method of producing ammonia (NH_3) from methane and atmospheric nitrogen. The "Haber-Bosch Process" created the fertilizer industry, which first set up shop right next to the natural gas wells.

When gas first emerged from Colonel Drake's first oil well, he had built a 5½-mile wooden pipeline to neighboring Titusville, to market it. Longer pipelines were impractical, however, until arc welding was developed in the 1920s. Then the country went on a pipeline-constructing binge and within a decade, natural gas from Texas and Oklahoma was making its way to Chicago and the Northeast. With the help of a massive advertising effort, Americans began abandoning their coal stoves and began "cooking with gas."

Before the pipelines arrived, however, "town gas" had fallen under the jurisdiction of municipal regulation. During the Progressive era, services such as water, electricity, and streetcars were dubbed "natural monopolies" and brought under a franchise system. The local government would grant an exclusive franchise—guaranteeing the company a customer base—in exchange for authority to regulate the price. Now as gas pipelines arrived in the cities, they also became targets of progressive reforms. As usually happens, it was the rival *coal* companies that supported the regulation of gas, since it gave them a chance to limit the competition. State governments also tried to regulate the interstate pipelines but the U.S. Supreme Court regularly turned them down.[3]

Then the New Deal arrived. By 1935 the Federal Trade Commission was expressing concern over the marketing power of electric and gas utilities. Congress adopted the Natural Gas Act of 1938, which turned regulation of interstate pipelines over to the new Federal Power Commission. Predictably, its first order was that *no new interstate pipelines* could be constructed in a market already served by another pipeline. "Cutthroat competition," after all, was believed to be the cause of the Great Depression.

Still, the new law extended only to gas *pipelines*, not to the producing wells, which were mostly in the Southwest. Soon this came under challenge. Attorneys general from several northern states began demanding the FPC impose price controls on gas *producers*. Finally in 1954 the Supreme Court succumbed. In Phillips Petroleum v. Wisconsin, it ruled that anyone who *sold* gas to an interstate pipeline was also part of the "monopoly" and therefore subject to federal regulation—even though independent drilling was one of the most highly atomized businesses in the country. What it amounted to was an attempt by northern consuming states to treat the producing states as economic colonies.

President Dwight Eisenhower, a native Kansan, immediately put legislation before Congress reversing the Supreme Court decision. The bill had widespread support and was about to pass when an executive from a small midwestern gas company dropped a $2,500 "campaign contribution" on the desk of a key Senator. The story hit the newspapers and scandal erupted. The bill passed Congress but by the time it reached Eisenhower's desk he felt compelled to veto it. And so the entire natural gas industry, one of the most widely diversified businesses in the nation, was put under federal price regulations—just as oil prices would later be regulated from 1973 to 1981.

The FPC made a game effort. By 1959 it had received 1,265 applications for price increases and acted on 240. Trying to get a handle on the task, the commission divided the market into five regions. By 1970 it had set prices for two. Meanwhile, the consuming states tried to control the regulation by bringing endless lawsuits, usually in the District of Columbia Court of Appeals. The court developed the "life of the field" doctrine, which said that once gas was submitted to a national pipeline it could not be withdrawn. Even if a producer went bankrupt, he must still keep selling his gas to northern states at what amounted to twenty-year-old prices.[4]

There was one escape, however. The FPC only had jurisdiction over *interstate* commerce. Well owners could still sell gas at unrestricted prices within their home state. And so by 1965, one-third of all gas in the country was being consumed in its state of origin. Texas and Louisiana were burning gas to produce more than half their electricity—which was universally regarded as a huge waste. Meanwhile, northern gas utilities were turning down new customers for home heating, its heat use.

The Arab Oil Boycott brought all this to the boiling point. As northern factories and homes desperately tried to switch from oil to gas, the whole system came crashing down. In February, 1977, only a month after President Carter had taken office, *Newsweek* ran a cover story, "Now, the Gas Crisis:"

> The harshest winter in decades held more than half of the United States in a choking grip last week, and brought with it a new and grave danger: a natural-gas crisis every bit as serious as the 1973 Arab oil embargo.... The combination of the chilling temperatures and the shortage of natural gas—which provides energy for half of the nation's homes and 40 percent of its industries—was devastating. Every school in Pennsylvania was closed, and 400,000 workers were laid off in Ohio alone. Nationwide, more than 2 million were already out of work, and the total mounted every day. Throughout the eastern half of the country, thousands of factories were cut back to "plant protection" gas levels (only enough fuel to prevent pipes from freezing) and had to shut down. Eight states—Pennsylvania, Ohio, Indiana Minnesota, Tennessee, Florida, New York and New Jersey—proclaimed states of emergency.... In office barely

> a week, Jimmy Carter was confronted by the first major challenge of his new Administration.[5]

As the new administration began sorting things out, it was appalled to discover that Texas and Louisiana were generating half their electricity with gas, while northern homeowners were freezing in their homes. There was immediate talk of more federal mandates. The producing states, on the other hand—tired of being exploited for their natural resources—rebelled. All over Texas and Louisiana bumper stickers proclaimed, "Let the Yankees Freeze in the Dark."

The battle was joined in Congress as part of the general debate over the Carter Energy Plan. Led by a trio of New York suburban Congressmen, the northern consumer states tussled with southern producing states over what was universally perceived as a diminishing resource. But the terms of the debate were changing. In the fall of 1978, David Stockman, a 32-year-old Congressman from Michigan, published an article in *The Public Interest* entitled, "The Wrong War: The Case Against National Energy Policy." Stockman argued that President Carter's government-led program of price controls and forced austerity had become the real source of the problem. Deregulate energy markets, and supplies would gush forth.[6]

As often happens, Congress compromised. The Natural Gas Policy Act of 1978 decontrolled prices but only ten years down the road. Meanwhile, the FPC's jurisdiction was extended to *intrastate* sales as well. Texans would be compelled to pay a lower price for their gas so that producers would sell more to northern markets. Meanwhile, the companion Power Plant and Industrial Fuel Use Act outlawed burning gas by electric utilities.

Stockman became President Reagan's Director of the Office of Management and Budget and ended oil price controls by administrative authority within a week of taking office, thus ending the "Oil Shortage." Deregulating gas, however, would not be so easy. Repealing the 1978 extension of price controls would mean another trip through Congress. So the matter was turned over to the Federal Energy Regulatory Commission (FERC), which had replaced the old FPC.

Surprisingly, the bureaucrats did a good job. Reviewing the transition in 1995, Robert Michaels, professor of economics at California State University at Fullerton, wrote:

> Although there were several credible candidates for the title, fifteen years ago natural gas was probably America's most misregulated industry.... Price controls on gas production had created shortage that exacerbated the energy crisis of the 1970s. Today, the controls are gone, and gas prices are about half those of a decade ago. The pipelines have been restructured into competitive entities by—of all people—federal regulators.[7]

FERC first made the pipelines common carriers, obligated to take anyone's gas. Then it cleverly set prices *too high* and allowed the parties negotiate down to a market price. The Maryland People's Counsel was soon in D.C. Court demanding that FERC set prices below market level, but this time the court

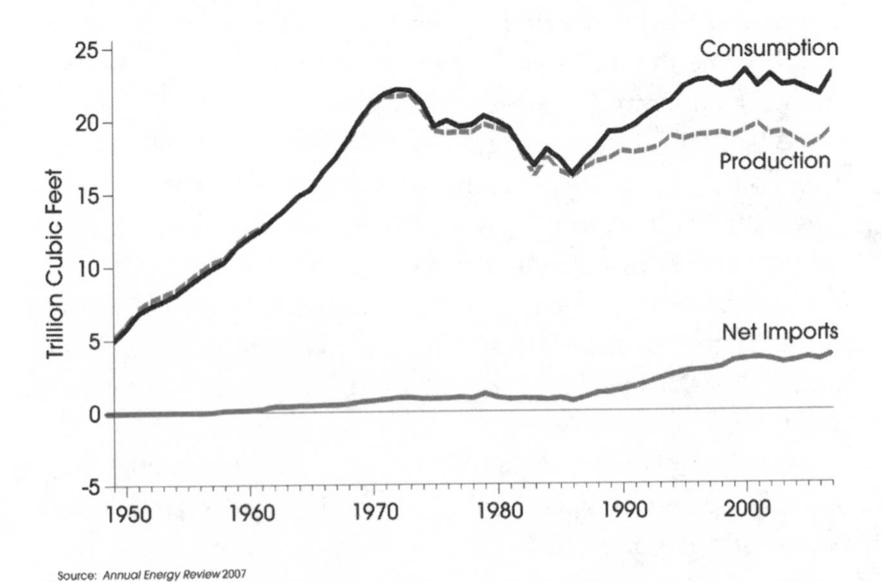

Natural Gas Overview, 1949–2007

said no. Judges Ruth Bader Ginsburg and Anthony Scalia, both future Supreme Court justices, voted with the majority.

Just as Stockman had predicted, deregulation prompted exploration in lands that had never felt a drill bit. The biggest discovery was the Overthrust Belt in the Rockies, which turned Denver into an energy capital. After years of decline, gas production turned upward again in a progression that would last almost two decades. By 1987 Congress felt so confident of gas supplies that it amended the Power Plant and Industrial Fuel Act, allowing natural gas once again to be burned in electrical boilers.

It was a fateful turn. Natural gas has enormous advantages for home use. It produces so little exhaust that it can be burned on open stovetops. The energy conversion in home heating is about 90 percent, which is as good as such transformations ever get. Natural gas is also a crucial feedstock for the drug, chemical, fertilizer, and plastics industries. With the passage of the 1990 cap-and-trade system, however, natural gas suddenly had another important use—reducing the sulfur emissions from coal.

In a market system, such conflicting demands are sorted out by price competition. Wherever a natural resource produces the most social and economic value, that is where it will be put to use. In this situation, however, the electric utilities had a decided advantage. They were still regulated monopolies. (Deregulation did not begin until late in the decade.) The system allowed them to pass through the cost of gas to their customers, whatever the price. As a result, electrical generation began crowding out other uses.

Compounding this was opposition to developing new supplies. Although environmental groups were wildly enthusiastic about natural gas' ability to "back out" old coal plants, they were far less enthusiastic about *drilling* for it. Land-based production in Texas and Louisiana had peaked in the 1970s and

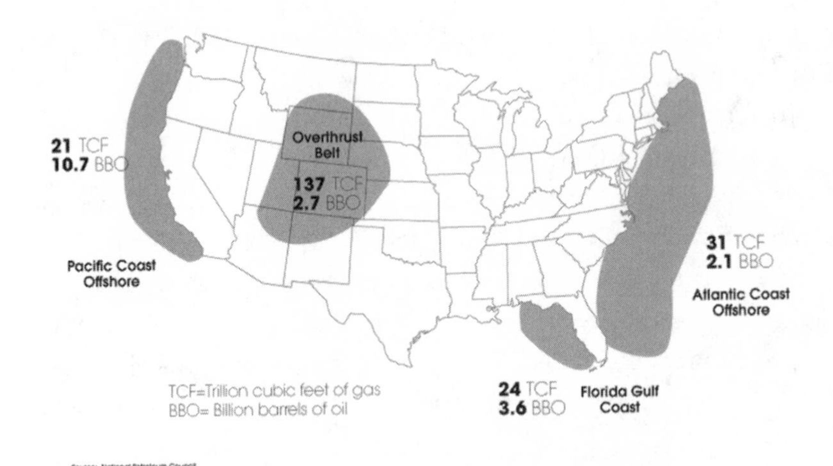

Areas Where Drilling Has Been Restricted

drillers moved offshore into the Gulf of Mexico. Huge deposits were also off the Atlantic and Pacific Coasts, but the 1969 Santa Barbara oil spill quickly put an end to that. Drilling off both the Atlantic and Pacific Coasts was suspended. In 1998, President Bill Clinton extended this moratorium until 2012, and President George Bush, Jr.'s subsequent efforts to lift it, failed. The National Petroleum Council estimates the continental shelves contain 16 billion barrels of oil and 75 trillion cubic feet of gas but opposition to tapping them remained strong until Newt Gingrich initiated his "Drill Here, Drill Now" campaign during the 2008 election. Without any new areas of operation, Texas, Louisiana, and the Gulf of Mexico have remained the workhorses of the nation.

The same scenario is being played out in the Overthrust Belt, which extends through the Rockies from New Mexico to Montana. There, deposits are believed to total more than 130 trillion cubic feet. For almost two decades, the Wilderness Society and other environmental groups have harassed drilling companies with lawsuits and regulatory opposition. President Clinton designated nineteen new National Monuments on 5.6 million acres in 2000, mainly to fend off oil and gas development. Drillers were particularly bitter about losing National Forest land around Glacier National Park, since Canadians have extracted huge amounts of gas and oil from Waterton National Park just across the border.

In January 2005, I shared a breakfast with a small group of independent oil and gas prospectors in Billings, Montana. All ran wildcat operations, doing their own geological work and drilling on state as well as federal land. All were extremely bitter at the restrictions put on their explorations in the Overthrust region.

"Here's the permit the state has me fill out," said Mac Clark, a transplanted Texan, showing me a small sheaf of about eight pages. "Now here's the EPA form." It was more than an inch thick. "They make you jump through every kind of hoop imaginable. You've got to do a geological survey, an archaeological survey, a botany survey, a wildlife survey, an endangered species study—and then when you're finished some Indian tribe will come in and say you're on sacred land."

The Wilderness Society, on the other hand, says these complaints are all a fraud. "The industry is always telling you about these horizontal drilling techniques where they don't even have to tread on the property," said Bob Ekey, Northern Rockies regional director. "But when it comes to using these new technologies, somehow they are never available."

Despite these disputes, wildcatters are still finding gas and oil. After my breakfast, I met Dick Findley, who was named "Explorer of the Year" by the American Association of Petroleum Geologists in 2005 after finding a 100 million barrel oil field previously overlooked in eastern Montana. "There's still oil and gas out there," he said. "You just have to look a little more carefully."

Beyond the Overthrust Belt lies Alaska, where huge quantities of natural gas are "stranded" at Prudhoe Bay. Nearly 30 trillion cubic feet of gas have been pumped back into the ground awaiting a pipeline to the lower 48 states. The Sierra Club and the oil industry have both enthusiastically embraced the pipeline but differ about where to put it. The gas industry wants to go through Canada but is having trouble with Native Canadian tribes and Canadian environmentalists. American environmentalists favor a pipeline next to the existing oil pipeline that would then follow the ALCAN Highway to the States. The industry says if it brings gas to southern Alaska it might as well sell to Japan, since Chicago and the East Coast will still be inaccessible. Either way, a pipeline will take ten years to build.

The Arctic National Wildlife Refuge (ANWR), which would also require a new pipeline, represents perhaps the last frontier in our conflict over natural resources. The oil industry argues it could add another 10 trillion cubic feet of gas to the nation's inventories. Environmentalists call the region "the nation's crown jewels." "The Bush Administration says drilling will help end our dependence on foreign oil," says the Natural Resources Defense Council. "In reality there is less than a year's supply of oil in the refuge and it would take ten years to access it."[8]

In truth, not even the upbeat National Petroleum Council is very confident that American and Canadian natural gas supplies can meet the demand looming ahead. In 2003 it reported:

> North America is moving to a period in its history in which it will no longer be self-reliant in meeting its growing natural gas needs; production from traditional U.S. and Canadian basins has plateaued.[9]

While North America is bumping up against limits, the rest of the world is discovering it is awash in natural gas. Total reserves are now estimated at 10 *quadrillion* cubic feet, enough to supply the entire planet for 200 years, at current rates of consumption. Once again, the Middle East is the epicenter, with 2.5 quadrillion cubic feet of reserves, while Russia is second with 2 quadrillion. Potential fields in Africa and East Asia are just being explored. Europe has already built two pipelines under the Mediterranean and is importing huge quantities from Russia—90 percent of consumption in some countries.

All this has led to more concerns about foreign dependence. When Russia cut off gas supplies to the Ukraine in 2006, demanding a five-fold price increase, visions of another Arab Oil Boycott sent shudders across the European continent. Although the situation was eventually resolved, a much more significant development has been Russia's announcement that it will build a $10 billion gas pipeline to China—locking up gas that Europe and Japan both covet. The U.S. has virtually no chance of accessing any of this gas.

And so, just as with oil, Americans find themselves falling further and further behind with access to natural resources. The only way we can join this world market is to build liquid natural gas (LNG) terminals that freeze the gas for ocean transport. The American Petroleum Institute is enthusiastic:

> LNG has a long and outstanding safety record with delivery across the oceans for 45 years, including more than 40,000 LNG cargo deliveries and more than 60 million miles traveled.... [A]ll of the gas consumed in Japan comes from LNG imports.
>
> LNG tankers are not likely to explode.... Over the last few decades, there have been only eight accidents with LNG tankers in which the cargo was spilled, and in no case was there explosion or fire. In almost all cases, when LNG is released, it regasifies and disperses, rising harmlessly into the atmosphere.[10]

Others are not so optimistic. Each tanker does pack the explosive power of a small hydrogen bomb and is far more dangerous than any nuclear reactor. Even without LNG technology, natural gas causes about half the explosive accidents in the country each year and about 1,000 people die. Ralph Nader's Public Citizen—a public interest group—also reminds us that some LNG tankers come from Algeria, where Al Qaeda members could sneak aboard.[11]

There was a flurry of LNG construction in the 1980s, giving the U.S. four terminals in Massachusetts, Maryland, Georgia, and Alabama. Three closed for lack of business but have since reopened. There are now proposals for a dozen new terminals along the East, West, and Gulf Coasts. All are mired in public opposition. Noting that the expansion of natural gas use has been spearheaded by environmental groups, *Washington Post* columnist Stephen Pearlstein wrote:

> Go to the Web site of the Los Angeles chapter of the Sierra Club...and you'll learn why any of three proposed terminals would be frighteningly dangerous and costly to electricity customers while making the nation even more dependent on foreign fuel.
>
> In Louisiana and Mississippi, the Sierra Club chapter warns of tens of thousands of innocents who would be burned to a crisp if there were ever an explosion and fire at any of the LNG terminals proposed for the Gulf Coast.
>
> On the East Coast, the Sierra Club's Delaware chapter has come out against BP's plan to build an LNG terminal in the Delaware River. In its April newsletter, the New York chapter lists eight environmental catastrophes that would befall the region if Shell were allowed to build a floating terminal 25 miles out in the Long Island Sound. And in Boston, the Sierra Club is leading the charge against a proposal by AES to build an LNG terminal on a small, unused and largely unusable island at the mouth of Boston Harbor.... It would appear that for the Sierra Club, LNG has become the energy source to be supported in principle but rarely in practice."[12]

Yet there remains a serious question of whether America will even be able to *afford* natural gas on the world market. Liquefying and un-liquefying natural gas for ocean transport doubles its price. For China and Europe all this is unnecessary. During recent winters half our LNG capacity sat idle because American companies couldn't outbid Europe and China for Russian and Middle Eastern supplies. By 2008 even the seeming abundance of world supplies was running up against increases in demand.[13]

Meanwhile, the continuing run-up in gas prices has devastated America's chemical and manufacturing industries. In 2003, Andrew Liveris, CEO of Dow Chemical, told Congress his company was moving its "center of gravity" to Europe and the Middle East where gas is more plentiful. "We need to declare a national crisis," he testified. "Dozens of plants around the country have closed their doors and gone away. They're never coming back." The fertilizer industry has been gutted, with more than half the manufacturing plants leaving for Mexico and Saudi Arabia. The American Chemistry Council estimates 100,000 job losses due to increasing gas prices. Dow alone has shut down twenty-three plants. Goodyear cut U.S. tire production 30 percent. After Hurricane Katrina, rescue workers in New Orleans had difficulty getting water to the victims, not because of a water shortage but a shortage of *plastic water bottles*.[14]

Utilities generate electricity to meet three kinds of demand—"base load," "mid-load," and "peak demand." Base load is the underlying demand for electricity that remains constant night and day. Mid-load is revved up to meet daytime levels, when consumption generally doubles. Finally, peaking power must be available to meet the highest periods of demand—usually hot summer days when air conditioning stretches the grid to its limits.

Base load is traditionally provided by hydroelectric dams (when they are available), coal and nuclear power. Mid-load is provided by coal or oil because they can be revved up to full speed with about an hour's notice. Since the 1960s, peaking power has been provided by gas turbines, which are essentially jet engines bolted to the ground. They are expensive but they are not producing steam and can be started and modulated almost instantaneously in order to follow peak demands.

During the 1990s, manufacturers developed a new type of

generator—the "combined-cycle" gas turbine. Combined-cycle uses two turbines, one driven by steam from the boiler, the other using the carbon dioxide exhaust from the natural gas. This can achieve a 60 percent conversion, as opposed to the usual 35–40 percent from traditional steam boilers. The greater efficiency made gas economical for both base and mid-load power—for a while. When prices took off after 2000, however, combined-cycle plants became uneconomical and many were idled. New "merchant" companies such as Calpine, which had made a business selling gas-produced electricity to the California grid, went bankrupt.

Natural gas grew from 13 percent of our generating capacity in 1990 to 39 percent today. Yet it still *produces* only 19 percent of our electricity. Nuclear, meanwhile, makes up 9 percent of our capacity but generates *20 percent* of our electric power. This is because nuclear reactors run all the time while gas plants have become ever more uneconomical.

Even if world gas supplies prove abundant, America is going to have a tough time accessing them—especially if we continue to lock up our own resources and oppose LNG terminals. This price crunch has ended the gas-electric boom and most new generating plants are now reverting to coal.

Is there any way out of this dilemma? The proposal of Larry Kazmerski at the National Renewable Energy Laboratory still looks appealing. Solar energy should substituted for gas in peaking power, and nuclear should replace coal for base and mid-load power. A carbon tax would make all this economical. The best way to break the status quo would be a nuclear-solar alliance. But that would require the leading opponents of nuclear power to give up their long-held illusion that the world can be run entirely on alternate and renewable energy. To that we will turn next.

Part Three
Solar and Renewables

Interlude

Snowmass, Colorado

A few days after my encounter with Thomas Friedman at the National Renewable Energy Laboratory, I am headed up through the snow-covered peaks of the Rockies to make my own pilgrimage to Snowmass, home of the Rocky Mountain Institute and Amory Lovins, perhaps the world's foremost opponent of nuclear power. It is a three hour trip from Denver and the roads are treacherous. Yesterday the state highway department set off an avalanche in a well-traveled pass where the overhang had become dangerous. Cars were backed up for three hours. Fortunately, I missed that one.

Lovins launched his career in 1976, only weeks before Jimmy Carter won the election, when the prestigious journal, *Foreign Affairs*, published an article by the then unknown 28-year-old Oxford don who was also the British representative for Friends of the Earth. The article eventually became the most highly requested reprint in *Foreign Affairs'* long, illustrious history—surpassing the previous record holder, John F. Kennan's "Mr. X" article of 1947, which proposed "containment" of the Soviet Union and established the ground rules of the Cold War.

In "Energy Strategy: The Road Not Taken," Lovins put forth the startling proposition that, even though we were facing an unprecedented Energy Crisis in which concern about coal pollution was limiting its use and the other fossil fuels—oil and natural gas—seemed to be running out, *nuclear power was not necessary*. In fact, it was an actual hindrance to our goal—energy independence. Centralized power plants were themselves inefficient because most of the steam went to waste. Instead, he argued, we should pursue a lower-technology, small-is-beautiful approach. (Eric Shumacher, also of Friends of the Earth, who had just published *Small Is Beautiful*, was his mentor.) This "soft path" would mean:

1) Much greater, even heroic, efforts at energy conservation
2) "Distributed power," where industries would once again generate their own electricity so they could use the steam for industrial purposes—called "cogeneration"
3) Movement toward our ultimate goal, a "solar economy," where gathering and storing energy flows from the sun would eventually power the entire world. This could happen by around 2050

At the time, Lovins' predictions seemed almost unimaginable. S. David Freeman had caused a huge controversy only three years before when he suggested that, with a heroic effort, we could limit our energy consumption to 85 quads instead of 100 by 1985. Now Lovins was saying we wouldn't reach 100 quads until 2000—when most projections said it would be 150 quads! Yet Lovins turned out to be exactly right. We barely nudged up to 100 quads in 2002.

Lovins' subsequent 1977 book, *Soft Energy Paths*, flashed like a bolt of lightening across the landscape. To that point, environmentalists had only been *against* things—hydroelectric dams, air pollution from coal, pumped storage facilities, and—when it seemed to be the only thing left—nuclear power as well. Now they could be *for* something—solar energy. The terms of the energy debate have not changed much since.

Lovins' work had enormous influence. When he visited Jimmy Carter in the Oval Office in 1977, he found a copy of *Soft Energy Paths* on the President's desk. Carter embraced the solar future, created the Solar Energy Research Institute (now NREL), and pushed subsidies for windmills, photovoltaics, and other forms of "renewable" energy. The entire biofuels effort can probably be traced to a single paragraph in *Soft Energy Paths,* where Lovins compared growing crops-for-fuel to the output of beer and wine industries. Jerry Brown embraced Lovins' vision and hired him as a consultant in putting California on the soft path in 1980—a route that led directly to the California Electrical Shortage of 2000.

Today it is hard to pick up an article on energy that doesn't mention Lovins' name. His curriculum vitae reads as follows:

> Lovins…has advised the energy and other industries for over 30 years, as well as the U.S. Departments of Energy and Defense. Published in 28 previous books and hundreds of papers, his work in ~50 countries has been recognized by the "Alternative Novel," Onassis, Nissan, Shingo, and Mitchell Prizes, a MacArthur Fellowship, the Happold Medal, nine honorary doctorates, and the Heinz, Lindbergh, Hero for the Planet, and World Technology Awards. He advises industries and governments worldwide, including major oil companies, and has briefed 18 heads of state. Since 1990, he has led the development of quintupled-efficiency, uncompromised, competitive automobiles and a profitable hydrogen transition strategy [i.e., his HypercarTM]…. Automobile magazine has called him the 22nd most powerful person in the global car industry; the *Wall Street Journal*, one of 39 people in the world most likely to change the course of business in the 1990s; *Newsweek*, "one of the Western world's most influential energy thinkers."[1]

In 2001, I wrote a cover story in *The Weekly Standard* called, "The Myth of Alternate Energy," calling into question Lovins' thesis, that solar energy can power the entire industrial economy. In particular, I was concerned with how much *land* solar energy technologies will occupy. It is my sense that, like any good American, Lovins has assumed that because our land resources are basically unlimited, we can "distribute" solar technologies everywhere without stressing land resources. Solar energy, after all, is extremely dilute. Fossil fuels are two to fifty times more concentrated than the various forms of solar energy and terrestrial energy is 2 million times more concentrated than fossil fuels. It's going to take a lot of land to collect and store all that solar energy in usable form. We are already covering the tops of mountains with giant windmills and a lot of people find that distressing.

It was the cover cartoon, however, by the gifted *Weekly Standard* illustrator Jason Seiler, that did me in. It showed Lovins atop a bicycle desperately trying to keep several windmills and solar collectors operating while pedaling out energy. I thought it was pretty funny, but needless to say, Lovins was not amused. He sent an eight-page, single-spaced letter critiquing my entire article, demanding all sorts of retractions.

Now, I've called ahead a couple of times to try to schedule an appointment, but without much luck. Lovins is off traveling somewhere in Europe and Japan. He may be back by the end of the week. The staffers on the phone are very gracious and invite me to come visit anyway. They probably don't recognize my name. In any case, it is doubtful Lovins will want to meet with me. If I had a camera, I could do one of those Michael Moore-type interviews where the surprised subject shouts, "Get that thing out of my face," but that wouldn't accomplish much.

What I'm really hoping is that Lovins will make some small gesture toward admitting nuclear can play a part in resolving global warming. I will raise my hand Indian-style and say, "I come in peace." Kaz Kazmerski at the National Renewable

Energy Laboratory has convinced me that the country is ready for a nuclear-solar alliance. Bernard Cohen proposed it a decade ago. Lovins has been the nation's leading opponent of nuclear power for twenty-five years. Isn't there room for compromise? Environmentalists have been preaching for decades that we have to change our lifestyles to deal with global warming. Isn't it just as possible they may also have to change *their* lifestyle and admit that nuclear is not the devil's work?

I am not optimistic.

The Rocky Mountain Institute is not as easy to find as I expected. After reading so much about it, I thought everyone in Snowmass would know where it is. Instead, I only get blank stares at the gas station on Route 82. Right next to it is Snowmass Creek Road, though, and somehow the name seems promising. I decide to give it a try. Five miles up the mountain I hit a lonely crossroads with a sign pointing to a monastery. I flag down a car and a stylish young woman with sunglasses tells me I should try the ski resort—that's down the hill again and five miles up Route 82, toward Aspen.

At the ski village no one has ever heard of Amory Lovins or RMI. With almost two feet of powder on the slopes, energy is the furthest thing from anyone's mind. Finally a bus driver tells me, yes, he knows where it is. It's right up on Snowmass Creek Road, at the first intersection. That's where I was half an hour ago! I didn't see anything. No, you can't miss it, he says. It's right there.

So it's back up the road again to the sign pointing to the monastery. I try both forks, left and right—still nothing in sight. In despair, I'm about to start down the mountain again when suddenly there it is, right in front of me, at the top of an uphill driveway, indistinguishable from all the other dream houses that dot the mountainside, except for the solar panels on the roof. It is here that Lovins claims to live comfortable without the benefit of centrally generated electricity.

The nation's most energy-efficient home is a U-shaped, one-story fieldstone structure with a large bay greenhouse window in the middle. The hand-carved wooden door bears a notice, "Please make sure both doors are shut." Inside the vestibule hangs a photo of Lovins sitting at a picnic table with an orangutan.

A sign explains that the staff is hard at work but visitors are welcome to take a self-guided tour. The inside is as quiet as the monastery up the street. A dim narrow hallway is lined with copies of Lovins' twenty-eight books. I start browsing when the overhead light suddenly goes off, leaving me in darkness. I realize it is motion-activated. I have to wave my arms every minute or so to turn it back on. In a recent special issue of *Scientific American* Lovins declared that, except for the Institute's office equipment, his house consumes "barely more electricity than a single 100-watt light bulb." I have the feeling I am standing under that one bulb.

The central room is about 30 degrees warmer than outside but has the fetid odor of a greenhouse. Two young interns sit at a long wooden table, their faces illuminated by laptops. They take no notice of my presence. "I'm working on oil," one finally says, apropos of nothing. He scrolls through long tables of numbers. I continue my self-guided tour into the tropical garden, where RMI has grown eight crops of bananas. On the far side of the garden is a more conventional office with desktop computers and copy machines. A white-haired older man shuffles silently among them. Lovins acknowledges these do not run on solar electricity.

The sign at one of the self-guiding stations says a hedgehog patrols the garden for insects. I ask an intern if it's still around. "He's spending the winter at somebody else's house," he responds. "He's hibernating." Tomorrow's guided tour has been cancelled. The guide is taking avalanche training. "Amory's in California and Washington this week," says the intern. "He may be back late Saturday but he leaves right away for Florida. He's going to a charrette."

What's a charrette?

I hang around Aspen for two more days, hoping to catch Lovins on the weekend. When I drive back late Saturday, however, there is only a teenage handyman shoveling snow away from the greenhouse window. He says everybody is gone until Monday. I browse the outdoor displays again and am about to leave when on impulse I ask, "Isn't the John Denver ranch around here somewhere?"

"Yeah, it's just up the road," he says. "It's part of the Institute. You'll see a sign."

Again, it takes some hunting. I miss the sign at first and drive five miles up the road before finally coming back and finding it tucked behind two hills at the end of a long dirt road. "Windstar" was the ranch from which Denver undertook his own efforts to save the planet before fatally crashing his homemade airplane in 1997.

The main building has the obligatory greenhouse along the front but there are no solar panels. The doors are all unlocked, so I step inside. I call out but no one seems to be around. There are several large conference rooms with plenty of three-pronged outlets in the walls. Utility wires crisscross the property. Lovins doesn't talk much about this place.

By the entrance is a wall-sized photographic collage of Denver with every celebrity in the world. On the lawn stands a life-size stature of him with an eagle landing on his arm. It was here (or maybe at his other ranch up the mountain?) that Denver famously installed two 500-gallon gas tanks on his property during the 1979 gas shortage—his contribution to solving the energy crisis.

The handyman and his mother arrive and start cleaning the conference rooms. I chat with them for a few minutes over the roar of the vacuum. Then I return to my fuel-sipping rented car and head out. It is amazing a place like this can have so much influence in the world. But the sun is shining brilliantly and there is a clear day ahead. All the way back to the highway, the electric utility poles follow me down the dirt road.

Chapter 9

Energy Conservation

In 2005, the Natural Resources Defense Council, one of the country's leading environmental groups, issued a 25-page report, "A Responsible Energy Plan for America," outlining how we could reduce carbon emissions by half while still continuing to grow as an economy. The report began:

> The cornerstone of NRDC's plan to secure America's energy future is increased energy efficiency. Not only is energy efficiency free of environmental impacts, but it is also by far the cheapest way of meeting our energy needs. The efficiency improvements we recommend do not rely on pie-in-the-sky, undeveloped technologies, but on readily available and cost-effective processes that allow us to gain more productivity out of less energy.[1]

An accompanying graph showed improved energy efficiency lowering carbon emissions 10 percent below projected levels by 2010 and 33 percent below projected levels by 2050. The addition of renewables plus carbon sequestration would cut emissions to one-half their current level. The report devoted only one paragraph to nuclear, saying it should receive no more government subsidies and is "saddled with proliferation risks and a costly, long-lived hazardous waste burden."

In *Natural Capitalism* (2000), written by Amory Lovins with his former wife Hunter, and Paul Hawken, the authors talk about the Factor Ten Club, which has declared that "within one generation, nations can achieve a ten-fold increase in the efficiency with which they use energy, natural resources and other materials."

> Austria, Sweden, and OECD environment ministers have urged the adoption of Factor Ten goals, as have the World Business Council for Sustainable Development and the United Nations Environment Program (UNEP).... Among all major industrial nations, the United States probably has the least familiarity with and understanding of these ideas.[2]

Arjun Makhijani, director of the Institute for Energy and Environmental Research and author of *The Nuclear Power Deception*, repeated this claim in 2004:

> Two-thirds of U.S. energy use per unit of economic output could be eliminated using available technology, while still maintaining all the functions present-day fuel use performs. With a sensible program of energy research and public policy, it is quite possible to achieve energy use per unit of economic output at *one-tenth* present levels within a few decades.[3]

Their claims that greater effieiency can achieve energy savings of upwards of 30 to 50 percent are often backed up by individual accomplishments. In 2001, for example, Texas Instruments was planning to move its newest chip fabrication plant to China, since energy costs in the United States had become too high. Then Paul Westbrook, sustainable design manager at TI, heard Lovins speak at an event called Sustainable Dallas.

Westbrook buttonholed him and offered him the opportunity to redesign the chip manufacturing plant. Several brainstorming sessions followed, in which TI engineers and Lovins' Rocky Mountain Institute redesigned the system for chilling water, cleaning air, scrubbing exhausts, and recycling heat at the new building in Leeds, Texas. Lovins had been developing these techniques—greater insulation, fat pipes, natural convection currents—for years.

"In the end, we came up with a building that saved 20 percent on energy costs and was cheaper to build as well," said Westbrook. "And we decided to build it in this country instead of abroad."

It's a great success for American enterprise, one that has occurred frequently over the past three decades and hopefully will be repeated even more in the future. But notice what's happened here. Did the new Texas Instruments fabricating plant *reduce* overall energy consumption in Leeds, Texas? No, it increased demand by a sizable chunk. The 20 percent savings came from *what was originally projected*. Thus, Lovins and the innovative designers at TI only "saved" what might have been a bigger increase in overall energy consumption.

This pattern is repeated over and over again when people talk of energy conservation. A sampling goes like this:

> "Energy efficiency has saved 70 quadrillion BTUs from 1972 to 1999." —Association of Energy Engineers

> "If not for energy efficiency and conservation measures taken since 1973, we would be using 40 percent more energy each year." —Alliance to Save Energy

> The 1 billion compact fluorescent lamps that have been put in use from 1988 to 1997 have saved the energy of 100 coal-fired electrical plants." —Rocky Mountain Institute

> "Our transportation technologies are projected to reduce oil imports by 2.3 million barrels a day by the year 2000, creating

a trade deficit reduction of $47 billion per year." —The Budget of the United States Government, Fiscal Year 1996

"With today's technology, we can achieve an average fuel economy of 40 miles per gallon; that would save us over 3 million barrels per day, more than we import from the Persian Gulf." —American Council for an Energy Efficient Economy

"The [2007 Energy Bill's] required boost in Corporate Average Fuel Economy standards—to 35 miles per gallon by 2020—will reduce national oil consumption by 1.2 million barrels per day, saving Mainers $140 million in 2020." —Environment Maine

After twenty years of this, it's often a wonder that we're still consuming any energy at all. In fact, our per capita consumption peaked in 1978–1979 and has declined 5½ percent since. Energy expended per dollar of GDP has done even better, declining steadily for half a century. It now stands at less than half its 1950 figure.

Yet for some reason, year after year, we still keep consuming more energy. Granted, it doesn't go up as fast as it once did or as fast as it might have—but it goes up nevertheless. Our consumption has risen steadily at about 10 quads per decade since the 1970s.

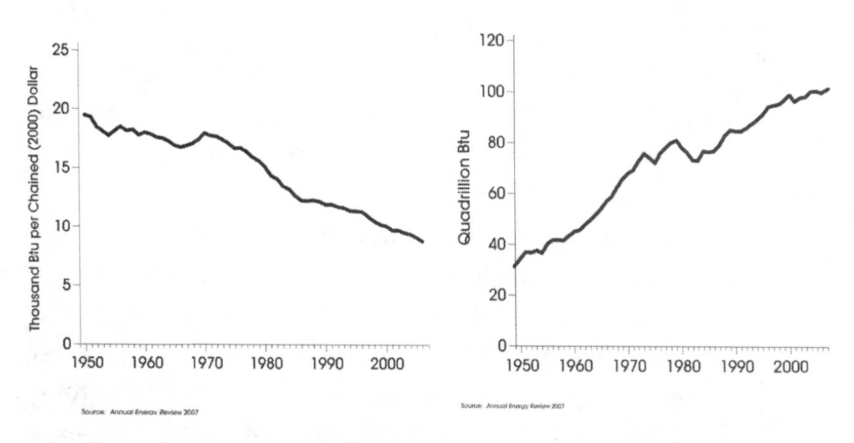

Energy Consumption per Real Dollar of Gross Domestic Product, 1949–2007

U.S. Overall Energy Consumption, 1949–2007

Only during three brief intervals since 1970 did our national energy consumption actually *decline* rather than increase. These were:

1) The two years immediately after the Arab Oil Embargo, when the economy went into a recession
2) From 1980 to 1983, during the worst downturn since the Great Depression
3) In 2001, right after September 11th, when the economy again went into a brief tailspin

In other words, when the economy turns bad, energy consumption goes down. At all other times, it keeps ascending. Even in a major downturn, the decrease is not a one-to-one ratio. During the depth of the Great Depression economic activity had fallen 33 percent but energy consumption only declined 15 percent.

The problem with claims about energy conservation is that they inevitably compare to *projected future* consumption. These savings also tend to be somewhat illusory, since the money saved is often put right back into consuming more energy. Herbert Inhaber likens it to the Dagwood cartoon where Blondie comes home and says, "Look, Dagwood, I bought this beautiful dress on sale and saved $25," and then adds, "Let's go out and have dinner with the $25 I just saved."

Why don't improvements in energy efficiency seem to lead to overall reductions in energy consumption? One of the reasons is Jevons' Paradox.

Stanley Jevons was a nineteenth century British mathematician often called the "father of quantitative economics." A child prodigy, he went to the University College of London after being rejected by Oxford and Cambridge for his Unitarian faith. Early on, Jevons became fascinated with the idea of solving everyday problems through mathematics. He refined Boolean algebra, the basis of computer logic, and in 1870 presented the Royal Society with a "Logic Piano," an early computer that did calculations at superhuman speed using keys, pulleys and switches. He also invented the concept of "marginal utility," a cornerstone of modern economic theory.

In 1865, Jevons published, *The Coal Question*, warning England that its coal resources would soon be exhausted. As with most such alarms, his prediction turned out to be quite premature. In the process, however, he introduced "Jevons Paradox," which says that *greater efficiency* in utilizing energy usually leads to *greater consumption*.

Jevons used Newcomen's steam engine as an example. The original engine, he noted, was extremely inefficient, converting only 0.75 percent of coal's potential energy into useful work. (Today's coal plants convert about 30 percent.) When James Watt's improvements raised thermal efficiency to 4 percent, the immediate impact was to reduce coal consumption by about one-third.

But that quickly reversed. Between 1830 and 1863, English coal consumption grew tenfold. What had happened? Watt's improvements in efficiency made coal useful for other tasks besides pumping water out of mines. It was soon powering factories, railroads, and steamships. After Jevons' writings, the trend only continued to grow. When electricity improved the application of energy to specific tasks, coal consumption soared again. The more efficient electrical appliances became, the more coal we consumed. After a century of improvements in efficiency, the world now burns a thousand times more coal than it did in the nineteenth century.

Now you may object, "These inventions weren't *saving* energy, they were finding ways to *use more* energy." But that's the whole point. There isn't any real way to differentiate. Doing the same amount of work with less energy is the same as doing more work with the same amount of energy. There is no prescribed limit to human demands. If it becomes easier to produce things by using less energy, then people will want more of it. As one commentator puts it, "If cars become more fuel-efficient, people will drive them over to the next county to save money at Walmart." Only in a completely static world would every improvement in energy efficiency lead to an actual reduction in consumption.

One of the most successful conservation measures to come out of President Carter's administration was the "Energy Star" program for improving efficiency in household appliances. Today's refrigerators, for example, are more than twice as efficient as those of 1980. Yet most homes now have two or more refrigerators. In 1980 only 27 percent of all homes had central air conditioning, today the figure is 59 percent. Microwave ovens were in 14 percent of homes in 1980, 88 percent today. Personal computers were unknown at the end of the 1970s, today 68 percent of homes have them.[4]

Energy consumption per square foot in American households has decreased significantly over the last quarter century. Yet the size of the average home has increased 50 percent.[5] That all our home appliances consume *less* energy than the model of 25 years ago is a remarkable achievement. But it doesn't reduce energy use. Overall consumption keeps steadily increasing.

Economists call this the "rebound effect." In 2001, the Congressional Research Service published a study showing the rebound can be as high as 50 percent for heating and air conditioning improvements, and 30 percent for improvements in gas mileage. "The rebound effect can increase the difficulty of projecting the reduction in greenhouse emissions from an improvement in energy efficiency," noted author Frank Gottron.[6]

Even if people reach the point of satisfaction in one area of energy consumption, they may spend their savings in another. Tom Wolfe loves to tell the story of how, if you have trouble reaching your plumber these days, it's probably because he's on a Caribbean cruise with gold chains hanging around his neck, dancing with his third wife. In the old days, plumbing contractors went fishing in a local creek. Now they take charter boats in Florida. Even the Internet, which was supposed to bring the world to your doorstep, has produced its rebound.

"The Internet has the funny effect of increasing the amount of travel," notes Vint Cerf, one of the original inventors of the Internet. "People use it to discover places to go and people they want to meet."[7]

The rebound has effectively eaten up most of the improvements in auto mileage. When the Corporate Average Fuel Efficiency (CAFE) standards were imposed in 1974, the U.S. auto fleet averaged 15.8 miles per gallon. Today it has climbed to 28 mpg. Yet our oil consumption has increased 33 percent, from 15 million to 20 million barrels per day. The main reason is that people travel more. Since 1980 the number of annual miles driven by the average motorist has climbed from 9,500 to 12,000. Total passenger mileage went from 1 trillion in 1970 to 2.3 trillion in 2004. Improvements in engine efficiency may improve gas mileage, but this also makes it easier for people to increase their driving.

Another rebound has occurred in the tendency for people to buy bigger cars. While improving efficiency has been achieved by making cars smaller and lighter, many people feel safer and more comfortable in SUVs. While SUVs barely existed before the fuel standards were imposed in 1979, they now constitute half the auto market. As a result, even though fuel efficiency improved in the regulated half of the market, overall mileage for the entire fleet has hardly improved at all. The 2007 Energy Act finally closed the SUV loophole, but even then the result might not be what is expected. If SUVs finally became more fuel-efficient, people may be even more eager to trade up to them, further shrinking the market for smaller, more fuel-efficient cars.

Even the presumed fuel savings from the new gas-electric hybrids have proved disappointing. Within two years, most of the new efficiencies had been redirected toward improved *performance* rather than lowering gas mileage. In an article titled "Pump My Prius," The *New York Times Magazine*'s "Style" section reported:

> This is the logic behind Detroit's new hybrid muscle: take a

> relative midpower gasoline engine, add electric power on top, and produce the illusion of V-8 strength. After all, if you're no longer worried about maximizing mileage, those electric motors can offer drag-race-style acceleration, giving a regular gas engine the appearance of far more torque. Last fall, Toyota engineers decided to prove this point by souping up a humble Prius to make it drive 130 mph in a dry lake bed in California.[8]

Thanks to improved technology, the average fuel-sipping American car now accelerates from zero to 60 in 9.9 seconds, faster than the "muscle cars" of the 1970s.[9] Efficiency can be used to build muscle as much as mileage.

Similar frustrations may come from the 2007 Energy Act's mandate that the nation switch to compact fluorescent light bulbs (CFLs) by 2012. A long a favorite of conservation enthusiasts, CFLs are claimed to be 80 percent more efficient than Edison's incandescent bulb, which will essentially become illegal to sell after the deadline. The mandate passed without much discussion, although it may set off a rebellion when the public finally becomes aware of the implications. CFLs can't be dimmed and give off a light that some people find unappealing. And they must be *kept on* for long periods of time, since switching them on and off shortens their life span considerably. Finally, the question remains, will CFLs save energy? In 1987, the municipal utility of Traer, Iowa held a Great Light Bulb Exchange, distributing 18,000 CFLs to customers. The outcome? Electricity consumption rose 8 percent. People kept them on longer because they didn't use as much energy.[10]

Even if conservation *does* stabilize the amount of energy we use, consumption will still grow because of two factors—population increase and growing affluence. The population of the United States grew 25 percent from 1980 to the present, from 235 million to 300 million. Over the same period, GDP per capita doubled from $21,000 to $42,000. All this translates into greater energy consumption. Even with the decline of consumption per capita and per unit of GNP, overall consumption will continue to increase.

The Alliance to Save Energy is a blue chip public interest group, founded by former Illinois Republican Senator Charles Percy and headed by Arkansas Democratic Senator Mark Pryor. Its board features Senators and Congressmen from both parties, plus an all-star cast from industry and environmental groups, including John Adams, president of the Natural Resources Defense Council, Stephen Brobeck, director of the Consumer Federation of America, Thomas Kuhn, president of the Edison Electric Institute, Ross Pillari, president of BP America, and S. David Freeman, one-time author of *A Time to Choose* and now president of the Hydrogen Car Company.

In 2005, ASE published "Vision 2010," a comprehensive plan to step up energy conservation on all fronts—transportation, buildings, industry, and electric utilities:

> Energy efficiency now contributes more than any single energy resource to meeting the country's needs, and is the quickest, cheapest, and cleanest way to meet the anticipated growth in energy demand in the United States. Energy efficiency and conservation since 1973 reduced America's energy consumption by 40 Quads.[11]

The plan envisions a broad raft of initiatives—instituting "feebates" that transfer money from people who buy big cars to people who buy small cars, increased funding for the EPA's Energy Star program, promoting industrial cogeneration (small power plants that utilize both electricity and steam), requiring facilities that produce more than 100,000 tons of carbon dioxide to report their emissions to the government, and instituting a five-state program to help electrical and gas utilities reduce consumption by 0.75 percent per year. "The Alliance estimates that the legislative provisions highlighted could reduce energy use by approximately 2.4 quads annually by 2010, adding up to 78 quads through 2020," concludes the report.

Yet even if *all* these measures take effect, the pace of increased consumption would only be *slowed* over the next 20 years. The Energy Information Administration found that in "Vision 2010's" *best*-case scenario, consumption would rise to 124 quads by 2025. The "moderate" path would result in 129 quads and "business as usual" in 133.[12] Most remarkable, because of declining American oil production, "Vision 2010's" best-case reduces imports only 1.5 percent in 2025, from 68.4 or our consumption to 67.3 percent.[13]

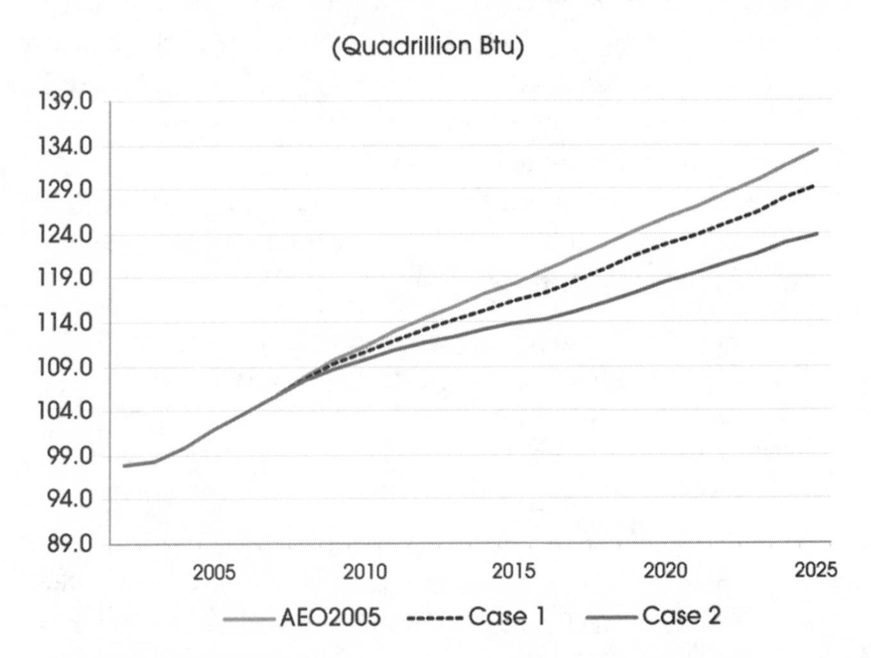

Source: Alliance to Save Energy

Three Projected Consumption Paths in "Vision 2010"

This is not the sort of thing environmental groups have in mind when they tell us we will be able to cut energy consumption or carbon emissions in half simply by looking for new ways to save energy. Energy conservation is worth pursuing on all fronts. It improves our lives and helps us meet the demands of growing consumption. But it is not going to give us a world where we no longer have to worry about producing more energy.

Chapter 10

Hydro

Hydropower is one of the many ways we can gather and store solar energy. The sun evaporates water and then returns it to the earth as rain. The water then flows downhill again to the oceans. While it is on its way, however, we can tap its kinetic energy or store large amounts behind dams. Our ability to draw on this energy and store it, however, is not unlimited and not always renewable. Nor is it necessarily enough to meet the needs of an industrial society.

The gristmills that sprung up during the Middle Ages operated on the modest flow of water in the "old mill stream." As power requirements increased, however, engineers began damming rivers and creating artificial waterfalls to power larger industrial complexes. The output of a waterfall is governed by the law of physics, which says that kinetic energy increases with the height of the drop. This creates an exponential advantage in building higher dams at natural waterfalls or in mountainous regions.

As with all forms of stored or direct solar energy, land requirements soon become huge. Hoover Dam, which produces 2,080 megawatts, backs up to Lake Mead, which is 247-square-miles. Lake Powell, behind Glen Canyon Dam, was originally also 250 square miles, producing 1,300 MW. But drought and silting have shrunk it to 131 square miles, reducing Glen Canyon's output to only 514 MW. Such facilities have worked in isolated desert areas largely because they add recreational value in the form of boating and fishing. But they can become a problem if the region has scenic or historic value, or if people have to be moved to make way to create the reservoir.

When electricity first made it possible to tap energy from remote places, Niagara Falls was one of the first sites to be developed. Dam construction combining flood control, recreation, and hydroelectricity became the proudest accomplishment of the conservation movement under Theodore Roosevelt and Gifford Pinchot in the early twentieth century. By 1935, nearly one-third of our electricity came from hydropower.

Unfortunately, nearly all the good dam sites have long since been employed and there is little room for expansion. The portion of our electricity provided hydropower has now dropped below 10 percent and is unlikely to rise again. Nonetheless, these dams constitute more than *90 percent* of what we call renewable energy. Remarkably, hydroelectric dams have also been long opposed by nature-oriented organizations such as the Sierra Club, which insist they favor renewable energy. This effort has now expanded into a campaign to *tear down* existing dams to restore free-flowing rivers.

The odd result is that environmental groups that propose solving energy shortages and global warming through conservation and solar energy are also often the biggest advocates of reducing this prime source of energy.

Water wheels were invented in ancient times. There is evidence that both the Egyptians and Mesopotamians experimented with the technology. Vitruvius, a Roman engineer of the Augustan Age (31 BC to 14 AD), mentions an undershot wheel but says it is a "machine rarely employed." One reason the Romans may not have pursued the technology is because they had slave labor.[1]

All this changed as the empire declined. By the fourth Century AD, the Romans had built a remarkable factory complex at Barbegal, in southern France, where sixteen overshot wheels turned millstones that ground grain for a village of 12,000 people.[2] In 537 AD, the Goths, who were besieging Rome, cut off the city's aqueducts, bringing the gristmills to a halt. On an inspiration, Belisarius, the Byzantine general defending the city, had barges with gristmills anchored beneath the Tiber bridges and suspended water wheels between them. This wartime improvisation became the "ship mill," which was soon a standard feature in cities around the Empire.

China also invented the waterwheel and by the first century AD was using it to blow bellows in iron smelting shops. In India the waterpower was employed only to run prayer wheels—a device where prayers were spun continuously, relieving Brahmins of their responsibility for constant prayer. (Eric Hoffer, the American longshoreman-philosopher of the 1960s, cited this as an example of how, in backward societies, inventions are usually hoarded by an elite, rather than bringing broad benefits to the masses.)

By the Middle Ages, rivers and streams had been harnessed to grind grain, husk rice, make paper, and husk sugarcane. In *De Re Metallica* (1556), Georgius Argicola illustrated a waterwheel driving an extensive series of camshafts and pistons for crushing rocks.[3] Running water was used to pump water from early mines. A wheel pumping municipal water supplies was installed on London Bridge.

Factories harnessing running water to weave cloth or grind grain were known as "mills" and soon expanded into "mill towns." In England, cities such as Manchester and Birmingham were powered by nearby rivers. In North America, Lowell and Lawrence, Massachusetts and Paterson, New Jersey were built from the ground up to take advantage of nearby dam sites. Even today, all over the East Coast of America, you can find beautiful old brick factories sitting beside the rivers and streams from which they originally drew their power. Many of these complexes are now being restored as shopping centers or incubators for small businesses. The publisher of this book is located in a restored historic mill in Maryland.

The steam engine eventually pushed waterpower aside, but hydro made a quick comeback generating electricity. Before Thomas Edison had opened his Pearl Street Station in 1882, H.F. Rogers, an Appleton, Wisconsin paper manufacturer, built a dam across the Fox River and introduced the world's first hydroelectric plant. The dam provided power to two factories and Rogers' home. Within weeks another hydroelectric plant went into commercial service in Minneapolis.[4]

In Holyoke, Massachusetts, a milldam built across the Connecticut River in the 1840s provided waterpower to an entire complex of paper and textile factories. In the 1880s the dam was converted to hydroelectricity and Holyoke formed its own municipal utility, the Holyoke Water Power Company. By the 1960s, it was still not hooked into the regional electric grid—and thus become the only town in the Northeast to miss the Great Blackout of 1965.

In the 1890s, dam building for electricity, flood control and recreational purposes all converged under the banner of the conservation movement, spearheaded by Theodore Roosevelt and Gifford Pinchot, head of the new National Forest Service. Conservation preached "wise use" of nature, bending it to human purpose while maintaining wildlife and scenic values. Dams fit this vision perfectly. On the rivers of the West Coast, they became the showcase of the conservation effort. When San Francisco was struck by an earthquake in 1906, plans were quickly made to dam the Hetch Hetchy Valley in Yosemite

National Park to provide drinking water and hydroelectricity to the beleaguered city.

One person objected, however—John Muir, the naturalist who had left his religious family in Ohio as a young man, walked to the Gulf of Mexico and ended up living alone for a decade in the High Sierras. Muir became a prolific writer, celebrator of wild nature, and founder of the Sierra Club. He and Roosevelt famously camped together in Yosemite at the dawn of the Conservation era.

Now Roosevelt and Muir found themselves painfully at odds over Hetch Hetchy. To Roosevelt, the dam was the ultimate wise use of natural resources. To Muir it was the ultimate sacrilege, defiling a "precious mountain temple." The Sierra Club drew support from around the nation and the battle raged for more than a decade before Congress finally approved Hetch Hetchy in 1913. When completed in 1923, the O'Shaughnessy Dam was the largest civil engineering project in the world except for the Panama Canal. Amazingly, the battle over Hetch Hetchy is not yet over, as the dam's old opponents still campaign to tear it down. The Sierra Club's Hetch-Hetchy Web site, titled "Time to Redeem a Historical Mistake," features a picture of Notre Dame Cathedral drowned in a valley and quotes a 1909 letter from Sierra Club secretary William Colby to Gifford Pinchot promising, "We have only begun to fight."[5]

Electrification grew rapidly in the 1920s but was limited mostly to urban areas. By the 1930s, 90 percent of city dwellers had electricity, while 90 percent of rural residents were still without. For farm families, life remained excruciatingly toilsome. As Vaclav Smil, the Canadian ecologist who grew up on a farm in Eastern Europe, writes:

> These burdens, falling largely on women, were much greater than the exertions of subsistence farmers in Africa or Latin America because…farmers tried to maintain a much higher standard of living and managed much larger farming operations. The word revolution is then no exaggeration to describe

the day when transmission lines reached the homes of such families.[6]

The Tennessee Valley Authority, perhaps the most ambitious undertaking of Franklin Roosevelt's New Deal, was launched in the First Hundred Days with a mission to extend the electricity to the rural Southeast. Private utilities had made some progress but the Great Depression brought the effort to a halt.[7] The TVA's motto was "Electricity for All."

From 1933 to 1944 the TVA built sixteen dams along the Tennessee River and its tributaries and converted two-dozen more existing dams to hydroelectricity. In 1935 the Rural Electrification Administration began organizing electrical cooperatives, run mostly by farmers. Next the TVA set up the Electric Home and Farm Authority to help farmers buy electric appliances. By 1939, the number of rural households across the country with electricity had jumped 25 percent.[8]

Inspired by the success of the TVA, Roosevelt moved on to the Northwest, where the Columbia River Basin promised similar multiple-use development. The State of Washington had planned the Grand Coulee Dam for more than a decade but was stalled by the Great Depression until Roosevelt provided funds in 1933. In 1937 Congress passed the Bonneville Power Authority Act, which began distributing electricity across the Northwest. The BPA paid Woody Guthrie the princely sum of $10 to celebrate the accomplishment in "Roll on Columbia:"

Now the world holds seven wonders that the travelers always tell.
Some gardens and some towers, I guess you know them well.
But now the greatest wonder is in Uncle Sam's fair land.
It's the big Columbia River and the big Grand Coulee dam.

Grand Coulee was so vast that it actually had trouble marketing its electricity until the power-hungry aluminum industry moved to the Northwest to take advantage during World War II. There are now 160 dams in the Columbia basin generating 15,000 MW, enough to supply 78 percent of Washington and

Oregon's electricity while exporting its excess to California during the summer months of peak demand.[9]

The Hoover Dam on the Colorado, which came to symbolize the dam construction of the New Deal, was actually begun under President Herbert Hoover. Ground was broken in 120-degree heat during the summer of 1931 while thousands of migrant workers and their families moved into adjacent "Ragtown." The dam was finished two years ahead of schedule and dedicated in 1936 by President Roosevelt—who changed the name to "Boulder Dam." (President Truman restored Hoover's name in 1947.) It provided 2,080 MW of electricity, mostly for Los Angeles, 250 miles away.

Inspired by this national mandate, the Army Corps of Engineers continued its march through the river valleys of the nation, building thousands of dams, large and small. Flooding on the Mississippi was tamed after a series of disasters, particularly the Great Mississippi Flood of 1927, which covered 27,000 square miles in seven states and killed 246 people. Southwest deserts bloomed under irrigation. Local politicians, municipalities and farmers welcomed the projects and were often given priority for cheap water rights.

The program finally hit a snag in the 1950s when the Corps of Engineers announced plans to dam Glen Canyon, just upstream from Grand Canyon. John Muir's Sierra Club vigorously objected. The Club had only tenuous legal standing, however, and the project rolled to completion in 1963. Lake Powell, 186 miles long with 1,963 miles of shoreline, took seventeen years to fill. The dam produced 1,320 MW, about the size of the largest nuclear or coal plant. It was at this point that the Sierra Club began proposing nuclear power as an alternative to dam projects.

Just as Glen Canyon was being completed, New York's Con Edison proposed another hydroelectric project—a pumped storage facility on the Hudson River at Storm King Mountain. Pumped storage was a new technology designed to solve the problem of storing electricity—still the major impediment to utilizing unpredictable energy flows such as wind and sunshine. Water is pumped to an upper reservoir at night, then released to meet the peak demands of the day. By storing off-peak electricity, a utility is able to smooth the extremes between daytime and nighttime demand and avoid new fuel-burning stations.

The project ran into opposition from upper crust residents in mountain retreats along the Hudson. This time, interveners were more successful. They formed the Scenic Hudson Preservation Committee, which testified before the Federal Power Commission. The FPC approved the project but a federal court overturned the decision, ruling that scenic values must be given weight. For the first time ever, environmental considerations had to be added to the equation. Within four years, Congress had adopted the National Environmental Protection Act, which required an "environmental impact statement" for every federal project.[10] The terms for building major power plants had changed completely.

Armed with this growing body of law, environmentalists were soon opposing dams everywhere. When the TVA proposed the Tellico Dam on the Little Tennessee River, a naturalist went out and discovered an obscure species of fish and named it the "snail darter." The new species was used to invoke the 1973 Endangered Species Act, halting the project. (The dam was eventually completed in 1979.) In one of his first acts as president in 1977, Jimmy Carter stunned Congress by cancelling appropriations for several western water projects—the first sign that the young environmental movement would play a significant role in his administration.

Dam building soon ground to a halt. There were few good sites left anyway—unless we consider damming Grand Canyon, which the Army Corps did propose at one point. Although the Corps still scours the countryside making proposals and drawing up plans, almost all meet fervent opposition and few are ever completed. In 2006, a proposal to drown several communities in eastern Nebraska with a dam on the Platte River was quickly scuttled when a group of high school students demonstrated at the state capitol.

Instead, all the talk now is about tearing down dams. In a 1998 issue of *Sierra* magazine, Reed McManus filed the following report:

> When the Sierra Club proposed draining Lake Powell [at Glen Canyon] last year, critics called it lunacy... But the assumption that dams are permanent fixtures has already been breached: around the country they're falling like dominoes in the name of river restoration.
>
> In December state and federal officials removed the Quaker Neck Dam on the Neuse River in North Carolina, making it the first large dam in the nation to be dismantled in an effort to rescue fisheries and renew a river.
>
> A month earlier, the Federal Energy Regulatory Commission (FERC) ordered the removal of the Edwards Dam on Maine's Kennebec River to restore fish habitats.
>
> In November the Western Canal Dam on a tributary of California's Sacramento River was demolished to let threatened chinook salmon run free for the first time since the 1920s.... The Newport No. 11 dam on the Clyde River in Vermont was removed in 1996, allowing salmon to return upstream for the first time in 40 years. And [the 1998] State of the Union address, President Clinton proposed spending $52 million to tear down the Glines Canyon Dam on the Elwha River in Washington State.
>
> Why all the sledgehammers? Some 550 dams will come up for relicensing by FERC in the next 15 years, and since 1986 the agency has been required to weigh a dam's impact on wildlife and recreation.
>
> At the same time, energy deregulation and low electricity prices caused in part by a glut of natural gas, more efficient natural-gas turbines, and conservation have turned many dams into economic dinosaurs.... While building dams has always demonstrated our ability to control nature, dismantling them represents our capacity to live in harmony with it.[11]

All this preceded the 2000 run-up in natural gas prices and the California electrical crisis, when electricity proved to be not so abundant after all. Yet the movement has not missed a beat. There is hardly a dam in the country now that does not have a local group trying to remove it and about 120 succumb every year. "Extinction is forever, dams are not," says the Northwest Energy Coalition, which is spearheading an effort to remove four dams from the Lower Snake River. "Together, the four dams only produce 5 percent of the Northwest's power. We can replace that power through common sense investments in energy conservation and renewable energy such as wind power."[12]

Hydroelectric dams constitute *90 percent* of the energy currently classified as "renewable." Inevitably, this contribution is lumped together with wind, photovoltaics, and biofuels to give the impression that renewables are making a hefty contribution.

In *Soft Energy Paths*, Amory Lovins tried to revive dam construction by proposing "low-head hydro"—small but beautiful dams putting out small amounts of power. As you will recall, however, a dam's output varies according to its height. Therefore, small dams provide even smaller amounts of electricity. In its 1990s effort to tread the soft path, California licensed dozens of low-head dams producing as little as 1.5 MW. These provide less than 1 percent of the state's electricity, however, and were of little help in resolving the 2000 California Electrical Crisis.

Surprisingly, even these small dams draw environmental opposition. On July 1, 1999, crowds cheered as crews began demolishing the Edwards Dam in Maine, a privately owned concrete structure built in 1837 to power mills along the Kennebec River, later converted to produce 3.5 MW of electricity (about the output of one giant windmill). It was the first dam in the country to be condemned by the federal government for environmental reasons. Two million alewives, striped bass, shad, sturgeon, and Atlantic salmon have returned to the Kennebec

River.[13] In 2008, a coalition of environmental groups, including the Atlantic Salmon Foundation and Trout Unlimited, raised $25 million to compensate the people for tearing down a series of small dams on the Penobscot River in Maine in an effort to restore Atlantic fisheries.

Major dam building is now reserved for other parts of the globe. The Chinese are currently completing the 12,000 MW Three Gorges Dam on the Yangtze River, which uprooted one million people. At the behest of American environmentalists, the U.S. Export-Import Bank refused to contribute funding. Canada has dozens of dam projects in the works, including the James Bay complex, which has flooded an area half the size of Lake Ontario and is now producing 15,000 MW—three times the capacity of Niagara Falls. Large dam projects are still being planned in India, Thailand, Afghanistan, and Australia. Still, in 2000 the World Commission on Dams issued a report saying there are barely 10,000 MW of good sites left in the world. Dam building is topping out across the globe, and there is certainly no place for expansion in Europe and the United States.[14]

Nor is hydropower eternally renewable, as lakes behind dams eventually silt up. Hoover Dam now often generates only 1,700 MW—down from its original 2,080 MW. Glen Canyon's loss of more than half its power has become one of the major arguments for tearing it down.

The amount of electricity we draw from dams—300 million megawatt-hours (MWh)—has not risen since 1970. Hydro's share, 23 percent at that point, has sunk to 8 percent today. In recent years, Western drought has cut production from several major dams by 30 percent.[15] Whatever possibility hydro has for providing us with energy has already been achieved. Rather than promising a solution to our energy problems, hydroelectricity illustrates the limitations of renewable energy.

Chapter 11

Wind

"Wind, the World's Fastest-Growing Renewable Energy Source is Coming of Age," says the cover of *E* magazine.[1]

"Wind Set to Become World's Leading Energy Source," says Lester Brown, president of the Earth Policy Institute.[2]

"Wind Power World's Fastest-Growing New Electricity Source," echoes the International Information Program of the United States Government.[3]

And indeed, the figures are exhilarating. Global capacity now stands at 60,000 MW, up from only 5,000 MW a decade ago. Germany has 18,000 MW generating 6 percent of its electricity. Denmark's 3,000 MW covers almost 20 percent of its grid. "America installed capacity just passed 20,000 MW and is growing an astounding 45 percent per year," said Christine Real de Azua, of the American Wind Energy Association in September 2008. At this rate we could reach 150,000 MW by 2020. This would give us 10 percent of our electrical capacity, although it's still only half of what the Department of Energy and Boone Pickens say is possible.

All this has conjured up visions of retiring coal plants and replacing large segments of foreign oil. In announcing his investment in a 4,000 MW wind farm in Texas in 2008, Pickens, perhaps the country's best known oil entrepreneur, said he wanted to use wind to replace the 20 percent of our electricity generated from natural gas so the gas could be moved to the transportation sector, replacing foreign oil. In 2008 the Department of Energy's Office of Energy Efficiency and Renewable Energy published a report, "20% Wind Energy by 2030," encouraging just that.

But wind has natural limitations. At best, windmills generate electricity only 30 percent of the time. The wind shifts unpredictably, and sometimes, doesn't blow at all. Electrical engineers say it will be impossible to rely on wind for more than 20–25 percent of an electrical grid's capacity. In addition, as with all forms of very dilute solar energy, wind farms will require staggering amounts of land. They are also likely to be sited on scenic mountain ridges, since that's where the wind blows the strongest. On the prairies, space requirements are mitigated in that wind can coexist with farmland.

What wind cannot supply is reliable peak load or base load electricity. The wind is not predictable enough for base load and—unlike solar electricity—does not peak at peak-loads during hot summer afternoons. In fact, those tend to be a period of doldrums. Since all utilities are required to carry an extra 20 percent of grid power as backup, wind's best role may be in adding to this "spinning reserve." This is an important contribution but one that will not eliminate the need for all the other energy sources.

Wind is essentially solar energy that is stored in the currents of the earth's atmosphere. The wind is created by the sun's action in heating different parts of the earth to different temperatures.

Where the sun shines hottest, warm air rises while cooler adjacent air flows in to replace it. This creates the wind.

The pattern takes place on a global level. Intense solar heat at the equator sends air aloft to be replaced by cooler air moving in from north and south. By the time this heated air reaches the 30th parallels north and south, it has cooled enough to descend again. The earth's rotation sweeps it westward, creating the "trade winds," which blow east to west around the Tropics of Cancer and Capricorn. This pattern is called the "Hadley Cell," after eighteenth century meteorologist George Hadley. The same pattern occurs at the poles, where a "polar cell" extends down to the 60th latitude and sends winds blowing in the same direction.

In the temperate zone, between the 30th and 60th parallels, air circulation is more complex. Called the "Ferrell Cell" after William Ferrell (1817–1891), it acts mostly in response to the Hadley and polar cells, and is often called the "ball bearing" between them. At these latitudes the weather is more turbulent, with prevailing winds generally blowing from west to east in opposition to the earth's rotation. In North America, the predominant pattern is the jet stream, which brings cold air down from Canada. Its strongest currents blow through the Dakotas and the upper Midwest—which is now being called the "Saudi Arabia of wind."

The use of wind power dates from ancient times. The Egyptians invented sailboats to cross the Nile, and sailing craft have been skimming the waters ever since. The ancient Babylonians built the first windmills around 2000 BC, using beveled vanes to create rotary motion. By the tenth century, windmills three stories high were grinding grain in Persia and Afghanistan. The technology remained in the Middle East until the Crusades brought it to Northern Europe.[4]

Wind soon joined water as Europe's primary source of non-animal power. The low countries, because they had few waterfalls, adopted windmills to grind their grain, saw their logs, and drain the Rhine Delta—so that the windmill became the symbol of Dutch nationality. Windmills also migrated south into Spain, where they had a famous encounter with a knight-errant named Don Quixote.

When Europeans crossed the Atlantic, they found ample wood and water on the East Coast of America, so windmills were rare. On the Great Plains, however, wind again became a prime source of energy. Postcards in the region still recall the days when "the wind pumped the water and the cows chopped the wood" (meaning windmills pumped wells while dried cow manure served as firewood in the treeless environment). Before electricity reached the high plains of Wyoming and Montana, cattle ranching would have been impossible without windmills, since cattle need water and there are few streams to be found. By 1890 there were seventy-seven factories manufacturing windmills in the United States and they had become a major export. Even today windmills still pump high plains cattle wells where electricity has not yet reached.[5]

The first windmill to generate electricity was built in Cleveland by inventor Charles F. Brush in 1888. A six-story wooden structure resembling a Dutch windmill, it turned very slowly, producing 12 kilowatts. Brush illuminated 350 incandescent lamps in his house, ran three small electric motors, and stored his surplus electricity in 12 lead-acid batteries in his basement. An 1890 article in *Scientific American* observed:

> The reader must not suppose that electric lighting by means of power supplied in this way is cheap because the wind costs nothing. On the contrary, the cost of the plant is so great as to more than offset the cheapness of the motive power. However, there is a great satisfaction in making use of one of nature's most unruly motive agents.
>
> [Nonetheless,] the amount of attention required to keep it in working condition is practically nothing. It has been in constant operation more than two years, and has proved in every respect a complete success.[6]

Brush's Electric Company was eventually bought by General Electric.

A few years later, Paul La Cour, a Dane, discovered that electrical output increases rapidly with turbine speed, so windmills were designed with smaller, faster rotating blades. By 1900, windswept Denmark had 2,500 towers operating pumps and mills, producing 30 MW of electricity. In 1930, a 1.25 MW windmill atop Grandpa's Knob in Castleton, Vermont became the world's largest.[7] While awaiting rural electrification, many American farms supplied their own electricity with windmills that could run a few light bulbs and perhaps a ham radio. Windmill brands such as Halladay, Star, Eclipse, Monitor, and Challenge were the favorites. Today they are collector's items.

Other methods of generation proved more convenient, however, and wind had almost been forgotten when the Energy Crisis struck in the 1970s. President Carter's Energy Policy Act of 1979 provided generous tax credits for wind power, which set off a small boomlet. By 1985, the nation had 1,000 MW of installed capacity. More than half of this was in a 580 MW wind farm at Altamont Pass, a narrow cleft in the Pacific Range 50 miles east of San Francisco. More than 1,000 ½-MW windmills stand scattered across 70 square miles of mountain slope, making the pass look eerily like Spindletop. (Even more eerily, many have now been abandoned since their tax credits expired.)

Altamont Pass revealed one of the big environmental problems of wind—their tendency to kill migrating birds. The carnage was brutal. More than 4,500 bald eagles, California condors and members of a dozen other species were butchered every year as they rode wind currents through the pass. The Audubon Society was so distraught that it became an opponent of wind farms, obstructing them in Los Angeles and other locations

As the Energy Crisis subsided in the 1980s, the Reagan Administration allowed the tax credit to expire and enthusiasm waned. What kept the wind industry alive, particularly in California, was the Public Utilities Regulatory Policy Act of

1978 (PURPA), which required the utilities to purchase electricity from "alternate" sources. California made particular use of this law—as we shall see in chapter 15. Research continued, however, and impressive strides were made. New materials allowed towers to rise 20 to 30 stories into the air, as tall as the Statue of Liberty. Researchers discovered that long, slow-moving blades, when combined with a gearbox, could raise rotational speed while cutting the carnage of birds. (Unfortunately, bats still seem unable to avoid them.) Soon giant windmills were producing 1 and 2 MW apiece.[8]

The real progress shifted to Europe, where Denmark and Germany adopted laws requiring utilities to buy wind energy at very high prices. With long Baltic coastlines near population centers, both countries have been able to build their facilities offshore, where the ocean offers no topographical resistance and the wind blows more steadily, providing a more reliable current.

Denmark now claims the most advanced wind economy in the world, with 3,500 MW of installed capacity, including a three-mile array along its Baltic shoreline. Across the rolling, gentle west Denmark countryside it is hard to find a place where windmills are not visible. The Danes now claim to get 25 percent of their electricity from wind, 90 percent of their low nighttime demand.

Germany has followed Denmark's footsteps, vowing to get 20 percent of its electricity from wind by 2015. The Germans have 22,000 turbines, producing 6 percent of their electricity. Spain has 11,000 MW—slightly ahead of the United States—providing 8 percent. The whole European Union is aiming for 50,000 MW by 2010.

Back in the U.S., Congress revived the 1.5 cents per-kilowatt-hour (kWh) tax credit in 1992 for any new windmills' first ten years of operation, making wind competitive with coal, which costs of 2.3 to 3.5 cents per kWh. Unfortunately, the tax credit was subject to annual renewal by Congress and underwent constant stops and starts, expiring every other year only to be revived again. Not until 2004 did Congress finally pass a seamless extension through 2008, leading to the recent burst of construction.

Wind energy is now growing in the U.S. at a brisk 25 percent a year. Surprisingly, most of the growth has been outside California, where capacity climbed only from 1,600 to 2,000 MW since 2000. On the Oregon-Washington border, a 300-MW wind farm will operate in conjunction with the hydroelectric dams of the Bonneville Power Authority. Iowa has added its own 1.5 cents tax credit, and Minnesota, Montana, New York, and several other states have taken similar initiatives to encourage development.

Beyond tax incentives, the favorite tactic has become the "renewable portfolio," whereby a state mandates that utilities must generate a specific percentage of their electricity from renewables by some future date. Minnesota has mandated 15 percent of the states energies be from renewable energy by 2015, Arizona 15 percent by 2025, Rhode Island 16 percent by 2020, and New York 24 percent by 2013 (the state already gets 19 percent of its electricity from hydro). Twenty-one states have now adopted such laws and a national renewable portfolio of 15 percent almost made it into the 2007 Federal Clean Energy Act. Such laws are relatively painless to legislatures, since utility companies and their customers will bear the costs.

New windmills are now being concentrated in three areas—farming communities, mountaintops, and offshore. In the Midwest, where America's wind currents are strongest, windmills can be situated on agricultural land without interfering with other uses. Farmers are happy to earn a few thousand dollars in rent each year, and gigantic 40-story structures are starting to dot corn and wheat fields across the Midwest.

With this expansion, however, has come the first opposition. Concern has grown over both bird kills and the scenic impact. In Flint Hills, Kansas, a group of local farmers has enlisted the Audubon Society and Nature Conservancy in opposing a windmill farm. "Wind is a renewable, virtually inexhaustible, source of energy," says Ron Klataske, director of the Kansas Audubon Society. "However, native prairie and prairie landscapes are not renewable and certainly not inexhaustible." "We have a bias in favor of wind energy," adds Greg Butcher, director of bird conservation at the National Audubon Society, "but the key is siting. We want to keep windmills away from important areas."[9]

In Leroy, Wisconsin farmers and environmentalists are locked in a duel over plans to put 133 40-story windmills on 50 square miles near the Horican Marsh, a stopover for migrating Canada geese. "These things are massive," complains Bill Volkert, a naturalist with the Wisconsin Department of Natural Resources. "And we won't know what's going to happen until it's too late."[10]

Offshore, there have been problems as well. Senator Edward Kennedy famously intervened to prevent the construction of a wind farm off Cape Cod that would be visible from the Kennedy compound in Hyannis Port. In New York, his nephew, Robert F. Kennedy, Jr., now a prominent environmental lawyer, was in the forefront of defeating a windmill complex in Long Island Sound.

The biggest opposition, however, has come from local communities concerned about noise and the aesthetic value of putting windmills along mountaintops and on otherwise rustic farmland. A whole gaggle of organizations have sprung up—"Windstop," "War Against Wind," "Vermonters With Vision," "Stop Ill Wind," "Save Our Allegheny Ridges," "Mountain Communities for Responsible Energy"—which have aggregated on Web sites such as "Wind-watch.org," "WindAction.org," and "AWEO.org (American Wind Energy Opposition)."

"I started out a strong environmentalist supportive of alternate energy and concerned about global warming," says Jon Boone, a western Maryland resident who has become a prominent opponent of wind farms on the East Coast. "But the more I looked into it, the more I realized how insubstantial the claims of wind advocates have become."

"I just do this because I have some experience in the electric industry and I know what isn't being said," adds Glenn R. Schleede, a semi-retired electrical engineer who has posted several Internet articles on the subject. "The things that people are claiming about wind just aren't true."

So is this opposition just another manifestation of the "not-in-my-backyard" syndrome, where people only oppose things because they are personally affected by them? Or are there deeper reasons for skepticism about wind? Unfortunately, there is strong reason for suspecting that the wind energy boom is less than meets the eye.

The important thing to recognize is that transmitting electricity through wires is not the same as shipping oil or gas through a pipeline. The "grid"—which has been called "the most complex machine ever made"—is actually a huge electrical circuit stretching across thousands of miles and millions and millions users and suppliers.[11] Once it is "up," the grid becomes a delicately balanced high-wire act where supply and demand must be kept in constant balance. (Recall how it usually takes more than twenty-four hours to restore the grid after a blackout.) Any sudden rise in resistance will cause current to sag and produce "blips" or brownouts. A sudden spurt in current will cause a "power surge" that can destroy appliances, wipe out data, or even start fires. That is why you have a "surge protector" between the wall outlet and your computer.

The voltage "drop" across the national grid—110–120 volts in America, 220–240 in Europe—must be maintained minute-by-minute, hour-by-hour, day-by-day. There is not a whole lot of room for error. The National Electrical Reliability Council estimates that voltage levels can vary about 5 percent before variations start damaging electrical equipment. Information technology is particularly susceptible. Computer geeks talk about the "high 9's," meaning that current flows must be maintained

99.9999 percent of the time in order to avoid losing data. In *Digital Power*, Peter Huber and Mark Mills report, "Some years ago, a Stanford computer center found its power fatally polluted by an arc furnace over one hundred miles away."[12] As *The Industry Standard* once put it: "Blips as brief as one-sixtieth of a second can zap computers and other electronic gear, and blackouts can be catastrophic."[13]

The problem with wind energy is that it is *always fluctuating*. The wind never blows steadily anywhere in the world. While coal, gas, and nuclear plants can maintain steady levels of power for days and years (in the case of nuclear), wind can fluctuate from minute to minute. Current output varies with the *cube* of the wind speed, exaggerating these power surges even more. A 20 percent increase in wind speed can *double* electrical output within a few minutes.

Under these circumstances, large numbers of windmills tend to be viewed by grid operators as a *liability* rather than an asset. A windmill is wildcard in the deck, functioning more like a consumer of energy than a supplier. Grid operators in Denmark, Ireland, and Japan have already declined to take any more wind on their systems due to problems of integrating its intermittent output. In fact, Denmark, the world's leading advocate of wind energy, built no new windmills in 2007.

Unfortunately, where the wind is predictable, it doesn't coordinate very well with demand. The wind tends to blow more at night than during the day, when demand is higher. It gusts stronger during spring and fall, rather than summer and winter, the periods of peak demand. Overall, though, the pattern tends to be random.

All this makes claims of wind's *installed* capacity very misleading. Most windmills produce electricity less than one-third of the time. Over the past decade, California's 1,500 MW of windmills have averaged 25 percent of their "nameplate" capacity. During peak summer demand it was only 9 percent. Germany has found its windmills producing only 6 percent of their nameplate capacity during hot summer days.[14] All this has led to restiveness about increasing wind energy among electrical engineers.

In November 2007, the entire issue of *Power and Energy*, the journal of the Institute of Electrical and Electronics Engineers (IEEE), was devoted to the subject, "Working with Wind: Integrating Wind into the Power System."[15] The authors—all dedicated to expanding wind power—included engineers from the National Renewable Energy Laboratory, the American Wind Energy Institute, the National Grid of Ireland, the Swedish Royal Institute of Technology, and a wide variety of consulting firms. In order to deal with wind's problems, they suggested the following recommendations:

1) *Dispersal of generation.* If wind farms are spread over a wide area of countryside and offshore locations, output variations in one place may compensate for variations in another.

2) *More integrated transmission systems.* The more the grid is broadened to include a wide variety of consumers and suppliers, the more wind's unavoidable fluctuations can be dampened.

3) *Better weather forecasting.* If wind speeds can be forecast accurately a day or more in advance, grid operators can make better plans for lining up supplies.

4) *Electrical storage.* The major problem with electricity is that it cannot be stored. It must be consumed as it is produced. Storage technologies have been pursued for decades. The most promising candidates are pumped storage plants, compressed air storage, giant chemical storage batteries, and hydrogen fuel cells (which we'll explore in chapter 14). Any of these would make wind power more "dispatchable," as the engineers like to say. Unfortunately, except for pumped storage, none have been developed to the point where they can handle large amounts of electricity.

All this will not be without considerable cost. Dispersing wind farms will mean covering entire mountain ranges with 30- and 40-story structures, since the wind blows strongest across mountaintops. Utility companies say transmission lines are now harder to site than power plants because of local opposition. (The 2005 Energy Act gave FERC the power to override this.) There are also technical limits. At 500,000 volts—the standard level for transmitting alternating current—10 percent of electrical energy dissipates as heat every 60 miles.[16] Even if California were connected to windmills in North Dakota, only 7 percent of the energy would ever arrive. Direct current lines at very high voltages could alleviate the problem, but that will mean building a new electrical infrastructure.

Some enthusiasts have even begun to ask whether the environmental benefits of wind power are that great. A study commissioned by Norway in 1998 found that wind power in Denmark had "serious environmental effects, insufficient production, and high production costs."[17] Reporting on the Danish experience to a British audience, Dr. V.C. Mason concluded:

> Although one-fifth of the electrical power produced annually in west Denmark is generated by its enormous capacity of wind turbines, only about 4 percent of the region's total power consumption is provided from this source. Most of the output of wind power is surplus to demand at the moment of generation and has to be exported at reduced prices to preserve the integrity of the domestic grid. Savings in carbon emissions are minimal.[18]

Windmills in the United States are now being built entirely for the tax credits, and to fulfill renewable portfolio requirements. A study of a wind farm proposed for Blairsburg, Iowa by Warren Buffet's MidAmerican Energy found that, with all the federal and state subsidies, the $323 million project could breakeven after only six years without ever producing a kilowatt of electricity.[19] Before its demise, the largest owner of windmills in the country, was Enron. The Marriott Corporation and other subsidy-hunters have since stepped into the breach. Boone Pickens is promising his investors a 25 percent return on a 4,000 MW windmill farm in Texas, based entirely on federal tax credits.

All these subsidies have to be absorbed somewhere and the usual end point is ratepayers. The Danes now pay the highest electrical rates in Europe—double many other countries. As we shall see, California achieved a similar outcome when it forced its utilities to buy alternate energy in the 1990s.

Perhaps the best role for wind would be to have it provide the "spinning reserve" required on all grids. Grid operators generally maintain a 20 percent margin of reserve capacity to be ready at a moment's notice in case a power plant shuts down unexpectedly or there is an unanticipated surge in demand. Wind power would fit this role ideally—especially since it requires it backup itself in case the wind suddenly dies down.

Countries such as Denmark and Germany have found that their ability to absorb wind is topping out around 20 percent. A recent study commissioned by the Minnesota legislature estimates it can go as high as 25 percent.[20] IEEE's special issue of *Power and Energy* also set 20–25 percent as a reasonable goal.

Wind energy is not inexhaustible. Windmills must be placed apart so they do not interfere with each other's flow. Boone Pickens' 4,000 MW project will cover 1,200 square miles. (He says he won't be putting any on his own 68,000-acre ranch because "they're too ugly.") Waiting in the wings are 6 MW windmills as tall as Manhattan's Trump Tower. There has even been talk of suspending windmills high in the atmosphere to capture the jet stream. But beyond this 25 percent, wind is unlikely to penetrate. It would be impossible—I-M-P-O-S-S-I-B-L-E—to run a contemporary electric grid solely on wind power. At bottom, it remains a medieval technology.

Chapter 12

Solar

Of all the forms of renewable energy, direct solar holds the greatest promise for helping to cut carbon emissions and provide alternate sources of energy. The amount of solar energy reaching the earth is about 8.2 *million* quads per year. Americans, you will recall, consume 100 quads and all of humanity 400. This means about eighteen thousand times our need reaches the earth each day.

Of course nearly all this energy already goes to heating and lighting the planet. Sunlight warms and illuminates the day. It also provides energy to plants, which use it to create organic molecules that we can eventually access as food. Remember, except for the terrestrial energy emerging from the earth, the sun has been the source of all energy throughout the planet's history.

When we talk about translating solar energy into *usable* forms of energy, however, the numbers begin to shrink rapidly.

Sunlight hits our outer atmosphere at a rate of 1,347 watts per square meter (W/m^2) and loses one-quarter of this filtering through the atmosphere. A square meter is about the size of a card table. The maximum that reaches the earth at the equator is 950 W/m^2, the equivalent of 9½ 100-watt lightbulbs. In the temperate zones, this drops to 750 W/m^2 in the Southwest and 500 W/m^2 at the 40-degree latitude of New York, Chicago, and San Francisco. For the entire country the average is 630 W/m^2.

That is a noontime maximum. Sunup-to-sundown, this reduces to about two-thirds or 490 W/m^2 in Phoenix and 320 W/m^2 in Columbus, Ohio, the population center of the country. All the solar energy falling on a large card table in Topeka, Kansas, for example, would power four 100-watt light bulbs. This would give us about the illumination of broad daylight—which makes sense, since most of the sun's energy that reaches the earth is in the form of visible light.

But 100 percent of this cannot be converted into usable energy. The best anyone has been able to achieve in turning sunlight into electricity through photovoltaics is about 15 percent, though optimists are hoping for as high as 25 percent. Boeing recently announced a solar cell that achieved 40 percent efficiency, but this was done by concentrating sunlight.[1] This means the very best technology anyone can imagine would power a single, 100-watt light bulb for every square meter in Topeka, Kansas. This would be a daytime average. At night, when lighting requirements are the greatest, we would get nothing.

Realistically, then, if we covered every rooftop in the country with solar panels, we would be able to provide the equivalent of our daytime indoor lighting. Lighting consumes about 9 percent of our electricity and rooftop electricity would probably be able to provide half of this. Once again, despite its ubiquity, solar energy does not measure up well against industrial-sized needs.

The limiting factor will always be the *land* required to

capture and store this energy. Windmill farms have to cover nearly a hundred square miles to generate as much electricity as one conventional power plant. Hydroelectric reservoirs cover hundreds and hundreds of square miles. Biofuels, as we shall see, have quickly run up against land limitations—and have crowded other land uses in the process.

Where solar electricity does have a great advantage over wind is that it *peaks* at the time it is needed most—hot summer afternoons. Since peaking loads can rise as high as 40 to 50 percent above the base load, covering the rooftop of every building in a city with solar panels would be extremely helpful in eliminating the need to build peaking plants, which, as already mentioned, are generally gas turbines running on extremely expensive natural gas.

The direct capture of the sun's rays to do useful work can be accomplished in three ways:

1) "Passive" solar energy for space heating in buildings
2) "Thermal" solar devices for providing heat and hot water
3) Electricity generated in one of two ways:
   a) Using solar heat to boil water, in the manner of conventional power plants; or
   b) Converting the sun's rays directly to electricity through photovoltaics (PV)

All three are now being employed.

Passive solar is basically the "greenhouse effect" used to capture solar energy and store it for heating buildings. An extreme example is Amory Lovins' Rocky Mountain Institute in Snowmass, as he described it in *Scientific American*:

> Consider my own house, built in 1984 in Snowmass, Colo., where winter temperatures can dip to -44 degrees Celsius [-14°F] and frost can occur any day of the year. The house has no conventional heating system; instead its roof is insulated

> with 20 to 30 centimeters of polyurethane foam, and its 40-centimeter-thick masonry walls sandwich another 10 centimeters of the material. The double-pane windows combine two or three transparent heat-reflecting films with insulating krypton gas, so that they block heat as well as 8 to 14 panes of glass. These features, along with heat recovery from the ventilated air, cut the house's heat losses to only about 1 percent more than the heat gained from sunlight, appliances and people inside the structure. I can offset this tiny loss by playing with my dog (who generates about 50 watts of heat, adjustable to 100 watts if you throw a ball to her) or by burning obsolete energy studies in a small woodstove on the coldest nights.[2]

Passive solar can be easily incorporated into new homes. At Texas Instruments, the Rocky Mountain Institute saved $30,000 annually in air conditioning costs by rotating the building 30 percent.[3] Unfortunately, these improvements are very difficult to retrofit into old buildings. There are 100 million households in the nation and only 160,000 new homes are built each year. Giving every one of them passive solar design would improve 0.16 percent of the housing stock. Many passive solar aspects have long been incorporated–picture windows, overhanging roofs, and trees that shade in the summer but lose their leaves in the winter.

Insulation—which is the key to passive solar—does save energy and Americans have spent more than $1 billion installing insulation since the 1970s. The North American Insulation Manufacturers Association estimates that the insulation in American homes and businesses saves more than 20 quads per year, the equivalent of *one-fifth* our consumption. Still, 65 percent of homes remain under-insulated.[4] The difficulty with insulation is that it tends to concentrate bad air. After the initial rush of enthusiasm in the 1970s, the EPA became very concerned about the accumulation of harmful gases in tightly insulated homes. The solution has been to clean the air through more extensive ventilation and air purification systems—which unfortunately run on electricity.

Beyond passive solar, the next step is to use the sun's energy for heating and hot water. Water is an efficient storage medium and can preserve solar heat for nighttime use. A solar hot-water heating system involves little more than carpentry and plumbing. Heat is trapped in a rooftop greenhouse and circulated throughout the building.

The system does not run by itself. It requires electric pumps plus care and maintenance. The unit sits outdoors, where it is subject to more wear and tear from the weather. Pipes can freeze or corrode. Leaves and snow must be cleared and the panels kept constantly clean. Even a thin film of dirt can reduce their effectiveness. Although the solar industry does not advertise it, labor-intensive maintenance is one of the most difficult aspects of the technology.

Another drawback is that it is much easier to raise water to 120° or 130°F using solar energy than 160° to 180°F, which is where most people take showers and wash dishes. Solar systems are good at producing *warm* water. In theory, it should be possible to trap enough solar heat to boil water (212°F) but inefficiencies set in. The pipes themselves act as conductors, sapping away some of the heat. As a result, most solar systems work best as "boosters" or "pre-heating" devices.

High-temperature devices went off the market after 1990 when much of the initial enthusiasm dissipated. Typical was this experience recounted in *Time* magazine:

> A quarter-century ago, the owners of a 13-story, 64-unit co-op at 924 West End Avenue on New York City's Upper West Side erected a steel framework on the rooftop, welded it to the building's steel beams and attached 117 solar-collector panels. Water heated by the sun flowed through pipes into a 5,000-gal. storage tank in the building's old coal bin and from there into the building's hot-water system. The project was funded in part with a $112,000 federal grant. Today the solar

> experiment is long gone. A building workman told TIME that the collectors behaved like sails, swaying back and forth so much that water leaked into apartments below. It cost several million dollars to repair the roof, he said.[5]

As oil prices came down and tax subsidies dried up, the solar heating industry gradually faded. In 1980 there were 233 manufacturers shipping 19,000 units of solar water heaters. By 1995 there were only forty-five manufacturers shipping 7,000 units. Even hot water systems have lost their appeal. The little bit of money saved was hardly worth the effort.

Solar-heated water has carved a niche in one market, however—heating swimming pools. Thermal solar's low-grade heat is perfect for keeping swimming pools at around 80 degrees. The market doubled in the 1990s and the industry now ships about 1 million square feet of collectors each year. Swimming pools are now 94 percent of the thermal market while residential hot water has shrunk to 4 percent and space heating to 2 percent. Most swimming pool sales are in the Sunbelt, with Florida constituting 51 percent of the market, California 23 percent and Arizona 7 percent.[6]

The technology that seems to promise the best long-range results is solar electricity. This can be generated in two ways: (1) Using the heat of the sun's rays to boil water as in a conventional power plant, or (2) converting the energy of sunlight directly into electricity through Larry Kazmerski's specialty—photovoltaics.

Thermal Electricity. The idea of concentrating the sun's rays for heat has been around since 213 BC, when Archimedes used an array of mirrors to burn holes in the Roman fleet at Syracuse. Twenty-two centuries later, the same concept was employed in the Mojave Desert near Barstow, California where the Sandia National Laboratory deployed hundreds of computer-controlled mirrors to focus the sun's rays on a 15-story Power Tower. Raising the temperature of the tower to 1500°C, the mirrors heated a synthetic oil called "therminol," which has a high boiling point but used its heat to boil water. The steam drove a turbine, producing 10 MW of electricity.

Solar One operated until 1988, when it was no longer deemed practical. The flow of electricity was always interrupted when the sun went behind a cloud. Over the next few years the facility developed a method for storing power in molten salts and reopened in 1996 as Solar Two, which sold electricity to the grid until going offline again in 1999. Solar Two occupied 137 acres (about one-fifth of a square mile) and generated 10 MW. To get to 1,000 MW—an average commercial plant—it would have to cover twenty square miles.

A second more successful experiment at Barstow is the Solar Electrical Generating System (SEGS), built in 1987 by the Luz Corporation. SEGS consists of 100 acres of 40-foot-high parabolic mirrors lined in long rows, north to south. The mirrors rotate to follow the sun, continually focusing its rays onto a small black tube running along their focal point. The tube contains therminol, which again is heated to 1500°C to produce steam. The system is more efficient and generates 354 MW.

SEGS' big problem, once again, was maintenance. The 10 million square feet of mirrors have to be washed every five days and scoured by high-pressure hoses once a month. The job is labor-intensive. Luz went bankrupt in 1991 but the project revived under California's PURPA effort, which requires utilities to buy power from alternative providers.

The system was plagued with fires. On January 10, 1990 a series of explosions rocked one of the cooling towers and ignited large quantities of therminol. Thirteen engine companies needed 1500 gallons of foam to quench the flames.Then on February 27, 1999 almost a million gallons of therminol caught fire, destroying considerable portion of the facility. The flames released toxic fumes and a half-square-mile area had to be evacuated. The Federal Aviation Administration also set up a no-fly zone around the facility.[7] When it reopened in 2000, SEGS still represented 90 percent of the world's solar thermal capacity.

Over the 1990s, research turned to photovoltaics as the more promising technology. The thermal idea was revived in 2007, however, when Spain built a new Power Tower in Seville to take advantage of a similar renewable portfolio law. A remarkably futuristic 30-story structure that looks like a giant carpenter's level stuck in the ground after arriving from outer space, the facility uses 136 acres to generate 11 MW—the exact same dimension as Solar One built in 1988. Nevertheless, it was on the cover of *U.S. News & World Report* in an article titled "The New Age of Solar Energy: Breakthrough Technology That Could Change the Future." Said the report:

> Solar energy may be poised to make the leap from the rooftop down to the floor of the desert—where some advocates say it needs to be if it's going to take its rightful place as a member of Big Energy.[8]

The Seville Power Tower, of course, is not a breakthrough but an exact duplication of a facility built in the 1980s. There is no chance of improving the technology, since it is already making maximum use of sunlight. The only way to grow is to use more land. Only the breathless reports in news magazines about "breakthroughs" are renewable, being recycled year after year.

Photovoltaics. In 1905, the same year he posited the Theory of Relativity, Albert Einstein wrote another paper identifying the "photoelectric effect," the ability of rays of light to knock an electron off an atom. It was the photoelectric effect paper that won Einstein the Nobel Prize in 1921. (The scientific establishment was much slower to accept the principle of $E=mc^2$.) Einstein's photoelectric effect is the basis of photovoltaics—generating electricity directly from sunlight.

Einstein's theory posited that radiation such as sunlight travels in discrete packets he called "photons." The process is very specific. The energy packet in the photon must match the energy level in the electron in order to dislodge it. So too, the light emitted from any element will radiate at a very specific wavelength. This is the basis of "spectrometry," where the elements in the stars can be identified by the wavelength of the light they emit.

In practical application, this means that sunlight falling on any substance will release electrons. If these electrons can be herded in the same direction, they form an electric current. Each element reacts with only one spectral line, however, which means that only a small portion of the total energy in sunlight can be transformed. The trick of photovoltaics is to find a substance that produces a strong electric current but is still cheap enough to manufacture. Silicon has been the substance of choice—not because it is particularly reactive to sunlight but because it is the second most common substance in the earth's crust. Germanium and a few other rare elements are more efficient at turning sunlight into electricity but are too expensive to be used in mass production. If solar cells are to be manufactured in industrial quantities, they will probably be made from silicon.

The semiconductor industry had made the same decision and for a while there was a mistaken impression that the two industries could follow the same trajectory. Moore's Law, first formulated in the 1960s, stated that the ability of computer chips to store information was doubling every eighteen months. The pattern has held true and now seems to be accelerating as researchers reach the quantum level. But solar panels are different. They are not *storing information*, they are *gathering energy*. In *The Earth in Balance*, Al Gore referred to solar cells as "small flat panels of silicon or similar materials that are designed to produce currents of electricity." There will be nothing small about solar arrays. They are being made *thinner*, which creates some cost savings, but there is no way technology can reduce the area requirements for PV electricity.

For this reason, the cost of photovoltaic electricity has not dropped significantly since the 1980s. When research first began, there was a flurry of progress that brought the price down by a factor of five. But in 1987 the technology hit a wall and has not improved much since. The major impediment to the spread of PV cells is that they remain very expensive.

The average American household uses 25 kilowatt-hours of electricity per day. A four- to six-panel solar system sold by Evergreen Solar of Waltham, Mass.—one of the leading suppliers—produces about one-tenth of this amount at an installation cost of $2,200. This quickly adds up to a large investment with a very long payback. California has allocated a $3.2 billion program to give rebates to homeowners and businesses to install solar systems, yet even with a rebate of $10,000 people are paying $16,000 to install modest systems. Recovering this investment through reduced electric bills takes an estimated 10–12 years.[9] "It is clear that the domestic market for PV cannot truly thrive without government incentives or a comprehensive requirement for utilities to buy back power," admits Matt Scanlon in *Mother Earth News*. "It's true, you have to be really green to buy one of these," agrees Paul Maycock, president of PV Energy Systems in Williamsburg, Virginia.[10]

In 2001, *Canadian Geographic* recounted the experience of Dr. Tom Touzel, an environmentally conscientious family doctor who lives with his wife and two daughters in a log cabin in rural Ontario. At a cost of $12,000, Touzel erected a vast array of PV panels in his backyard generating 600 watts—about two-thirds the consumption of an average American home—at peak hours of sunlight on a sunny day. He was still connected to the grid but sells electricity back to the utility during sunny hours. By practicing draconian conservation, he has been able to make the system break even. Still, he was forced to mull long and hard before buying his daughter an aquarium. The oxygen pump would put too much strain on his system.[11]

The Oak Ridge National Laboratory in Tennessee is now the site of some of the most advanced solar research. At the town's sizable library, solar panels cover the entire front pavilion, stretching almost the length of a football field. They are stacked across the portico overhang and cover the first floor roof as well. The display informs visitors that this vast rooftop installation is sufficient to provide the electricity for *one* suburban home.

As with the Power Tower, constructing commercial-sized power plants using photovoltaic cells will consume huge amounts of land. A 1,000 MW facility will occupy at least 50 square miles. Rather than building vast complexes in the desert, a much better idea has been to put PV cells on walls, rooftops, and every exposed surface available. In Japan, an entire high-rise apartment has been built with a photovoltaic skin.

In 1997 President Clinton launched a "Million Rooftop Initiative," vowing that the United States would spend $100 million a year to put photovoltaic panels on a million roofs by 2010. The Department of Energy offered grants to commercial and industrial buildings but not to homeowners. Instead, DOE formed partnerships with cities and counties in the hope that local governments would provide residential funding. The entire project was designed to produce 3,000 MW of installed capacity. Since solar devices run at 30 percent of capacity, this would equal the output of one conventional power plant. By 2005, about 500,000 roofs had been retrofitted when Congress quietly diverted the appropriation to homeland security.[12]

In the middle of the 2000 California Electricity Crisis, Los Angeles announced its intentions of becoming the "Solar Capital of the World." The Los Angeles Department of Water and Power offered to reimburse ratepayers half the cost for installing rooftop photovoltaic cells—about $10,000 for what is generally a $20,000 investment. After three years, however, the *Daily News of Los Angeles* reported that only 500 units had been installed and they were delivering only 45 percent of the expected power. "The losses are extremely high," said Henry Martinez, DWP assistant general manager for power generation. "Somewhere along the line, we would have expected the energy we generate from these systems would be a lot more than what we're getting."[13]

Perhaps the most successful effort to date has been at the U.S. Postal Service's processing plant eleven miles southwest of downtown Los Angeles. With great fanfare, the Environmental Protection Agency announced that it had covered the entire roof, the size of a football field, with PV panels. The array would provide the one-story building with *one-tenth* of its electrical requirements. The entire project cost $225,000 at the commonly cited price of 25–30 cents per kilowatt-hour. (In 2008, the average price of electricity off the grid was 8.55 cents per kilowatt-hour [kWh].)[14]

In trying to become the "Solar Capital of the World," Los Angeles would be following in the footsteps of Gardner, Massachusetts, a mill town of 20,000 on Route 2, just 20 miles north of Worcester. In 1985, the Massachusetts Electric Company approached the citizens of Gardner with a proposition. How would they like to put photovoltaic panels on their roofs? The company was beset with problems of siting power plants and responded to the romance of the technology. Going solar, company officials told the residents, would save them up to 40 percent on their electric bill.

About 30 families responded. Five commercial buildings also put up solar collectors, including the town library and a Burger King. Soon Gardner was being visited by engineers, politicians, and solar enthusiasts from all over the planet. The experiment is still going on today.

The results have been almost identical with other experiments. The first problem—although hardly anyone ever mentions it—is that PV cells produce direct current (DC) whereas the grid operates on alternating current (AC). You need to buy an inverter, which, along with a storage battery and other safety devices, pushes the installation costs to $15,000. Mass Electric agreed to pay the costs and upkeep.

Leon Rice, a meat cutter in the local supermarket, had a typical experience. He was happy to have the new-fangled contraptions but found that they were only saving him $5 a month on electricity. Then the inverter failed. The electric company replaced it, but then the system stopped working altogether. Rice went on his roof and found that water was leaking into the panels, corroding the cells.

This time the utility didn't respond. Months passed before workers finally showed up and lowered the ninety-pound panels off his roof, stacking them against a backyard fence. They promised to come back with new ones. Months later Rice finally gave the company an ultimatum: "Either they're outta here or I'm going to have my son throw them in the trash," he told them. Two men finally showed up and hauled them away. He never heard from the utility again.

Only four of the thirty participants gave up completely, but others have seen their system fall into disrepair. No one ever saved more than about $15 a month on electricity. "It really did look like a breakthrough twenty-five years ago," said Joe Broyles, a New Hampshire state energy planner. "But most of the buzz you heard did not talk about cost."[15]

Amory Lovins claims to be doing much better with photovoltaics, although you have to read him carefully.

> Eliminating the need for a heating system [with passive solar] reduced construction costs by $1,100 (in 1983 dollars). I then reinvested this money, plus another $4,800 [i.e., $12,500 total in 2008 dollars], into equipment that saved half the water, 99 percent of the water-heating energy and 90 percent of the household electricity. The 4,000-square-foot structure—which also houses the original headquarters of Rocky Mountain Institute (RMI), the nonprofit group I co-founded in 1982—consumes barely more electricity than a single 100-watt lightbulb. (This amount excludes the power used by the institute's office equipment.) Solar cells generate five to six times that much electricity, which I sell back to the utility. Together all the efficiency investments repaid their cost in 10 months with 1983 technologies; today's are better and cheaper.[16]

The RMI headquarters has six rooftop panels, each covering about two square meters. This produces, as Lovins reports, enough electricity to run his "household electricity," a few light bulbs and a small refrigerator. The institute's office equipment—computers and copy machines—is excluded from the equation, so "selling back to the utility" probably involves no more than a small monthly deduction from his overall bill. He is still connected to the grid.

So there you have it. The best technology in one of the most advanced solar homes in the country provides indoor lighting and not much else. There is no thought of running major appliances. (Lovins hangs his laundry out to dry and recommends everyone else do it as well.) An energy-efficient dishwasher (1,500 watts), coffee maker (1,000 watts), clothes dryer (2,500 watts), dehumidifier (750 watts), microwave oven (950 watts), or even an aquarium oxygen pump (600 watts) would overwhelm a home photovoltaic system.

Reluctant to dish out any more subsidies, twenty-two states have started dumping the costs onto the utilities through "renewable energy portfolios." These "unfunded mandates" allow the government to tell utilities they must get a fixed percentage of their electricity—usually around 10 percent—from renewable sources. The utilities must bear the costs. California tried this throughout the 1980s and 1990s with very unsatisfactory results—as we shall see in a later chapter.

Sometimes these mandates are mere grandstanding. New York has adopted a renewable portfolio of 25 percent by 2013 but it already gets 17 percent from hydroelectric dams, mostly in Canada. Maine has adopted a standard of 30 percent by 2010—but the state already gets 40 percent of its electricity from hydro! (Good old hydro—where would renewable energy be without it?)

Stuck with this high-priced electricity, utilities have begun trying to off-load the risks onto the providers. Southern California Edison has announced an agreement with Stirling Energy

Systems to buy 500 MW of electricity from a solar plant to be built in the desert 80 miles northwest of Los Angeles. Stirling Energy will use four-story mirrors to heat hydrogen that will run a Stirling engine, an internal combustion engine invented in the nineteenth century by Robert Stirling, a Scottish minister. The array will cover seven square miles.

Does the project represent the long-awaited breakthrough on solar energy? It is not likely. Once again, such facilities almost never differentiate between *capacity* and *output*. As with windmills, solar electric facilities produce only 20 to 30 percent of their nameplate, since the sun doesn't always shine. The Stirling technology has never been tried on a large scale. If it does not deliver, the Phoenix company will take the loss.

The biggest impediment to advancing solar electricity is that although it provides its best power at the most crucial time of day, when air conditioning puts the biggest strain on the system, there is currently no way this advantage can be translated into the price. The way to do this would be through a system called "time-of-day" pricing.

Of the 8,760 hours in the year, 19 percent are considered "on-peak"—generally 4 PM to 10 PM during winter months and 10 AM to 6 PM from April to September. Household customers consume 25 percent of their electricity during these hours. Utilities would love to get homeowners to shift 5 percent to off-peak—by washing clothes or running dryers at night or cutting down on lighting or air conditioning during the day. That would remove the necessity of building expensive peaking plants that sit idle most of the time.[17]

To date, time-of-day pricing has been advocated as a way of promoting energy conservation. But its real strength may lie in promoting solar energy. If customers are paying higher rates for on-peak hours, then rooftop collectors become more economical. If homeowners are then able to sell excess electricity back to the utility at peak prices, the package becomes even more attractive.

The problem with solar electricity is that it is now being oversold as the solution to *all* our energy problems. `In January 2008, *Scientific American* featured yet another cover story, "A Grand Plan for Solar Energy," in which three prominent researchers argued that we can power our entire economy by erecting photovoltaic collectors on 30,000 square miles of desert in the American Southwest (about one-quarter of the area of New Mexico).

> Studies by the National Renewable Energy Laboratory in Golden, Colo., show that more than enough land in the Southwest is available without requiring use of environmentally sensitive areas, population centers or difficult terrain.[18]

Such a facility, of course, would require tremendous backup and storage capacity, since one cloudy day would shut down the whole country.

Such efforts to oversell solar energy are bound to lead to disillusionment, both with individual investors and the general public. A much more sensible strategy would be to take solar's strengths—zero carbon emissions and strong peaking capacity—and incorporate them into the price. This would mean two things:

a) a carbon tax or marketable rights system reflecting the long-term environmental damage of fossil fuels, and
b) a time-of-day pricing system that would properly reward solar's peaking power.

But a carbon tax or marketable rights system also emphasizes the environmental strength of terrestrial energy. Once again, the best strategy for pushing these things through Congress or the state legislatures would be a solar-nuclear alliance.

Chapter 13

Geothermal, Waves, Biofuels

We will now consider three other forms of energy that fall under the category of "renewables"—geothermal, wave and tidal energy, and biofuels.

Geothermal is a good source of heat and steam for generating electricity. Tidal and wave energy taps the vast forces that move the ocean and will probably be converted to electricity as well. Biofuels have been promoted as a way of replacing foreign oil and—almost as an afterthought—a way of reducing carbon emissions.

As terrestrial energy, geothermal is *not* trying to capture large amounts of energy from the sun and therefore does not require large amounts of land. Instead, the energy is concentrated deep in the earth. The limiting factor is how close terrestrial heat comes to the surface. Where magma—liquid rock—breaks through the earth's crust, geothermal heat is readily accessible.

That is why Iceland, a volcanic island, is planning to extract 80 percent of its power from terrestrial energy.

Tidal energy is really "lunar energy," since it is driven by the gravitational pull of the moon in its orbit. Waves, on the other hand, are driven by the wind, which derives from the sun. In either instance, the kinetic energy of tide and waves faces the inexorable restrictions of gathering solar energy–huge requirements of surface area, in this case the ocean. With tidal and wave energy, no one has even begun to explore the consequences, but they will inevitably be great.

Unfortunately, we have discovered the land requirements of biofuels and they have turned out to be far greater than ever anticipated. The effort to distill ethanol from corn now consumes *one-third* of the American crop, yet this offsets only 3 percent or our oil consumption. Even worse, biofuels have ended up competing with conventional agriculture and are now being held responsible for the doubling of world food prices. The UN attributes food riots in more than a dozen developing countries, in part at least, to American and European attempts to grow biofuels.

Environmentalists have long rejoiced that biofuels were enlisting farmers to their cause. Now that the adverse consequences have been revealed, however, it may prove impossible to get rid of them. No American political constituency is more firmly entrenched than Midwestern farmers. No matter how bad the outcome, biofuels subsidies may prove as difficult to end as farm subsidies in general, which have been around since the 1920s.

It is a lamentable lesson. Energy problems cannot be solved by forcing a technology to do what it cannot do. Government cannot simply "pick winners" and then subsidize them into success. By overriding market signals, we create economic and environmental destruction—and in this case a world food crisis. The biofuels fiasco is a clear warning that solar technology has its limits. It is not sitting around "free" in "renewable" and "sustainable" quantities. Unless it has long since been distilled into more compact form and buried in the earth—i.e., the fossil fuels—the amount of land required to capture and store solar energy is staggering. Land, after all, is also a limited resource.

Geothermal. Geothermal energy, as we have said, is terrestrial energy, produced when the decay heat of radioactive uranium and thorium encounters groundwater in the earth's crust. The total aggregate of this heat is extraordinary. The earth's surface constantly radiates 16 watts of energy per square meter, about 1/200 the strength of bright sunshine. Over the entire planet this accumulates to huge quantities. One second's output is enough to provide all of America's energy needs for four years.[1] As with any naturally occurring source of energy, however, the big problem is to concentrate it and harness it for work.

With geothermal vents, nature has done this already by heating rock and groundwater enough to produce hot water or steam. Historically, these sites were "hot springs" that attracted people for their health effects. Often these sites became resorts such as Saratoga Springs, N.Y., Hot Springs, Ark., and Marienbad and Baden-Baden in France and Germany. For centuries, it was thought that the warm waters and their sulfur content was the source of healing power. Only recently has it been recognized that the radioactivity in the rocks may also have a healing effect—as we shall see in a later chapter.

Geothermal sites have long been tapped for energy. In China, geothermal waters are used in agriculture and aquaculture to prevent damaging frosts and extend the growing season. In Romania, geothermal waters are used to heat 3,000 homes and 50 acres of greenhouses, sanitary hot water for 16,000 dwellings, and industrial hot water for 10 factories. Geothermal heat is being used for timber drying in New Zealand and for other agricultural purposes in Greece, Israel, Guatemala, and Mexico. Through heat-pump technology, it can also be employed for air conditioning and refrigeration.[2]

Most prized, however, are sites that produce enough steam

to generate electricity. These tend to be located on volcanic islands or along geological faults. Iceland has so many geothermal sites that the country is planning to tap them for 80 percent of its energy, building several large electric plants and producing hydrogen for auto fuel. Indonesia has extensive resources and the Philippines already gets 27 percent of its electricity from geothermal vents. In the United States, Hawaii gets 25 percent of its electricity from geothermal energy, Nevada 10 percent and California 6 percent. Altogether twenty nations derive a total of 12,000 MW for heating and 8,000 MW for electricity.[3]

In a startling new departure, Australia has discovered a huge formation of heated granite two miles below the surface in the remote interior desert, far from any fault lines or volcanoes. The radioactive heat has apparently become trapped beneath a sedimentary cap rock—just as do oil and gas. Although far from population centers, the huge reservoir promises to supply the continent with a considerable portion of its electrical power. The discovery has sent geologists scurrying to find for similar formations in other parts of the globe.[4]

In only a few places, however, does geothermal steam vent to the surface in a way that it can be directed through turbines. At the vast majority of sites, steam must be reached by cutting through geological layers, just like drilling for oil. The most common technology is the "single flash power plant" where superheated subterranean water is brought to the surface while still under pressure. It enters a "separator" where it flashes into steam. The steam drives a turbine, then runs through a condenser and a cooling tower. The process is identical to a boiling water nuclear reactor, the first kind developed. From the outside it looks just like a nuclear plant. The hot water may also be siphoned off for heating purposes.

In the more complex "binary" plant, geothermal hot water transfers its heat to an organic fluid that has a low boiling point. The organic fluid vaporizes, driving a turbine. The design is identical to a pressurized-water nuclear reactor, where water superheated by the nuclear material passes its heat on to a secondary fluid, which evaporates, driving a turbine.

Geothermal plants do have their problems. The hot water or steam often contains high levels of poisonous hydrogen sulfide that must be diverted. Chemical residues also build up and pipes can corrode. Sometimes reservoir pressure is not enough to carry hot water to the surface. Pumps must be then employed—just as oil must be pumped out of the ground when reservoir pressure begins to fall. Underground aquifers are by no means inexhaustible and usually it is the water that plays out first.

After twenty-five years, The Geysers, the world's largest site, in northern California, experienced a sharp decline in reservoir pressure. In 1997 a 29-mile pipeline was built to carry sewage wastewater from neighboring Lake County. This reinjected effluent has been enough to restore the site's 70 MW capacity. At other sites, however, injected water has disappeared into the earth, never to be seen again.[5]

Although volcanic islands such as Iceland, Hawaii and New Zealand may be able to tap geothermal energy for large portions of their energy budget, large continental masses such as the United States and Europe will have only limited production. California, the nation's leading earthquake zone, has 1,800 MW of installed capacity, providing 5 percent of its electricity. However, there are few other regions as favorable in the United States.

Geothermal is pure terrestrial energy, simply tapping the earth's eternal heat. In places where the earth's hot magma has intruded close to the surface—usually where continental plates are rubbing together—it can be readily accessed. On the continental plates themselves, however, it can be necessary to drill ten to fifty miles or more to reach magma. (The deepest oil wells now go three miles.) A much easier way to gather terrestrial energy may be to mine its source—the uranium and thorium atoms—and reproduce the process in a nuclear reactor.

Tidal Energy. Waves are caused by the wind blowing across the surface of the ocean. In that way, they are solar energy. The tides, however, are driven by a combination of the earth's rotation and lunar gravity and they are properly called "lunar energy." Both waves and tides are "renewable" and "sustainable" in that the motions of wind and tide are eternal and seemingly unlimited for our purposes. Once again, however, the big requirement will be the huge quantities of ocean surface needed to gather them.

The flow of the tides is strong enough to produce commercial quantities of electricity only ten of the twenty-four hours in a day. The tides are perfectly predictable but they keep shifting. Because of the moon's month-long journey around the earth, the ten-hour period moves forward about fifty minutes a day. (The sun's gravity also has an impact, making the schedule a little irregular.) Thus lunar energy is constantly moving in and out of phase with periods of high demand.

Nonetheless, the search for new sources has led inventors all over the world to try harnessing energy in the oceans as a useful form of power. The world's first and only tidal electric plant is a 240 MW "barrage" built across the Rance estuary near St. Malo, France. The concrete structure stretches across the inlet and its turbines spin as the tide moves in and out. There are a few dozen other sites around the globe where geography offers similar opportunities. Nova Scotia's Bay of Fundy, which has the highest tides in the world, has been studied for two decades although nothing has yet been attempted. One problem is that tidal estuaries are amongst the world's most productive ecosystems. Building concrete barriers and slowing the tidal flow could possibly disrupt ecological cycles.

Another possible strategy is to capture the energy potential of ocean waves, which are created through a combination of ocean winds and tidal ebb and flow. Seawater is 832 times as dense as air, which makes an 8 knot current the equivalent of a 240 mph wind. Several inventors have made the attempt, although no large-scale efforts have yet emerged.

In 1974, Professor Stephen Salter of the University of Edinburgh introduced a small, free-floating device that turned 80 percent of wave motion into electricity in pilot experiments. The device has ever after been known as "Salter's Duck." For a while Britain had a UK Wave Energy program that conducted experiments with the duck. A large-scale project was finally scuttled at a 1982 committee meeting that is still the subject of conspiracy theories around the world. Drawing on this power would involve covering the ocean with thousands of ducks and connecting them by a web of electrical wires. No one has yet attempted this.[6]

Another larger device developed by the Japanese is called the "Mighty Whale."[7] Unlike the duck, the whale and other similar devices must be anchored to the ocean floor. This creates problems, since sands are constantly shifting and corrosion occurs rapidly in salt air and water. Blue Energy, a Canadian company, has invented a "tidal bridge" that can extend for miles and miles, generating thousands of megawatts of power, at least in theory. A joint study by NREL and the Electric Power Research Institute concluded that wave energy might be economical in Hawaii, Washington, Oregon, Northern California, and on Cape Cod but would require considerable investment.

Portugal has decided to become the "Denmark of tidal energy" and is experimenting with Pelamis, a device invented in Scotland that resembles a sea snake. A company called Ocean Power Delivery has built three 750-kilowatt installations, each the length of a football field, for a total of 2.25 MW. The company is now planning a 5 MW system off the coast of England.[8]

Alexander Gorlov, a 73-year-old Russian mechanical engineer, has developed a Helical Turbine that sits in any body of water of more than three feet and turns no matter which way the current is moving. Gorlov's device has received the Thomas Edison Patent Award. South Korea has installed a 15-foot turbine in the Uldolmok Strait and is providing 1 MW of electricity on Jindo Island. New York City is planning to install six smaller turbines in the East River. Gorlov envisions "farms" of thousands of turbines operating in the Gulf Stream and sending electricity and hydrogen to power North America.[9]

Once again, though, we are faced with the twin dilemmas of any form of solar energy. It is highly dilute and gathering it leaves a huge environmental footprint. The United Kingdom's Department of Trade and Industry estimates the global potential for ocean currents is 3,000 gigawatts—enough to power the current world economy—but only 90 GW are recoverable under current technologies. This is almost precisely what we get from our fleet of 104 nuclear reactors. Harvesting this energy would mean covering dozens of straits, bays and river channels around the globe with infrastructure, perhaps extending to the Gulf Stream.

How would these thousands of rapidly spinning underwater turbines affect marine life? Would they repeat the bird kills of land-based windmills? Would such vast networks interfere with boat traffic? Could electricity be transmitted over these vast ocean distances? What would all this cost? Answers to these questions—and the large-scale development of ocean power—probably lie far in the future.

Biofuels. Beginning with the Carter Administration, biofuels rapidly gained favor as a strategy for replacing foreign oil. Alternate energy enthusiasts have long recommended them and President George W. Bush adopted the cause in his 2008 State of the Union Address, even mentioning switchgrass, which is a kind of shibboleth among biofuels enthusiasts.

The premise is that growing our own fuel will make us more energy independent. As an afterthought, biofuels were also an advanced way of dealing with global warming. The argument is that they are "carbon neutral" because they only put carbon into the atmosphere that was taken out last year.

All this involved some dubious accounting, since carbon that is not incinerated immediately will remain in one of the various "sinks"—the soil, plant life, animal—for decades and more.

What was totally ignored, however, was the amount of *land* that would be needed to make any dent in our oil consumption. It turns out to be staggering. Our current effort is already driving up food prices both domestically and on the world market as farmers switch to corn to take advantage of the subsidies. In the tropics, projects to clear rainforests to make way for biofuels plantations are already considered an environmental disaster. As land becomes a limiting factor, we are suddenly faced with the ugly confrontation of affluent people buying up crops for fuel while people in the developing world go hungry.

How did biofuels go almost overnight from being the darling of the environmental movement—the centerpiece of Amory Lovins' collaboration with the Defense Department in *Winning the Oil Endgame*—to being radioactive, as the saying goes? It's a long story.

Like most forms of alternate energy, the idea of distilling crops for fuels was first suggested in Lovins' *Soft Energy Paths*. In two short paragraphs, he introduced the concept as follows:

> "[E]xciting developments in the conversion of agricultural, forestry, and urban wastes to methanol and other liquid and gaseous fuels now offer practical, economically interesting technologies sufficient to run an efficient U.S. transport sector.
>
> The required scale of organic conversion can be estimated. Each year the U.S. beer and wine industry, for example, microbiologically produces 5 percent as many gallons (not all alcohol, of course) as the U.S. oil industry produces gasoline. Gasoline has 1.5 to 2 times the fuel value of alcohol per gallon. Thus a conversion industry roughly ten to fourteen times the physical scale (in gallons of fluid output per year) of U.S. cellars and breweries, albeit using different processes, would produce roughly one-third of the present gasohol requirements of the United States.... The scale of effort required does not seem unreasonable...[10]

Biofuels were immediately labeled "renewable," "sustainable," and "solar," and by the late 1970s had tremendous momentum. The 1978 Energy Tax Act gave gasoline blended with 10 percent ethanol an exemption from the 4-cents-per-gallon federal excise tax. Over the years this exemption has grown to 52 cents per gallon. In addition, auto manufacturers were allowed to take credit on their CAFE fuel standards for making cars that could operate on E85, a blend of 85 percent gasoline and 15 percent alcohol.

Distilling corn into fuel quickly became the franchise of Archer Daniels Midland (ADM), the agricultural giant headquartered in Decatur, Illinois, that already owned 250 processing plants worldwide. ADM had a product on the market in a matter of months and within a year was selling 175 million gallons. When oil prices dropped rapidly after 1981, the company assembled a contingent of Midwestern Congressmen to increase the subsidy and keep the program alive.

From the beginning, biofuels advocates argued the program would benefit farmers and they were right. Land was shifted out of wheat and soybeans and into corn to take advantage of the incentives. Farmers planted fencepost to fencepost and land prices soared. Ethanol distilleries sprung up all over the Midwest, reviving small towns. The Iowa Presidential primaries became a contest of who could lavish the most praise on ethanol. The program exploded from 3 billion to 7 billion gallons a year between 2004 and 2008 and President Bush—always a few years behind the curve—gave it his endorsement in his 2006 State of the Union. Ethanol now consumes more than 30 percent of the 13-billion-bushel corn crop. Still, this continent-sized effort only replaces 3 percent of our oil consumption.

Now notice that while Lovins considered the *industrial* effort required to produce biofuels, he never estimated the amount of *land*. That's easy enough to calculate. The hops and vineyard industries—the raw materials for beer and wine—occupy 40 million acres. Using Lovins' estimate of "roughly ten to fourteen times the scale" of the beer and wine industries, we can multiply by twelve, giving us 480 million acres—about half the cropland in the United States.

But remember that beer and wine are not all alcohol. Beer is only 5 percent and wine 12 percent, so we can average that at 7 percent, since there is much more beer than wine. This means we must again multiply the 480 million acres by a factor of fourteen, leaving us with 6.5 billion acres—*three times the area of the United States, including Alaska*—in order to produce *one-third* of our transportation needs in 1977.

Those numbers have hardly changed. Writing in the *Washington Post* in 2006, James Jordan and James Powell, two research professors and former biofuels enthusiasts at the Polytechnic University of New York, noted:

> It's difficult to understand how advocates of biofuels can believe they are a real solution to kicking our oil addiction...[T]he entire U.S. corn crop would supply only 3.7 percent of our auto and truck transport demands. Using the entire 300 million acres of U.S. cropland for corn-based ethanol production would meet about 15 percent of demand.... And the effects on land and agriculture would be devastating.[11]

The question that has always dogged corn ethanol is whether there is any energy savings at all. Ordinarily, prices would tell us, since energy costs money, but the subsidies overwhelm this information. Instead, the battle is joined in typical Washington fashion–with competing academic studies. Early examinations showed little or no energy gain, since growing corn requires large amounts of fuel and fertilizer. New methods have improved the balance so that some experts now claim an energy gain of 34 percent, although these figures remain under dispute. In 2000, a DOE study concluded, "The production of ethanol from corn is a mature technology that is not likely to see significant reductions in production costs."[12]

Sugar yields eight times more energy per acre than corn but does not grow well in America's colder climate. Brazil has had great success with sugarcane. Brazilian cars are designed to run on any grade of gasoline-and-alcohol mix and the country now gets 30 percent of its auto fuel from ethanol, although the entire transport sector is only 10 billion gallons—compared to 170 billion in the U.S. This created a volatile issue—should we abolish the import quotas that keep the price of domestic sugar at 21 cents a pound while the world price is 4 cents? As late as 2007, urban representatives such as New York's Senator Charles Schumer were recommending it, while farmers and sugar lobbyists argued it would ruin domestic industries. This was before it was noticed that Brazil's effort was cutting down huge swathes of Amazon forest. Lifting import quotas is now not likely.

The holy grail of ethanol has always been to produce it from cellulose, the more complex carbon molecules that make up a plant's fibrous tissue. Distillation only works on simple sugars and starches in roots and seeds. Anything made from cellulose—stalks, leaves, and woody parts of plants—cannot be digested by humans nor distilled into alcohol. Cows and other ruminants have bacteria in their stomachs that break down cellulose allowing them to eat grass. Termites also host bacteria that digest wood. For several decades, biofuels scientists have been trying to domesticate these microorganisms and deploy them on a commercial scale. Researchers occasionally claim success in the laboratory, but nothing has yet made it into mass production.[13]

The gold standard for domestic ethanol is switchgrass, mentioned in President Bush's State of the Union Address. A narrow-leafed plant that once covered the American plains, switchgrass grows ten feet high with thick stems as strong as pencils. It takes root almost anywhere and thrives with little water or fertilizer, returning nutrients to exhausted soils. As a perennial, it does not have to be replanted. It can be harvested with standard farm equipment. The Department of Energy is trying to develop even more hardy varieties at its Biofuels

Feedstock Development Program at Oak Ridge National Laboratories.[14]

In "Growing Energy," an 85-page report issued in 2005, the Natural Resources Defense Council's Nathanael Greene estimated that growing half our fuel from switchgrass by 2050 would take 114 million acres—about 25 percent more than the 90 million acres we now plant in corn. But this would be contingent on cars getting 50 miles per gallon.[15] In *Winning the Oil Endgame*, Amory Lovins convinced the Defense Department we could provide *all* the nation's transportation needs by planting all of North and South Dakota with switchgrass—90 million acres. "Would we rather be dependent on the Middle East or the Middle West?" he exuded.[16] All this, however, was contingent on automobiles getting 140 miles per gallon. It also depends—it should be repeated—on finding a commercial way to convert cellulose to ethanol, which has not yet been accomplished.

"Biodiesel" is yet another biofuels technique, based on Rudolf Diesel's versatile diesel engine, invented in 1892. The diesel is a two-stroke engine with no spark plugs that can run on much lower grades of fuel. (Diesel originally wanted to burn coal dust.) Combustion is less complete than gasoline, which makes diesel exhaust much dirtier—as anybody can testify who has ever become trapped behind a bus in traffic. Nonetheless, Europe has adopted diesel for 40 percent of its auto market, mainly for fuel-economy reasons. Although biodiesel emits the same carbon dioxide, carbon monoxide and particulate matter as oil-based diesel fuels, enthusiasts claim its exhaust is "sweeter" and has a faintly organic aroma.

Seeking new sources of "green" energy, biodiesel enthusiasts have proposed tapping all kinds of sources—cooking grease, food scraps, crop wastes, anything organic. Typically, someone will design a car that runs on some organic waste—turkey droppings, hayseed, coconut oil—and drive it around until it attracts press attention. Then they will announce they have solved the world's energy problems. A few years ago a group of Hampshire College students drove cross-country in a car that burned discarded restaurant grease. When Jane Fonda announced she would visit eight cities on an anti-Iraq-War tour in 2005, she said she would travel in a bus powered by vegetable oil. (The trip was later scaled down to two cities.) In May 2007, Staten Island Congressman Vito Fossella introduced legislation to double the tax credit for turning restaurant grease into biodiesel. ""From cooking fried calamari to powering trucks, restaurant grease represents a viable energy source for our nation," he announced.[17] Former Secretary of Energy Spencer Abraham says he once attended a national energy conference in which the first half-hour was given to a woman who had started a small company recycling cooking oil.

What is completely ignored in all these efforts is any element of *quantity*. So let's look at some numbers. In the summer of 2008 gasoline had dropped to $3.50 from $4 a gallon and was still considered very expensive. At the local supermarket, the house brand of vegetable oil sells for $4 a gallon. Brand name canola, sunflower peanut and corn oils range between $5 and $7.50 a gallon. Don't even ask the price of virgin olive oil. All these are well-established commodities in mature industries. Millions of acres are devoted to growing these crops. Yet the entire cooking-oil industry produces only 60 million barrels a year—three days' worth of oil. Think of how often you go through a gallon of cooking oil. Now think of how often you go through a gallon of gasoline. How are we going to substitute one for the other?

Or consider restaurant waste. A recent *Wall Street Journal* story recounted the experience of a British company named Green Fuels that sells European farmers recycling kits enabling them to burn cooking grease in their tractors. With oil prices rising, sales had climbed from eight to 45 units a month. At that point, company president James Hygate was stunned to discover his customers were encountering a cooking grease shortage! There were only a few restaurants in town and all had run out. The biggest problem for biodiesel vehicles comes when other people start adopting the technology.[18]

There are 15,000 McDonald's restaurants in 79 countries across 6 continents. To make a wildly generous estimate, say each produces five barrels of kitchen grease per day. If all this were converted to biodiesel, it would produce 75,000 barrels per day—approximately .004 percent of America's 18-million bbd oil consumption. Our daily oil consumption would cover 250 football fields ten feet high. All the McDonald's kitchen grease in the world would cover three-fourths of one field to the same height. The EPA estimates that all U.S. restaurants produce 300 million gallons of waste oil per year—exactly one gallon for every American.

Facing a shortage of cooking oil, biofuel enthusiasts have expanded their horizons. Sawdust, rapeseed, castor oil, soybeans, palm oil, hemp, animal fats, agricultural wastes—anything is regarded as fair game. Yet claiming these wastes for incineration assumes they are not being recycled already. Look at any bag of garden fertilizer and you will find it has agricultural origin. Cow manures, turkey droppings, pig carcasses—all are put to some use. Sawdust and timber wastes are used to make fiberboard. Most excess vegetable oil and animal fats are already recycled as soap.

American farms and industries are far more efficient at recycling than is generally acknowledged. Much organic material is plowed back into the soil as compost. This is the principle of "organic farming"—a subject that once interested environmentalists before they became experts on alternate energy. Converting these supposed wastes to biodiesel means putting carbon back into the atmosphere that would otherwise remain in plant and soil for years or decades.

Most surprising is supporters' continued insistence that burning crops is "carbon neutral" and will therefore help prevent global warming. It's not like burning fossil fuels, the argument goes, whose carbon has been stored for millions and millions of years. With biofuels we are only putting back carbon into the atmosphere that was just taken out. Therefore the net addition is zero.

But all this assumes that *nothing was growing on this ground beforehand*. If an acre of grass is left fallow, most of the carbon will be incorporated into the soil, where it may remain for thousands of years. Except for coal and the deep oceans, soils are the world's biggest carbon sink. Even if the grass is fed to cattle or humans, it will be a long, long time before it finds its way back into the atmosphere. If crops are burned as biofuels, however, the cycle is truncated and carbon is immediately converted back to carbon dioxide. Only if we ignore everything that could grow in their stead can biofuel crops be called "carbon neutral."

All this finally came to a head in February 2008 when *Science*, the highly respected publication of the American Association for the Advancement of Science, simultaneously published two studies questioning the entire biofuels effort. One, headed by Environmental Economist Timothy Searchinger of Princeton, concluded that growing biofuels almost anywhere will result in clearing land somewhere else for food or fuel. This, of course, is generally called "deforestation." The other study, headed by Joseph Fargione of the Nature Conservancy, finally asked the question, "What was growing there before?" Factoring this into the equation, the researches concluded that clearing grasslands for biofuels increases carbon emissions *93 times* in the first year. "So for the next 93 years you're making climate change worse," concluded Fargione.[19]

Suddenly, the roof was crashing in on biofuels. *Time*, which in February 2008 had celebrated Richard Branson's conversion of one of Virgin Airways' jets to biofuels, ran an April cover story calling biofuels "environmentally disastrous" and a "catastrophe." "Hyped as an eco-friendly fuel, ethanol increases global warming, destroys forests and inflates food prices," said

Michael Grunwald, in "The Clean Energy Scam." "So why are we subsidizing it?"[20]

While biofuels' adverse calculus had been disguised by subsidies in the Midwest, the consequences have become ruthlessly clear in the tropics. South American and Asian countries are clearing forests to make way for sugarcane and palm oil plantations that sell biofuels to Europe and America. The results have been devastating. Burning rainforests for palm oil plantations had already helped make Indonesia the world's third largest emitter of carbon dioxide, behind the United States and China.[21]

Friends of the Earth, the home organization of Amory Lovins, David Brower, and *Small is Beautiful* author E.F. Schumacher, first sounded the alarm in 2005 with a study called "The Oil for Ape Scandal:"

> Between 1985 and 2000 the development of oil-palm plantations was responsible for an estimated 87 percent of deforestation in Malaysia.... In Sumatra and Borneo, land clearing for palm oil is threatening the last redoubts of the orangutan and other endangered species.... Unless we act now, the orangutan—Asia's only great ape and one of humankind's closest living relatives—will become extinct.[22]

In response to these studies, the European Union was talking about banning certain imported biofuels.[23]

The biggest threat, however, was to world food supplies. With more than 30 percent of the U.S. corn crop now going into gas tanks and other countries joining the crowd, food prices had already more than doubled between 2005 and 2008. The International Monetary Fund blamed the worldwide surge on biofuels:

> [M]ore recently food prices have jumped sharply, at least in part because of an attempt to encourage the use of so-called biofuels.... The surge in corn prices over the past two years has been remarkable—prices have roughly doubled both in the United States and worldwide.... This then has knock-on

> effects on other crops, as land is switched from wheat on the margin, for example, into corn or, as has been most marked in Europe, out of dairy production and into crops used for biodiesel (for example, rapeseed, whose prices have also increased sharply). In the IMF staff's assessment, a significant part of the latest jump in food prices can be traced directly to biofuels policy.[24]

Mexico had "tortilla riots" in 2007 over rising corn prices. In the next twelve months similar disturbances occurred in Morocco, Egypt, Cote d'Ivoire, Guinea, Mauritania, Cameroon, Senegal, Uzbekistan, and Yemen. In April 2008, the prime minister of Haiti was deposed by hungry mobs demonstrating over food prices.

Suddenly, the condemnation was universal. UN food expert Jean Zeigler labeled biofuels a "crime against humanity" and called for a five-year moratorium.[25] Lester Brown, former head of the Worldwatch Institute and one of American's most revered environmentalists, published a critique arguing, "The grain required to fill a 25-gallon SUV gas tank with ethanol will feed one person for a year. The grain to fill the tank every two weeks over a year will feed 26 people."[26] In 2007, C. Ford Runge and Benjamin Senauer of the Center for International Food and Agricultural Policy at the University of Minnesota published an article entitled, "How Biofuels Could Starve the Poor" in which they revised a previous prediction that there would still be 625 million hungry people in the world by 2025.

> [I]f, all other things being equal, the prices of staple foods increased because of demand for biofuels.... [t]hat means that 1.2 billion people could be chronically hungry by 2025—600 million more than previously predicted.[27]

The article appeared in *Foreign Affairs*, where Amory Lovins first proposed biofuels in 1976. The idea had come full circle.

Biofuels brings to mind the legend of Jubilation T. Cornpone, the hapless Civil War general of the Li'l Abner cartoon. As the song in the Broadway play put it:

With our ammunition gone and faced with utter defeat,
Who was it that burned the crops and left us nothing to eat?

Some day someone should erect a statue of General Cornpone in front of the Department of Energy to commemorate the effort.

It would be hard to find another example of a national policy entered so heedlessly. Markets tell us whether an energy crop is producing any useful energy because energy costs money. But biofuels have been exempted from this calculus. Instead, they have been overlain with so many subsidies and mandates that it is impossible to know whether they are accomplishing anything at all. And none of this is factored in the long-term effects on the agricultural environment.

Perhaps E.F. Schumacher said it best in *Small is Beautiful*: "In our time, the main danger to the soil, and therewith not only to agriculture but to civilization as a whole, stems from the townsman's determination to apply to agriculture the principles of industry."

Chapter 14

Hydrogen and Electric Cars

It is often said that nuclear power won't do anything to replace oil because it can't run automobiles. This is true under present technologies but will not be so under emerging technologies. In fact, the most promising use for terrestrial energy may be that it will allow us to move the transportation sector onto a much-expanded electrical grid. This would reduce our oil consumption and carbon emissions all at once.

The two technologies that hold the most promise are hydrogen-powered vehicles and electric cars. Both have been in the experimental stage for decades. There have been plenty of stops and starts—wild promises, investment bubbles, and total collapses. California has volunteered to be a laboratory for the country, forcing the technology through various draconian mandates. The results have not been very successful. The electric car turned out to be the wrong technology at the wrong time—giving rise to conspiracy theories in the movie, *Who Killed the Electric Car?*

The original electric car failed because it didn't work very well. Slowly but surely, however, other alternatives to the all-gasoline engine are emerging. Hydrogen vehicles and gas-electric hybrids have both gained a foothold. The duel is still being fought out before the California Air Resources Board, which has now become the focus of intense lobbying from all sides. One way or another, however, politics will eventually fall aside and market forces will prevail.

The important thing to remember is that neither hydrogen nor electricity is a *source* of energy. Both are simply a way of transferring energy from other sources into vehicles. If successful, both these technologies could replace oil but will not reduce carbon emissions. Instead, they will require either the creation of a new hydrogen industry or the near doubling of the electrical grid. In either case, the choice will remain the same: "How do we generate this much new energy—with fossil fuels, solar, or terrestrial energy?"

In Jules Verne's 1874 science fiction, *The Mysterious Island*, three Union soldiers escape from a Confederate prison in a hot-air balloon. As they are blown off course toward the Mysterious Island, they find themselves discussing what will happen after the world's coal resources are used up.

> "They will discover something else," said Herbert.
> "But what will they find?" asked Pencroft.
> "Water," replied Harding.
> "Water!" cried Pencroft, "water as fuel for steamers and engines! Water to heat water!"
> "Yes, but water decomposed into its primitive elements," replied Cyrus Harding, "and decomposed doubtless, by electricity, which will then have become a powerful and manageable force.... Yes, my friends, I believe that water will one day be employed as fuel, that hydrogen and oxygen which constitute it, used singly or together, will furnish an inexhaustible source of heat and light, of an intensity of which coal is not capable.... Water will be the coal of the future."

In recent years, visionaries have touted hydrogen as the fuel of the future: a clean, benign combustion that produced warm water as its exhaust. As early as 1976, the Stanford Research Institute published a study called: "A Preliminary Technology Assessment." Hermann Scheer, Green Party member of the German parliament, president of EuroSolar and a "Hero of the Green Century" according to *Time* magazine, saw hydrogen as the chief means of storing solar energy in *The Solar Economy* (1999). Jeremy Rifkin, the peripatetic prophet of futuristic technologies, had promised in *The Hydrogen Economy* that hydrogen would mean "the creation of the worldwide energy web and the redistribution of power on earth." Amory Lovins had even gone so far as to *build* a hydrogen vehicle—the "Hypercar"—raising millions for the effort. Lovins' vision extended even further into the future. He believed the Hypercar could serve as a network of "mini-power plants" that would spell the "end of the car, oil, steel, aluminum, nuclear, coal, and electricity industries."[1]

In 2003, President George Bush brought all this mainstream in his State of the Union Address by announcing a $1.2 billion "Freedom Fuel" initiative to make hydrogen cars commercial by 2020. Bush proclaimed that the hydrogen car would "free us from dependence on foreign oil while producing no pollution and no greenhouse gases."

Yet by that time the reaction among environmentalists was surprisingly negative. "The FreedomCAR is really about Bush's freedom to do nothing about cars today," complained Ashok Gupta of the Natural Resources Defense Council. "The President should be tightening CAFE [Corporate Average Fuel Economy] standards instead." Other groups lamented that $300 million of the $1.2 billion would come out of funds already committed to solar and renewable research. "President Bush is merely playing a shell game," said Patricia Monahan of the Union of Concerned Scientists.

What had happened to hydrogen along the way? Once

again, as with all "green" and "renewable" forms of energy, it is important to understand the technological limits.

Jules Verne was not pulling ideas out of the thin air when he envisioned his hydrogen economy. The fuel cell—the central device of such a system—was invented in 1839 by a Welsh lawyer named Robert William Grove.

In 1800, William Nicholson, an English chemist, had discovered the electrolysis of water. A polymath who invented several industrial machines, Nicholson stuck two electrically charged plates in a container of water and found gas bubbles collecting at each. He identified the gases as hydrogen and oxygen. Within a few years, Sir Humphrey Davey, Britain's outstanding chemist, had used the electrolytic process to isolate a host of other elements as well.

Three decades later, Grove, who supplemented his legal practice with some amateur inventing, began wondering if the process could be reversed. Could combining hydrogen and oxygen back into water produce an electric current? After a long trail of experiments, he finally took two platinum strips, one surrounded by pure hydrogen, the other by pure oxygen, and submerged them in a bath of weak sulfuric acid. A small current flowed between them. Grove realized the commercial possibilities and speculated that hydrogen would one day replace coal and wood as fuel. In 1872, he was knighted for his efforts.

In 1800, Allesandro Volta had created the "electric pile" by stacking alternate layers of zinc and copper, separated by cardboard soaked in brine, which generated weak sulfuric acid (H_2SO_4). The hydrogen ions migrated to the copper, while the sulfate attached to the zinc, creating an electric charge—ever after measured in "volts." A wire connecting the two plates produced a current, creating the first electric battery. Grove now called his new device a "gas voltaic battery," since it generated the same current using hydrogen and oxygen. In 1889, two German scientists, Ludwig Mond and Charles Langer modified Grove's system by pairing hydrogen-heavy coal gas with atmospheric oxygen. They called their device the "fuel cell."

Although a laboratory curiosity, fuel cells faced monstrous challenges before they could compete with turbines in generating electricity on a commercial scale. Little progress was made until Francis T. Bacon, a direct descendent of the great Elizabethan scientist, decided to devote his career to the task in 1932. By 1959, after more than a quarter-century of effort, he had developed a 5 kW fuel cell that could power a welding machine. In the same year, Allis-Chalmers Manufacturing demonstrated a fuel cell that could run a 20-horsepower tractor.

Still, fuel cells remained hopelessly uneconomical. Hydrogen was scarce and the same electric current could be obtained by sticking a plug in the wall. Then came the space program. Searching for an onboard power source, NASA found miniature nuclear reactors too complex and solar panels too bulky. Fuel cells seemed ideal. NASA funded more than 200 research contracts and ended up choosing Bacon's entry. Manufactured by General Electric, "Bacon's Cell" produced electricity aboard ten Gemini missions and the seven Apollo moon voyages. On the space shuttle, astronauts drank the exhaust water.[2]

Providing electricity for a spaceship with hydrogen was a marvelous accomplishment but other people had already envisioned greater possibilities. In a famous lecture given at Cambridge University in 1923, British scientist John Haldane, who would later become an outstanding geneticist, put flesh on Jules Verne's dream by outlining the basics of a "hydrogen economy." Solar energy, Haldane pointed out, was ubiquitous but diffuse. It could not substitute for fossil fuels. But hydrogen offered a means of concentration and storage. Electricity from windmills could split water, producing hydrogen in large quantities, which could then be widely distributed. An entire country could be powered by such a system.[3]

Nonetheless, success in space did not easily transform back to earth. Mass-producing Bacon's fuel cell with platinum would be prohibitively expensive while fitting it into an automobile would require shrinking its parts. Still, General Electric tackled the problem and soon developed a "proton exchange membrane" (PEM), where a thin polymer film replaced the bulky electrolytes. A hydrogen atom is split and the proton passes through the membrane, attracted by an oxygen ion on the other side. The electron is repulsed by the membrane but travels around it via a wire, creating an electric current. It finally reunites with the proton on the other side to reform hydrogen, which combines with the oxygen, producing water.

The promise of fuel cells is their thermodynamic efficiency. While producing electricity with turbines wastes two-thirds of the energy on steam, fuel cells can achieve conversions of 60–70 percent. Their low temperature and lack of exhaust would make it possible to "distribute" them at specific sites, matching power to the task at hand while eliminating the need for transmission wires. Still, by 1980 even the best compact fuel cells could produce only 300 watts, enough to power three 100-watt light bulbs. The average home absorbs 5,000–20,000 watts (5–20 kilowatts) and the average car needs 90 kW.

At this point a petroleum geologist named Geoffrey Ballard entered the picture. A Canadian-born entrepreneur who had worked on energy conservation with the U.S. government under Jimmy Carter, Ballard became convinced that the solution lay on the supply side. In the late 1970s, he and two partners founded a Vancouver company that tried to develop a lightweight rechargeable battery for electric cars. Then in 1983, the Canadian government sought proposals for improving the fuel cell. Without much knowledge of the technology, Ballard submitted a bid and won.

Ballard Power Systems hired a team of engineers and began applying good old Canadian ingenuity, convinced that nobody had ever done much with the technology. The group quickly started making progress. At one point they produced such a surge of electricity that it melted a cable.[4] Soon rumors of Ballard's advances were circulating through the auto and oil industries. The response was hostile. A few American companies took out ads ridiculing the fuel cell and denounced it at auto shows. The reception abroad was more welcoming, however, and in 1993, Daimler-Benz, the German automaker, agreed to a partnership.

Commercial prototypes followed immediately, although the technology was not overwhelming. Daimler Benz's NECAR, introduced in 1994, was so crammed with fuel cells and hydrogen tanks that that it had room for only two passengers. It could do only 60 mph and ran out of hydrogen after 55 miles. NECAR II, introduced in 1996, was an improvement. It could carry six passengers, reach 70 mph, and had a range of 150 miles. (The average gasoline-powered car has a range of 450 miles.) In 1997, Toyota brought out its own prototype.

Meanwhile, Amory Lovins, determined to be more than a theorist, launched his own effort. In 1994, the Rocky Mountain Institute spun off a Hypercar Center with the intention of reconceptualizing the entire automobile. The researchers discovered ultra-strong and lightweight carbon fibers were being used in Formula 1 racecars as well as experimental military jets, and decided they could be applied to ordinary vehicles. With improved aerodynamics, the team came up with an ultra-lightweight car that would require only one-third to one-fifth as much fuel per mile. This would fit the limitations of hydrogen perfectly.

Lovins took his design to Detroit, where auto executives were impressed but declined to participate. (Ralph Nader told him he was foolish to expect any different.) So in 1998, RMI formed its own for-profit auto-manufacturing company. The Hypercar quickly attracted $500,000 from Amoco and Sun Microsystems, plus $1 million from Sam Wyly, a solar-minded Texas billionaire. A year later, the company rolled out a prototype and Lovins began asking for $20 million for full-scale production. Meanwhile, the auto companies started blending carbon fibers into some of their high-end models, although they still insisted a pure carbon vehicle would cost an additional $10,000–$20,000 per car.

Not satisfied simply to produce a new form of transportation, Lovins soon proclaimed that a fleet of Hypercars could grow into a vast network of "micro-power" stations that would revolutionize the entire electrical grid:

> Drive your ultraclean car, then park it and plug it into the grid. It just became a 20- to 40-kilowatt mobile power plant on wheels, selling back electricity at your workplace, when and where it's most valuable.... The U.S. car fleet ultimately will have five to ten times as much generating capacity as all power companies now own. This scenario, plausible before a nuclear plant ordered today could be built, could replace all U.S. power plants many times over and profitably could protect the climate.[5]

This vision was reiterated in forums as diverse as *Business Week*, *Time International*, the *New Zealand Herald*, *The Japan Times*, and the *Ottawa Citizen*. In a 3,500-word profile in *Fortune* entitled, "Can This Man Solve the Energy Crisis?" Lovins again promised the Hypercar would "end the car, oil, steel, aluminum, nuclear, coal, and electricity industries."[6] Even *Public Utilities Fortnightly* became enthusiastic, conjecturing that "a full Hypercar fleet in, say, the United States would represent about five or 10 times the generating capacity that all power companies now own. So, it doesn't take many people accepting the value proposition [...] to put at least all the central stations out of business."[7]

All this generated quite a buzz in Washington. In 1996, Congress funded research through the Hydrogen Future Act. President Bill Clinton carved out a new hydrogen division in the Department of Energy, appointing Joseph J. Romm, a former Lovins' staffer, to oversee it. By 2000, the federal government had spent more than $1 billion on hydrogen-and-fuel-cell research.

As the millennium approached and the dot-com boom mushroomed, hydrogen loomed as the next big thing. Ballard went public and raised $370 million. Daimler merged with Chrysler in 1998 and together they joined Ford to produce TH!NK, a four-door hydrogen-powered family sedan. *Time* called GM's Hy-Wire, another hydrogen vehicle, "The Car of the Future." "It has no engine, no steering column and no brake pedal. It requires no gasoline, emits no pollution (just a little water vapor) and yet handles like a high-performance Porsche," said the editors.[8]

Small firms popped up everywhere. Plug Power, an upstate New York start-up, claimed to have a compact fuel cell that would enable homeowners to generate their own electricity from natural gas. In 1998, the company joined with General Electric to create GE Fuel Cell Systems. Plug Power agreed to deliver a product by 2001 while GE would use its vast marketing network to distribute it worldwide.

When the dot-com bubble burst in 2001, however, hydrogen went down with it. Indeed, there were already creeping doubts about the technology. Ballard's stock took a huge hit, falling from $140 a share to $6. When the 2001 deadline arrived, Plug Power found it didn't have a home generating package after all. General Electric went to court for breach of contract and a blizzard of lawsuits followed. Two-thirds of the new fuel cell companies went under. From "the fuel of the future," the buzz became, "Who put the hype in hydrogen?"

So things stood when President Bush made his State of the Union Address in 2003. Behind the curve as usual, the President was indeed coming more than a little late to the party. Enthusiasm was already waning and gas-electric hybrids were overtaking hydrogen as the "Car of the Future." The most telling blow came a year later when Joseph J. Romm, President Clinton's former head of hydrogen research, published *The Hype About Hydrogen*. After half-a-dozen years in the field, Romm had decided hydrogen wasn't so wonderful after all.[9]

The problems, he pointed out, were always known but too easily ignored. Hydrogen is the smallest atom and escapes from any pipeline or storage container. Losses are 2 percent over 100 miles and could quickly mount to 10–20 percent. A group of Caltech engineers even speculated that escaped hydrogen could damage the ozone layer, although others have disputed this.[10]

Storing hydrogen aboard a car presents huge problems. First

it must be liquefied at very low temperatures and high pressure. Even then, hydrogen has only about one-fourth the energy density of gasoline, meaning fuel tanks must be four times as large. Romm believes it is unrealistic to expect hydrogen cars to be commercially available before 2025.

Still, there may be a market for *stationary* fuel cells. Data storage centers and server farms drink up so much electricity they can often support their own generating stations. (When the Lawrence Livermore National Laboratory first turned on its huge 8 MW IBM "ASCI Purple" computer, customers all over San Francisco saw their lights dim.)[11] Such on-site generation, now provided by noxious diesel generators, could easily be provided by efficient fuel cells.

Yet before we go any further in talking about the large-scale deployment of fuel cells, we must return to the question: "Where do you get the hydrogen?"

Although it is rarely clarified, hydrogen is not a *natural* resource. There is no hydrogen sitting around waiting to be mined or harvested. Enthusiasts talk about hydrogen being "the most common element in the universe" but this is only true on a cosmic scale. On earth, oxygen and silicon are more abundant. While oxygen is heavy enough to remain in the atmosphere, all the free hydrogen floated long ago into space—just as a child's balloon will float high into the sky if it slips out of his or her hand. What hydrogen remains on earth is locked up in chemical compounds, mainly water (H_2O) and methane (CH_4). Freeing hydrogen means extracting it from other resources. This requires *energy*.

As a result, hydrogen cannot serve as a *source* of energy. It can only be a *carrier* of other energy resources, terrestrial or solar. In this way, it is identical to electricity, which is only a way of transporting energy generated by coal, gas, wind, nuclear, or hydro to other uses. Hydrogen's only advantage is that it may be fluid and compact enough to power automobiles. It will not *replace* terrestrial or solar energy. All it can do is transfer them to other places.

The main source of commercial hydrogen is now natural gas (CH_4), which is stripped of its H_2 through a process called "reforming." This leaves the carbon, which forms carbon dioxide. Thus we are back to greenhouse gases. On the gigantic scale required to power our transportation sector, hydrogen would be much better manufactured through electrolysis—which means electricity. A hydrogen transport system would probably double the burden on our national grid. A more direct way of using electricity to run vehicles would be the electric car itself. This has now been attempted with some success.

The Electric Car. As previously related, electric cars were common at the turn of the twentieth century. For awhile they competed head-to-head with gasoline and steam. There had been electric trains since the 1830s but these ran on contact with "third rails." Powering a freestanding vehicle was something different. From the beginning—and right up until the present—the problem has always been the life and power of the battery.

Early batteries were lead-acid, but in 1899, Waldemar Jungner, a Swedish scientist, invented the nickel-cadmium battery that eventually evolved into nickel-iron. Thomas Edison took note and proclaimed it perfect for an electric car. Soon there were several companies—Detroit Electric, Columbia Electric, Baker Electric—selling a variety of models, most of them made to run around town. Edison purchased a Baker for $850 in 1902, his first car. Electrics generally made 10–12 mph, although experimental models set speed records at 120 mph. The limiting factor was the car's *range*—only around 80–100 miles. Even so, a pair of wayfaring adventurers were able to drive a Columbia Electric 250 miles from Boston to New York in 1903 on a single charge. The trip took 23 hours, averaging 11 mph.

As recounted earlier, the advantages of the internal combustion engine—particularly its greater range—soon won out and electric cars faded. In 1910, Detroit Electric was still selling 2000 cars a year but Columbia was sold to a gasoline-car company and Baker went under in 1916. There was a brief revival during World War I when gas became scarce, but after 1918 the Model T swept everything in its path. By the 1950s, the only person still driving an electric around town was Grandma Duck.

With the Energy Crisis of the 1970s interest revived. The auto companies renewed research and in 1990 General Motors announced it had plans for a new electric, EV1. The vehicle might have evolved from there, but the California Air Resources Board leaped upon the news and *mandated* the auto companies produce "zero-emissions vehicles" (ZEVs) by 1998. ZEVs would have to be 2 percent of California sales the first year, 5 percent by 2001, and 10 percent in 2003. Stiff fines or exclusion from the California market were the penalties. At the time, lead-acid batteries still had a range of only 80 miles and required four hours to recharge. Just for opening its mouth, General Motors, plus all the others, would be required to rush cars into production.

The auto companies worked furiously to improve their batteries but in 1996 they told the CARB they would be unable to meet the deadline. The CARB relaxed the 1998 and 2001 mandates but insisted the manufacturers market 1,800 ZEVs in California by 2000. Then, in a surprise move, Toyota introduced the Prius to Japan in 1997. A gas-electric hybrid, the Prius draws some inertia off the wheels during braking and uses it to produce an electric current that recharges the battery. Unfortunately, the Prius still emitted some exhausts and therefore could not be sold as a ZEV. It would not help for Toyota in California.

In 2000 the majors all brought out electric vehicles in the Golden State. Because of an obscure federal safety regulation, the cars could only be leased instead of sold outright. In any case, they attracted few customers. Despite a massive advertising effort, Toyota was only able to lease 300 of its RAV4-Ev model. Most electrics were dumped onto car rental agencies where they sat quietly on the back lots.

In 2002, Vijay V. Vaitheeswaran, the energy correspondent for *The Economist*, came to Los Angeles on a story. At the rental agency, the sales clerks persuaded him to try an electric, telling him it had a range of 100 miles, would recharge "pretty quickly" and allow him to drive in HOV lanes. Reckoning this as part of his research, Vaitheeswaran took up the offer. As he recounted a year later in *Power to the People*:

> The vehicle proved to have a much shorter range than I thought it would—closer to 50 miles than a 100. The fact that I sped along at 80 mph in those empty HOV lanes might have drained the battery faster, but only certain highways had that lane; more often, I was crawling along in traffic like everyone else. And most of the time, I was going nowhere at all, since my vehicle kept running out of power. Charging proved the biggest nightmare. There were plenty of chargers around, but some were of the wrong sort; others were locked or nonfunctional. And rather than the "pretty quick" recharge, my useless battery took more than five hours for a full charge. As a result, my entire visit turned into a fiasco of delayed or missed appointments, apologetic cell-phone calls, and panicky exits from the highway to obscure malls and commuter-rail stations in search of a charger.[12]

In 2001, GM and the other manufacturers sued to block the implementation of the 2003 standards, arguing that California was usurping the federal authority to regulate fuel efficiency. The U.S Justice Department joined the suit. In 2003, the CARB relented and switched to a cap-and-trade system where the manufacturers could earn credits by marketing a variety of low-emission vehicles—hybrids, hydrogen cars, natural-gas powered vehicles, and gasoline engines with advanced emissions controls. The CARB itself would decide how the credits were awarded.

Unfortunately, this made the process even more political. The companies now spend their time lobbying CARB, since how the credits are awarded determines the fate of each vehicle. Meanwhile, the manufacturers gathered all their old electrics off the back lots of leasing agencies and crushed them into scrap.

All this became fodder for *Who Killed the Electric Car?*, the 2006 documentary that made the Sundance Film Festival. Rounding up a handful of the eighty-five people who had leased electrics, the filmmakers recorded their testimony on how much they liked them. Then the electric's demise was attributed to a conspiracy among the oil companies, the auto companies, and the Bush Administration. Most interesting was that the hydrogen car was also portrayed as a villain—something that would have interested Ballard, Lovins, and all the other pioneers. In any event, the images of dozens of electric cars being crushed and shredded gave the indelible impression of sinister forces loose in the world.

Choosing the original EV1 as one of "The 50 Worst Cars of All Time" in 2007, *Time* wrote the following obituary:

> The EV1 was a marvel of engineering, absolutely the best electric vehicle anyone had ever seen...quick, fun, and reliable. It held out the promise that soon electric cars—charged from the grid with all sorts of groovy power sources, like wind and solar—could replace the smelly old internal-combustion vehicle.... In fact, battery technology at the time was nowhere near ready to replace the piston-powered engine. The early car's lead-acid batteries, and even the later nickel-metal hydride batteries, couldn't supply the range or durability required by the mass market. The car itself was a tiny, super-light two-seater, not exactly what American consumers were looking for. And the EV1 was horrifically expensive to build, which was why GM's execs terminated the program—handing detractors yet another stick to beat them with. GM, the company that had done more to advance EV technology than any other, became the company that "killed the electric car.[13]

Meanwhile in 2004, Toyota introduced the Prius to Americans. There was a rush of orders and by November the company was forced to step up production. Released in January 2005, the Prius became the "Motor Trend Car of the Year," with waiting lists stretched out for six months. Strangely, the American version couldn't be plugged into the wall for recharging, although the Japanese version could. Toyota explained. The EV1 experience had given electric cars such a bad name that the company did not want the Prius associated with them in buyer's minds. Undaunted, hot-rodders began hacking their Priuses and recharging them overnight. In 2007 Toyota finally relented and made the plug-in a standard feature.

Hydrogen got a new boost in 2004 when California Governor Arnold Schwarzenegger promised a "Hydrogen Highway"—a network of 200 fueling stations across California by 2010.[14] The Big Six—Toyota, GM, Ford, Nissan, DaimlerChrysler, and Honda—all revitalized their hydrogen programs and were once again rushing to meet yet another mandate from the CARB that they introduce 250 fuel-cell cars by 2008.

The California Air Resources Board continues to muddle the situation by micromanaging technology. Gas-electric hybrids get only one-fourth the credit of pure electrics and one-sixteenth the credit of hydrogen cars, which are still the CARB's favorite. The obvious standard is *how reluctant consumers are to buy the car*. Hybrids sell easily so they require no mandates, but few people want hydrogen. Meanwhile, Phoenix Motorcars, of Rancho Cucamonga, announced it would bring out a lithium-equipped pickup truck that recharges in ten minutes. Unfortunately such a hefty charge is only available at utility substations. Tesla Motors, of San Carlos, unveiled an all-electric sports car that goes from 0 to 60 in 3.6 seconds, has a range of 220 miles, and costs only 2 cents per mile to operate. Its price tag, however, is $70,000. Desperately trying to improve its finances, both Phoenix and Tesla were petitioning the CARB to tip the credit system back in electrics' favor.[15]

Finally, GM, the "company that killed the electric car," is jumping back in the fray with the Volt, a fully electric hybrid unveiled as a "concept car" in 2007 and scheduled for 2010 production in 2010. While the Prius switches back and forth between gas and electric motors, the Volt will be all electric, using its three-cylinder, on-board gasoline engine only to recharge the battery. A full charge will carry the Volt 40 miles with on-board recharging extending the range to 300 miles—in theory. The lithium-ion batteries are still in development and everyone in Detroit is holding their collective breath. "I understand what the Volt means to the company and, quite frankly, to the industry and to our country," says Denise Gray, a graduate of Detroit's Cass Technical High School who is overseeing the project.[16] If the Volt doesn't succeed, the documentary filmmakers will be circling the waters, ready to chronicle the latest episode in the great auto-oil-company conspiracy.

Whatever mistakes can be made in forcing technology, California seems determined to make them. Along with coming close to killing the gas-electric hybrid, the state had created a Third-World-type shortage on its electrical grid by forcing alternates and renewables. In the end, though, the rest of the country has probably benefited—if only in learning that "soft energy" can't be mandated to carry more than its natural share.

Whatever new mode of transportation emerges—hydrogen or electric—the fundamental question remains the same: "What is the *primary* source of energy?" Transportation currently consumes 28 quads of power while the electrical grid produces 39. Moving transportation onto the grid would probably only require 30 percent new capacity, since most hydrogen or additional electricity would be produced off-peak. Still, that's a lot of new power plants. Says David Crane, president of NRG Energy of Princeton, N.J., "In our industry, electric cars would be the best thing that's happened since air conditioning." That is why in 2007 NRG became the first company in thirty years to propose a new nuclear reactor before the Nuclear Regulatory Commission.

The Idaho National Laboratory has developed a Very High Temperature Nuclear Reactor that can turn water directly into hydrogen. Amory Lovins proposes covering all of North and South Dakota with windmills—essentially a resurrection of John Haldane's 1920s vision. Whichever way we turn, powering cars and trucks with electricity or hydrogen will not reduce our energy consumption. It will only mean a favorable shift from domestic and imported oil to some other form of terrestrial or solar energy.

Chapter 15

The California Electrical Crisis

The premise of Amory Lovins' 1977 book *Soft Energy Paths,* was that central electric generation was inefficient and should be replaced by more efficient "cogeneration" plants distributed throughout the country. Because these smaller industrial facilities would utilize their waste steam, they would save enormous amounts of energy.

Combined with the development of "soft technology"—solar panels, windmills, geothermal sites, low-head hydroelectric dams, and other innovations—this "distributed" system would eliminate the need for new power plants, coal or nuclear.

California Governor Jerry Brown adopted the "soft path" in 1980, right after the enabling legislation emerged from Washington. From 1980 to 2000 California did not add a single large power plant—except, ironically, the Diablo Canyon Unit 2 Nuclear Reactor, which was commissioned in the 1960s and not completely until 1986. Although the state undertook heroic efforts in conservation—producing the lowest per-capita rate of electrical use in the country—and although hundreds of alternate "soft" facilities came online, the state found itself in the midst of electricity shortage in 2000—something that no other advanced nation has ever experienced.

For a state so enamored with alternate energy, however, it is difficult to admit that such an experiment might be to blame. And so the legend has arisen that the California Energy Shortage was caused by several other factors. One target is California's electrical deregulation, which was done very poorly but only aggravated a bad situation and was not the underlying cause. Another myth is that Enron and several energy trading companies were to blame. These companies did try to circumvent the system of price controls during the crisis and did make a few windfalls, but they were not the *cause* of the problem.

California's electrical shortage occurred because for twenty years the state decided it did not need new power plants. Instead, it followed the conservation-and-renewables script, pushing demand as low as possible and trying to make up the gap with small solar and renewable facilities, the vast majority contributing no more than 1–5 MW. It didn't work. The problem was not solved until the state went on a building binge and constructed 12,000 MW of new capacity, all of it burning natural gas.

There are many lessons from the California experience, but the main one is this: while alternate and renewable energy sources can make contributions to our energy supply, they cannot *replace* coal and nuclear as the base load of electric power. It is a lesson the whole country would do well to learn. Otherwise the California drama of 2000 could be repeated on a national scale.

On a sunny afternoon in May 1986, officials of the Pacific Gas & Electric Company cut the ribbon on the Unit II of the Diablo Canyon Nuclear Power Station, halfway between San

Francisco and Los Angeles, on the Pacific Ocean. First proposed in the mid-1960s, the plant went through two decades of regulatory review before finally moving to completion. "It was a nightmare to build but it's been a dream to operate," said Jeff Lewis, a plant spokesman. Purring along at close to 95 percent capacity, Diablo Canyon produces 2,160 MW, about 5 percent of the state's 45,000 MW capacity.

That spring morning of 1986 was the only time between 1980 and 2000 that what was then the world's sixth largest economy added a major generating station to its electrical grid.

In 1980, Governor Jerry Brown, a former Zen monk and frequent escort of pop singer Linda Ronstadt, was finely attuned to every crosswind of the counterculture. Not at all surprisingly, he was the country's leading political opponent of nuclear power and advocate of solar energy. "We saw solar as the future," said Brown, who is now attorney general of California. "It's a vision I still embrace. As Mayor of Oakland [1998–2006], I pledged to add 5 MW of solar electrical power by the time I left office. In August 2005, I threw the switch on a 1 MW solar installation that will provide 80 percent of the electricity for Federal Express's hub at Oakland International Airport. It was another important step in the struggle to reduce our energy independence and greenhouse gas reduction."[1]

In the midst of the Energy Crisis of the 1970s, seeking to find alternatives to fossil fuels and nuclear, Congress adopted the Public Utilities Regulatory Policies Act of 1978 (PURPA). The bill ended the utilities' monopoly on electrical generation and opened up the grid to alternate producers. Anyone making electricity with a windmill or a hand crank could sell it to the utilities, under conditions set by the states. The system was voluntary and no one was obligated to implement it. No state embraced it more enthusiastically than California.

The California Public Utilities Commission (PUC) set up rules for "qualifying facilities" (QFs)—independent generators qualified to sell to the grid. The standard was "avoided costs"—the amount it would supposedly cost the utility to provide the same amount of power. This number was very elastic and by setting it high, the state could oblige the state's major utilities—Pacific Gas & Electric, Southern California Electric, and San Diego Electric—to buy alternate power from just about anybody. This is what the PUC did.

"There was a gold rush into renewables," says Karl Stahlkopf, vice president of the Electric Power Research Institute in Palo Alto. "People came in with every sort of technology—cogeneration, windmills, small dams, methane-from-garbage, solar collectors, every conceivable form of alternate energy."[2]

On the demand side, the PUC undertook the most ambitious conservation effort in the country. California residents were inundated with billboard, leaflets, radio, and television advertisements urging them to insulate their homes and buy efficient appliances. The campaign cost $200 million in 1993 alone. At one point the *San Jose Mercury* calculated that the state had spent more money urging people to install insulation in low-income housing than it would have cost to install the insulation itself.[3]

Hiring Amory Lovins as a consultant, the state began a "best buy first" program, which required utilities to exhaust all conservation efforts before adding new power plants. These included free energy audits and zero-interest loans for energy-saving appliances. In some circumstances, the utilities were told to go ahead and install the energy-saving equipment themselves and put the costs in their "rate base"—the sunken investment on which they were guaranteed a profit. From 1980 to 2000, Pacific Gas & Electric spent $1.3 billion on conservation, displacing the need for 2,300 MW of new power. Between 1983 and 1995, electrical consumption in the Golden State grew only .5 percent per year, the lowest rate in the country. Customers were said to have saved $1 billion in electrical bills. By 2000 California's per-capita rate of electrical consumption was *37 percent below* the national average.[4]

On the supply side, California stopped building coal and nuclear (except for Diablo Canyon) and actually shut plants down. In 1991, the venerable S. David Freeman, architect of President Carter's Energy Plan, became CEO of the Sacramento Municipal Utility District and quickly closed the 914 MW Rancho Seco Nuclear Station, perhaps the worst run reactor in the country. Around the shuttered facility Freeman planted twenty acres of photovoltaic panels, which, during the summer months, could produce 45 MW of electricity.[5] As a result of these efforts, California also ranked forty-ninth in per-capita electrical *generation*.

In place of conventional power plants, the PUC encouraged every sort of alternate provider. Between 1980 and 2000 it commissioned more than 250 cogeneration plants, ranging from the Claremont Tennis Club (0.1 MW), the Episcopal Home (0.2 MW), St. John's Hospital and Health Center (1.08 MW), Sea World (2.56 MW), and Hershey's Chocolate (6 MW) to Rockwell International (28 MW), Exxon (49 MW) Shell (60 MW), Chevron (99 MW), Procter & Gamble (171 MW), and ARCO (385 MW). Together these added 5,000 MW to the state's 45,000 MW capacity. Campbell Soups' 158 MW cogeneration facility, completed in 1994, was the last sizable addition to the grid before the 2000 crisis.

Beyond cogeneration came renewables. The state licensed 200 sites to burn biomass, half of them at landfills, the remainder incinerating everything from agricultural and animal wastes to walnut shells and old tires. Most generated around 25–50 MW. Other projects involved siphoning methane from landfills for electricity. By 2000 the state had generators operating at almost every landfill, each producing about 1–5 MW.

The PUC commissioned 600 low-head hydro plants, ranging from 1.25 to 125 MW. Sites were soon exhausted, however, and by 2000 there were only four new dams totaling 13 MW in the pipeline. The state also approved 76 geothermal sites, 12 of them producing more than 100 MW. Windmills constituted another

232 QFs, the smallest generated 1.5 MW, the largest 580 MW at Altamont Pass. Finally, the state licensed the Power Tower (10 MW), the Luz Corporation's parabolic mirrors (350 MW), and four photovoltaic plants (3.83 MW), making California the world leader in solar electricity. By 2000, *non*-large-hydro renewables delivered *11 percent* of the state's electricity, as opposed to the less than ½ percent in the rest of the country.

Meanwhile, faced with universal opposition to power plants, the state's utilities had shifted their biggest generators out of state. Southern California Edison took a 48 percent interest in two 740-MW coal plants at Four Corners, New Mexico, a 56 percent interest in the Mojave coal plant in Nevada, and a 16 percent interest in the Palo Verde Nuclear Reactor near Phoenix. The Modesto Irrigation District and several municipal utilities bought into out-of-state coal plants, as did the Los Angeles Department of Water and Power. If California consumers were producing any pollution, it would happen somewhere else.

Seeking to solidify the system, the PUC required utilities to enter 20-year contracts with QF's. "The utilities had to buy so much electricity from the QF's that they ended up substituting it for their own cheaper power," said Stahlkopf, of EPRI. As a result, California electricity became extraordinarily expensive. "By 1991, we had a reserve margin well above the industry standard of 20 percent," said Stahlkopf. "But our electrical rates were double the surrounding states."[6]

Manufacturers began to balk and move out of state. By 1990, unemployment stood at 10 percent and the state government faced a $10 billion deficit. And so, the California Chamber of Commerce and Business Roundtable began campaigning for electrical deregulation, arguing that competition was the only way left to lower the cost of electricity.

Nationally, electrical deregulation was an idea whose time had come. Airlines, trucking, telephones, and the securities industry had all deregulated in the 1980s, setting off competition that improved service and lowered costs. Electrical deregulation, it was argued, would do the same thing. In 1992, the Energy Policy Act encouraged the Federal Energy Regulatory Commission (FERC) to end price-fixing in interstate markets. In 1995 there were 25,000 transactions over the national grid. By 2000 there were 2 million. An unregulated marketplace for electricity was suddenly developing.

Now it was the states' turn. Flushed with success, FERC began encouraging them to deregulate at the retail level, abolishing utility franchises and opening the grid to all comers. Homeowners and businesses would choose their own utility. Competition would drive down prices—just as MCI and Sprint had challenged AT&T, and Delta and Southwest had toppled the old giants, Pan Am and TWA. Rather than generating their own power, utilities would become middlemen, buying from a host of suppliers and competing for retail customers. In order to achieve this restructuring, utility companies would be encouraged to sell their generating plants. Most states made it an option. Only California mandated it.

There was one rub. Long protected from competition, the utilities had accumulated a few uneconomical investments, called "stranded costs." In most states this meant nuclear reactors. Still paying off huge cost overruns from the 1980s, the nation's reactors were plagued by shutdowns and lumbering along at 60 percent capacity. Many were approaching the end of their forty-year licenses. Decommissioning costs would be astronomical. Most of the industry regarded them as white elephants. If nothing else, electrical deregulation would mean the end of nuclear power.

Oddly, this didn't apply to California. Rancho Seco was shut down but Diablo Canyon and San Onofre were running well. Instead, the "stranded costs" were the long-term contracts with the QFs. The legislature decided the utilities must sell half of their generating capacity, then buy the remainder form the new merchant energy companies—Calpine, Duke Power, Entergy, Texaco Global Gas and Power, and Enron. The utilities asked permission to sign long-term contracts from their providers but the legislature—somehow fearful that the utilities would outmaneuver the merchants—wouldn't allow it. Instead, the utilities would buy power day-to-day through the California Power Exchange, a state-appointed body. Since the Power Exchange had no financial credit, it could not enter long-term contracts either. By stripping the utilities of half their generating capacity *and* forbidding them from entering long-term contracts, the legislature left them at the mercy of the spot market.[7]

As deregulation approached, the retail price of electricity stood at $70 per MWh. Fearing competition would drive that price *down*, the utilities asked the PUC to set a *floor* price of $65 per MWh. The regulators agreed, giving the companies three years to recover their stranded costs before they could enter the unrestricted market. The state's eleven municipal utilities—Los Angeles, Sacramento and various small irrigation districts—were exempted from deregulation. Disastrously, the $65 retail price floor soon became a ceiling. Within two years, wholesale prices on the spot market were $360 per MWh—occasionally rising to $1,000. Yet the PUC, with a rigid hand on the controls, refused to relinquish the fixed price. Soon the utilities were paying $300 for electricity that could be resold for only $65 to their customers.

Electrical consumption began climbing for the first time since the 1970s. With the Internet booming, the rate of annual increase jumped from 2 to 8 percent in 1998—12 percent in Silicon Valley, the fastest growing economy in the nation. New supplies, meanwhile, were nowhere to be found. The utilities had given up building plants and the merchant companies were waiting to see how the market developed. In 1996, the first year of deregulation, *nothing* was built. The merchants soon saw demand rising and by 1999 had 5,000 MW in the pipeline, but none of it would be available for several years. As the 2000 summer peak approached, California's electrical consumption was running 21 percent ahead of 1999 with no immediate possibilities for new capacity.

The spark that set off the firestorm was natural gas. After being flat for almost a decade, gas prices suddenly leaped from $3 to $5 per million British thermal units (BTUs) in *one month,* May 2000. At the same time, a drought had taken hold in the Northwest and hydroelectric dams were operating at less than half-capacity. The Bonneville Authority announced it would be unable to ship its usual summer allotment to California. The perfect storm was taking shape.

There are certain politicians in whom regulated utilities bring out their sadistic impulses—just as tethered elephants bring out the sadistic impulses in small boys. Sacremento was full of such politicians. In September, wholesale electricity climbed to $100 per MWh, $35 above the fixed retail price. The utilities begged the PUC to let them enter long-term contracts with the merchant providers. Unnerved, the PUC relented, but the state legislature—still obsessed with the utilities' presumed market power—vetoed it.

Within a month, the state's Independent System Operator (ISO) was buying electricity at spot-market prices of $1,000 per MWh—*15 times* the retail price. "We hired the very best system reliability operators, the people who know how to keep the lights on," admitted Kellan Fluckiger, chief operations officer of the ISO. "But in terms of matching wits with some M.B.A. who's got a Ph.D. in chaos theory, who's working on the derivative of whatever, the answer is no way. We couldn't do that."[8] Forced to rely on inexperienced public servants, the utilities lost $12 billion in six months—almost their entire net worth. Meanwhile, California consumers—protected by the price freeze—went on consuming as if nothing was happening.

By December 2000, things were coming apart. Even though this was the low point in the annual cycle, the state was far short of meeting demand. While being interviewed by *The Wall Street Journal,* a PSE&G executive looked ruefully out his window at San Francisco Christmas lights and lamented, "We're losing money on every one of those bulbs."[9] Natural gas prices, aggravated by a lack of pipelines, hit $15 per million BTUs—up from $5 the month before—and briefly spiked to $60. The state began to suffer rolling blackouts. In San Francisco there wasn't enough electricity to run traffic lights.

In January 2001, the year's low point in demand, brownouts and blackouts became common. PG&E and Southern California Edison had stopped paying creditors and were teetering on the verge of bankruptcy. Governor Davis demanded that FERC require all western utilities to ship electricity to California, but the feds refused, saying it was California's problem. In any case, there was no guarantee anyone would ever be paid.

For a while, San Diego Gas & Electric avoided the debacle. With fewer QF's, SDE&G had retired its stranded costs in 1999 and escaped the price freeze. At first the utility lowered rates to $35 per MWh. When natural gas prices took off, it had raised them to $128. By September SDE&G was charging retail customers $179—nearly triple the floor price. The response was exactly what supply-and-demand theory would predict—San Diego customers cut their consumption by 10 percent and the city avoided the brownouts that plagued the rest of the state.

To Governor Gray Davis and the legislature, however, the price increases in San Diego were as much a crisis as the brownouts and blackouts everywhere else. The governor called a press conference to lament both. The legislature couldn't do anything about brownouts but decided to solve the San Diego situation by bringing SDG&E back under price controls. San Diego consumers immediately responded by going back to their old ways, and demand soared. In four months SDG&E lost $605 million. "The PUC guaranteed us recovery but it didn't say when or how," lamented Ed Larson, a utility spokesman.

At this point, electricity was so short and wholesale prices so astronomical that aluminum plants in Washington with long-term contracts with the Bonneville Power Authority started closing down and selling their electricity to California. It was more profitable than manufacturing aluminum. Meanwhile, factories and retail stores in California began practicing "distributed generation" by installing $250,000 diesel generators. Air pollution got worse but environmental authorities were fearful of cracking down because of the power shortage. Cisco Systems—which was opposing a new natural gas utility plant across from its headquarters in San Jose—installed twelve diesels just to keep its operations going.

To Governor Davis and the legislature, the blame lay with the merchant companies, particularly those from out of state. Texas's Enron and its CEO Ken Lay became targets of particular venom. At one point, State Attorney General Bill Lockyer told *The Wall Street Journal*, "I would love to personally escort Lay to an 8 by 10 cell that he could share with a tattooed dude who says, `Hi, my name is Spike, honey.'"[10] Although Lay avoided this particular fate, he later put himself in the arms of the law by bankrupting his company and was facing a long prison sentence when he died in 2006.

Throughout the crisis, Governor Davis refused to acknowledge that price controls were having any affect. A mere 10 percent cut in consumption would have solved the problem—as had already happened in San Diego—but Davis refused to deregulate. At one point he said famously, "Believe me, if I wanted to raise rates I could solve this problem in twenty minutes. But I am not going to ask ratepayers to accept a disproportionate burden."[11] And so Californians paid as taxpayers instead.

With fearful suppliers refusing to sell electricity to the nearly bankrupt utilities, the legislature appropriated $1 billion in January to enable the California Water Authority to buy electricity from out of state. To everyone's astonishment, the money was gone in six weeks. In February, the legislature adopted a "comprehensive solution"—a $10 billion bond issue, the largest municipal borrowing in history—to enter long-term contracts with the merchant companies. In 2011, Californians would still be paying their electric bills from 2001. With virtually no bargaining experience, the state negotiators paid top-of-the-market prices—although the state later reneged and overturned the contracts in court. Meanwhile, the legislature appropriated more money for renewable energy.

Publicly denouncing deregulation, free markets, and anything else to do with the private enterprise, Governor Davis made plans to take over the state's two major utilities. The government would restore them financially but had to demand something in return. Davis suggested their transmission lines. John Burton, president pro tem of the California Senate, learning the basics of trade, expressed it succinctly: "I give you a dollar, you give me a hot dog." In the meantime, Governor Davis created the California Consumer Power and Conservation Financing Authority to prepare for the state takeover.

Selected to head the Authority was the venerable S. David Freeman, the zelig of America's long energy odyssey. Now director of the Los Angeles Department of Water and Power Department (LADWP), the 74-year-old Freeman had emerged a hero of the crisis, keeping rates low in Los Angeles and even offering to send spare power to San Francisco—except there were no transmission lines available to carry it. On television, Freeman announced: "The California crisis may signal the end of the age of big power—nuclear power plants and dams and coal-burning generators. The future is micro-turbines and fuel cells and clean sources for power." The press hailed him as a "wise old pro."

How did Freeman's LADWP managed to survive the crisis? First, as a municipal utility, it was not required to deregulate. This meant it got to keep its power plants. These plants included:

1) Coal plants in Utah, Nevada, and Arizona (providing 50 percent of LADWP's power)
2) Natural gas plants around Los Angeles, many of them 40 to 50 years old (30 percent)

3) An interest in the Palo Verde nuclear plant near Phoenix (10 percent)
4) An allotment from federally owned Hoover Dam (the final 10 percent)

Because of its heavy reliance on out-of-state coal, the LADWP has been called "one of the dirtiest public-owned utilities in the country."[12]

What happened to all the windmills, cogeneration plants, methane-from-garbage plants, small-head hydro, and solar collectors? "Because we're a municipal, the PUC did us a favor and set our avoided costs very low," explained Eric Tharp, director of public affairs at LADWP, in a telephone interview. "That meant we didn't have to buy any power from the QFs." With nuclear and coal plants churning out electricity in the desert and no renewables cluttering its portfolio, the LADWP easily weathered the perfect storm.

Freeman celebrated his success by announcing "Green Power for a Green Los Angeles," which "marries renewable energy sources with energy efficiency measures." Within months the program had won plaudits from former Clinton Secretary of Energy Bill Richardson, Ralph Nader, and Robert Redford. Then, after heading the state power authority for three years, he moved on to become chairman of the Hydrogen Car Company, which introduced an internal-combustion hydrogen-driven car in 2004. In 2007, he summarized his experience by writing *Winning Our Energy Independence*, in which he named "the Three Poisons—dirty coal, imported oil, and dangerous nuclear power" as the source of our energy problems. The road to energy independence, he said, led through conservation and renewables.[13]

California eventually solved its energy crisis the old fashioned way—by building power plants. Construction was accelerated and by 2001 the state had 12,000 MW of new natural gas generating stations online or in the pipeline—more wattage than it has constructed since Jerry Brown embarked on the soft path in 1980.

Meanwhile, California public officials continued to blame the crisis on Enron. For a long while FERC dismissed this interpretation, but finally succumbed to extreme political pressures and commissioned a study. The investigation found Enron and other merchant companies had indeed sold power back and forth to itself a few times to circumvent the regulations and get higher prices. Thus was born the urban legend that Enron caused the California Electrical Crisis. "Ken Lay play[ed] his role using his Enron smoke and mirrors tactics to create a faux energy crisis in California," *New York Times* columnist Nicholas Kristof was still telling readers in 2007.[14]

To anyone who looked at the situation, however, the answer was obvious. As Daniel Yergin, Pulitzer-Prize-winning author of *The Prize: The Epic Quest for Oil, Money, and Power*, founder of Cambridge Energy Research, and winner of the 1997 United States Energy Award for "lifelong achievements in energy," put it: "The power simply wasn't there."

The outcome of twenty-five years of pursuing conservation-and-renewables is that California now gets 41 percent of its electricity from natural gas—twice the national average. As gas prices rise, this will clearly become unsustainable. Calpine, which constructed almost half the new gas plants, went bankrupt in 2005 trying to pay high gas prices and didn't emerge until 2008. With drilling off the Pacific Coast and in the Overthrust Belt off-limits, the only solution was to build LNG terminals and start importing. Yet this only produced more opposition. "We don't need more natural gas in California because renewable energy and energy efficiency can meet future energy demands," announced Renewable Energy Access, one of dozens of groups opposing LNG terminals in the courts. "Opting for LNG may divert attention from renewable energy and energy efficiency—the far more preferable alternatives."[15]

And so the gas merchants followed the coal and nuclear

industries and moved across the border—this time into Mexico. In 2003, ChevronTexaco announced plans to build the Costa Azul LNG terminal *and the companion power plant* on the Baja Peninsula. Electricity will be pumped across the border. In only a short time, however, the first plant had run into opposition—not from Mexicans but from Americans who with second homes in the area. A delegation of a hundred surfers petitioned Governor Arnold Schwarzenegger to block the terminal because it is near Harry's, a popular surfing spot.[16]

All this is clearly unsustainable. Almost a decade after its Electrical Shortage, California continues to pay the highest electrical prices in the nation while Google, Cisco and other major employers continue to move their data farms and manufacturing facilities to Texas and Oregon. While continuing to preach conservation-and-renewables, California now imports 40 percent of its electricity—more than any other state.

So just out of curiosity, what do you suppose will happen if Mexico elects a populist government—say along the lines of Venezuelan President Hugo Chavez—and decides to nationalize that LNG facility and divert its electricity to Mexico City? Will we send in the California National Guard? Or maybe the surfers.

Chapter 16

The Solar Utopia

By now you're probably wondering what happened to the plan where the Hypercars are going to replace all the coal and nuclear power plants in America? Good question. We'll get to it in a moment.

First let's tackle the issue of whether electricity generation is inherently inefficient. Lovins began *Soft Energy Paths* with this argument:

> The laws of physics require, broadly speaking, that a power station change three units of fuel into two units of almost useless waste heat plus one unit of electricity.... At least half the energy growth never reaches the consumer because it is lost in elaborate conversions in an increasingly inefficient fuel chain dominated by electrical generation.[1]

Much of this "end use," Lovins pointed out, was residential heat and/or industrial steam. The chain of events in centalized electricity consists of raising temperatures in a nuclear reactor to nearly 800°F in order to produce steam at 212°F in order to spin a turbine to push electrons across the landscape to raise the temperature in your living room to 78°F. In the process, about 80 percent of the original energy was wasted. "It's like cutting butter with a chainsaw," Lovins said famously.

Solar energy and other "appropriate technologies" on the other hand, were much better suited to the end task. They could do the job on a local or distributed bases, while leaving a much smaller environmental footprint.

> Some 8 percent of all U.S. energy end use, and similarly little abroad, requires electricity for purposes other than low temperature heating and cooling. Yet since we actually use electricity for many such low grade purposes, it now meets 13 percent of U.S. end-use needs—and its generation consumes 29 percent of U.S. fossil fuels.... [B]y applying careful technical fixes, we could reduce this 8 percent total to about 5 percent (mainly by reducing commercial overlighting), whereupon we could probably cover all those needs with present U.S. hydroelectric capacity plus the cogeneration capacity available in the mid to late 1980s. Thus an affluent industrial economy could advantageously operate with no central power stations at all![2]

As you can see, this is the "soft path" that California undertook.

Nuclear power, on the other hand, represented gigantism. Its arcane mysteries would require a "technological priesthood" (the phrase borrowed from Alvin Weinberg). In a passage that would prove at least partially prophetic, Lovins argued:

> The scale and complexity of centralized grids...increase the likelihood and size of malfunctions, mistakes, and deliberate disruptions. A small fault or a few discontented people become able to turn off a country. Even a single rifleman can probably black out a typical city instantaneously. Societies may therefore be tempted to discourage disruption through stringent controls

> akin to a garrison state. In times of social stress, when grids become a likely target for dissidents, the sector may be paramilitarized and further isolated from grassroots politics.
>
> If the technology used, like nuclear power, is subject to technical surprises and unique psychological handicaps, prudence, or public clamor may require generic shutdowns in case of an unexpected type of malfunction: one may have to choose between turning off a country and persisting in potentially unsafe operation. Indeed, though many in the $100 billion quasi-civilian nuclear industry agree that it could be politically destroyed if a major accident occurred soon [Three Mile Island was only three years away], few have considered the economic or political implications of putting at risk such a large fraction of society capital.
>
> For all these reasons, if nuclear power were clean, safe, economic, assured of ample fuel, and socially benign per se, it would still be unattractive because of the political implications of the kind of energy economy it would lock us into.[3]

The important point was this: the systems were so different, we had to choose between one or the other.

> It is important to recognize that the two paths are mutually exclusive. Because commitments to the first may foreclose the second, we must soon choose one or the other—before failure to stop nuclear proliferation has foreclosed both.[4]

What was wrong with this analysis? As already noted, Lovins' prognostications for energy conservation turned out to be amazingly prescient. Alone among energy experts, he predicted that overall energy consumption could be held under 100 quads by the turn of the twenty-first century.

At the same time, "distributed energy" has been widely practiced. California used PURPA to distribute just about everything built between 1980 and 2000. Other states have not opened the grid as aggressively, but the utilities' monopoly on generation—which Lovins rightly criticized—has long been laid to rest.

Yet beyond this, nothing has happened in the way Lovins predicted. In fact, it has been almost the opposite. Lovins was convinced that decentralization would reduce the demand for electricity. "Electricity-specific's needs are already met by present capacity with a good deal left over," he told *Electrical Week* in 1982.[5] "The long-run supply curve for electricity is as flat as the Kansas horizon," he announced two years later in *Business Week*.[6]

Instead, as the economy revived in the 1980s, a whole new generation of electronic products reached the market—microwave ovens, VCRs, sophisticated sound equipment, electronic tools, electronic scanning machines in medicine. Then came personal computers. By the 1990s, this new world of electronic communication had produced the Internet. Instead of remaining flat as a Kansas prairie, the "electrification" of the economy has advanced steadily, so that electricity now constitutes 40 percent of our energy consumption, as opposed to only 29 percent 1976. It is still climbing.

Has all this just added waste? As it turned out, there was something else at work. No one quite managed to put a finger on it until the publication of *The Bottomless Well* (2005) by energy specialists Peter Huber and Mark Mills.[7]

Soft Energy Paths was talking about "First-Law" energy efficiency, named after the First Law of Thermodynamics. The First Law, you will recall, says that energy is never lost or gained. It can change form, from kinetic to heat to light and back again, but the sum total of energy always remains the same (unless energy changes into matter). Some of this energy is squandered, however, when it finds no use—like the steam vented from a power plant. We can quantify the First Law efficiency of energy transformations by dividing the energy that is available for end use by the energy with which we started:

First Law Efficiency = End-use energy / Starting energy

By this measure, electricity is relatively inefficient. The amount of heat lost in generating electricity in America is now *double* the useful work done by electricity—and almost the equivalent of all the energy derived from oil and natural gas. Yet still we persist in using electricity. Why?

The answer lies in what Huber and Mills describe as "Second Law Efficiency." The Second Law of Thermodynamics, you will recall, describes the level of disorder or "entropy" in a closed system. The great advantage of electricity is that it can be used to achieve a much higher level of order than can be achieved in other energy systems. More and more, the work being done in our economy requires such a high degree of order.Storing data on server farms, for instance, involves creating an extraordinarily high level of complex order that—as noted earlier—can be destroyed within only a fraction of a second if power is interrupted. Having electronic robots assemble automobiles in factories goes far beyond any kind of order that could be achieved with mechanical motion or industrial heat.

The great advantage of electricity is that it can *concentrate* huge amounts of energy on a single task. Lasers, x-rays, MRI's—all operate at levels of order that are unobtainable through any form of mechanical or heat energy. More and more, these tasks are becoming part of our everyday life.

Second Law Efficiency, Huber and Mills noted, can be measured by taking the *final level of order achieved* and dividing it by the *energy wasted in the system.*

Second Law Efficiency = Final order level / Waste heat

By this measure, electricity proves to be surprisingly *more* efficient than any other technology. Nothing else allows us to concentrate so much energy on a single task. You can use a magnifying glass to concentrate solar energy so that it will burn through a piece of paper, but not through steel. You can use the wind to spin a turbine, but not at the speed of a dentist's electric drill. You can use mirrors to concentrate sunlight on a single Power Tower, producing close to 1,000 degrees of heat, but until you turn that heat into electricity it is not much use. Low-grade energy systems can only get us so far. As the late Dixy Lee Ray, who chaired the Atomic Energy Commission from 1973 to 1975, used to put it, "You can hold a match under a pot of water *forever* and it won't boil."[8]

As Huber and Mills have brilliantly argued, electricity may be relatively inefficient at generating and transmitting energy, but it can do things that are *inconceivable* with other energy systems. You can't make the new steel alloys in old-fashioned coke ovens—they're not hot enough, you need an electric-arc furnace. You can't weld buildings together with hot irons—you need an electric welding gun. You can't operate a microwave oven with rooftop photovoltaic panels. (You can barely run an oxygen pump in your aquarium.) Ironically, one of the biggest beneficiaries of high-quality electricity has been energy conservation. Huge advances have come from using information technology to match consumption with production, eliminating waste. These advances are possible only with information delivered electronically.

Electricity is no longer just a better way of powering the nineteenth century technology that was run by the Corliss engine. It has created a whole new electronic world that demands an energy *quality* that can only be achieved by electricity. That's why the world of "soft energy" never materialized. Lovins saw the inefficiencies of cutting butter with a chainsaw. He didn't see the futility of trying to cut steel with a butter knife.

So what happened to the vision of the Hypercars that were going to eliminate the steel, auto, coal, and nuclear industries? That was another alternate energy utopia put forth in the 1990s. It went like this:

The Hypercars would become the ultimate form of distributed energy. Each could produce 20–40 kilowatts. With 150 million of them—the size of the American fleet—they could sit all day generating power for the electrical grid. Here's the way it is described on the Rocky Mountain Institute Web site:

> You drive your Hypercar® vehicle to work.... As you park your car, you plug it in both to the electricity grid and to a little snap-on pipe that brings surplus hydrogen out from the reformer in the building.
>
> But you're not plugging in to recharge your car: quite the contrary, while you sit at your desk, your power-plant-on-wheels is silently sending kilowatts back to the grid.
>
> It wouldn't take many people taking advantage of this deal to jeopardize remaining coal and nuclear power plants. If the entire U.S. light vehicle fleet consisted of Hypercar® vehicles, it would collectively have about five times the generating capacity of the national grid.[9]

Of course if you had a fleet of 150 million gasoline-driven vehicles, you could attach their drive shafts to small electric generators and power the grid until then as well. The question then, of course, would be, "Where do you get the gasoline?" As usual, however, Lovins' revelation was greeted with hosannas among anti-nuclear groups, who celebrated that he had once again seen the way to a world without power plants. Jeremy Rifkin, another futurist, repeated this vision in his book, *The Hydrogen Economy*.

> In the new hydrogen fuel-cell era, even the automoblie itself is a "power station on wheels".... Commuters drives their cars to work, then plug them into the hydrogen line coming out of the natural gas reformer installed as part of the building's fuel cell. While they worked, their cars would produce electricity, which they could then sell back to the grid. The car, instead of simply occupying space, would become a profit center. "It does not take many people doing this to put the rest of the coal and nuclear plants out of business," says [Amory] Lovins, who's been trying to do just that for decades.[10]

If you've read this far, I'm sure you are already asking the right question: *"Where do you get the hydrogen?"* In other writings,

Lovins said there were two possible sources—natural gas or the "ubiquitous electric grid." In this vision, natural gas would be reformed into hydrogen in the basement of every building or the grid would be tapped to split water into hydrogen.

But why bother? Reforming natural gas means shipping it all over the country to create hydrogen to feed into individual fuel cells. Why not just burn it at a power plant to begin with? The energy lost in the hydrogen transition would be enormous. The second scenario is even more ludicrous. You are using electricity to create hydrogen to make it produce electricity? What's the point?

In 2001, I wrote a cover story for *The Weekly Standard*, "The Myth of Alternate Energy," challenging Amory Lovins' thesis that solar and renewable energies could replace coal and nuclear power. Lovins' system seemed to say you could use power plants to generate hydrogen and when you were done you would have so much hydrogen you wouldn't need the power plants. This would be a perpetual motion machine, a violation of the Second Law of Thermodynamics.[11]

A year later, Lovins came to the New York Academy of Sciences to give an address. My wife at the time was director of media and publishing at the Academy and was his host for the evening. Halfway through dinner she had to tell him that it was her husband who had written the article in *The Weekly Standard* and that I would be attending lecture. That ruined dinner.

After his speech I once again posed the question: Where do you get the hydrogen to run the Hypercars? By that time natural gas prices had skyrocketed and there was no longer the illusion that gas is an unlimited resource. Electricity was the only alternative. Where would that come from?

Lovins had the answer—windmills. He would cover all of North and South Dakota with windmills. (This is in addition to covering North and South Dakota with switchgrass.) I have read dozens of versions of Lovins' vision for replacing all the coal and nuclear plants in the country with the Hypercars, yet

I have never found in them any mention of windmills. However, there it is.

So Lovins' vision is essentially an updated version of John Haldane's idea in the 1920s—wind, hydro, and solar power stored as hydrogen. It is not an unreasonable vision. Of course moving the hydrogen around the country to tens of millions of plug-in stations might create problems, since hydrogen does not transport easily. Why not just transport it to a few hundred *power plants* and generate the electricity there? Wouldn't that be more efficient and less expensive? But no, it would puncture the myth that windmills and hydrogen and Hyper Cars are somehow "small and beautiful" even when the windmills occupy whole states.

And now that we mention it, why not just run those hundred-or-so power plants with nuclear power and avoid all the inefficiencies of creating and transporting hydrogen in the first place? But that obviously is not the point. There is something infinitely more appealng about a world run on windmills and solar panels than by the dark and forbidding forces of terrestrial energy. And that is the subject to which we shall turn at last.

Lovins, S. David Freeman, the scientists who want to cover 34,000 square miles of New Mexico with solar collectors, the Union of Concerned Scientists who say it would only take a few hundred square miles of Nevada—all have the same thing in common. They are quintessentially American, in that they all live with the historical memory that somewhere out there beyond the edge of the frontier is a world of endless terrain waiting to be painted with our dreams. They are the last of the Westward pioneers, the "People of Plenty" David Potter wrote about in 1956, still imbued with the sense that America is blessed with infinite resources—specifically, land. It is only necessary to make use of it.

Of course this is no longer true. It hasn't been for some time, although solar enthusiasts do not seem to have noticed. It certainly doesn't apply to the rest of the world. Europe, India, and Japan hardly have 34,000 square miles to spare for solar collectors—although somebody will eventually propose covering the Sahara or Gobi Desert. For the present, however, the image of a solar utopia remains a uniquely pastoral vision of America—a land running gracefully on picturesque windmills and solar collectors requiring little or no effort—certainly preferable to the sinister mysteries of nuclear technology.

And this is the ultimate appeal of alternate energy—the promise of a simpler life. As Lovins put it in *Soft Energy Paths:*

> In an electrical world, your lifeline comes not from an understandable neighborhood technology run by people you know who are at your own social level, but rather from an alien, remote, and perhaps humiliatingly uncontrollable technology run by a faraway, bureaucratized, technical elite who have probably never heard of you.
>
> The reason electrical grids are designed to such exemplary—and expensive—standards of reliability is that they must be, because so many people depend on them that a failure could be a social catastrophe. If your solar system fails (which, of course, it should not do, as there should not be much to go wrong with it), you can put on a sweater or go next door.[12]

This pastoral vision has a long and honored tradition in America. Perhaps its most complete expression was in *I'll Take My Stand*, an agrarian manifesto published in 1930 by the Fugitive Poets clustered around Vanderbilt University. Seeking to defend the Old South from the encroachments of northern industrialism, John Crowe Ransom wrote:

> In most societies man has adapted himself to environment with plenty of intelligence to secure easily his material necessities from the graceful bounty of nature. And then, ordinarily, he concludes a truce with nature, and he and nature seem to live on terms of mutual respect and amity...But the latter-day societies have been seized—none quite so violently as our American one—with the strange idea that the human

> destiny is not to secure an honorable peace with nature, but to wage an unrelenting war.... Our vast industrial machine, with its laboratory centers of experimentation, and its far-flung organs of mass production, is like a Prussianized state which is organized strictly for war.[13]

Seeking to avoid these depredations, the Fugitives also recommended a return to the simpler life:

> To avoid the dire consequences and to maintain a farming life in an industrial imperialism...the farmer...must deny himself the articles the industrialists offer for sale. It is not so impossible as it may seem at first, for, after all, the necessities they machine–factured were once manufactured on the land, and as for the bric-a-brac, let it rot in their hands. Do what we did after the war and the Reconstruction: return to our looms, our handcrafts, our reproducing stock. Forsake the movies for the play-parties and the square dances. Throw out the radio and take that old fiddle down from the wall.[14]

This appeal is made over and over in the shelves full of books warning "the party is over" and that energy shortages and man-made climate change will soon be overtaking our civilization.

Ted Trainer in *The Final Energy Crisis*:

> We need to convert our neighborhoods, suburbs and towns into small, thriving local economies which produce most of the goods and services they need, using local resources wherever possible. Every suburb would have many small productive enterprises such as farms, dairies, local bakeries, and potteries. Many existing economic entities would remain but their operations would be decentralized as much as possible, with workers living close to their place of work, enabling most of us to get to work by bicycle or on foot. Many farms could be backyard and hobby businesses.... We would...retain some mass production facilities, but many items of general necessity such as furniture and crockery could in the main be produced through craft-working. It is far more satisfying to produce things using craft-working than in factories.[15]

Julian Darley in *High Noon for Natural Gas*:

There are lots of ways to have low-energy fun, especially with other people. Learn how to make things—our villages and towns once made all the things we used and wore in life. If you can play an acoustic orchestral instrument, join or form an orchestra. If you can sing, join a choir. Dance to music played by live musicians using no electronic amplification whatsoever. Write, read, and perform plays. Tell stories. Observe nature, and tell others what you observe.[16]

Or Amory Lovins in *Soft Energy Paths*:

Real people generally want to understand their own world and feel responsible for their own destinies, not be mere economic cogs. That is why, for example, even the most materialistic among us complain about impersonal and shoddy "service," arrogantly paternalistic utilities, incompetent auto mechanics, outrageous bills for simple repairs, and petty bureaucracy. That is why many increasingly cherish the small corner shop, feel more guilty than satisfied at eating hamburgers that are a by-product of petrochemicals manufacture, and try to persuade the dentist to explain, as one adult to another, exactly what is wrong with that tooth. That is why creative personal activities of all kinds are flourishing—from gardening and canning to weaving and do-it-yourself carpentry (and arguably citizen's band radio). People from all walks of life are changing their own fuses, making their own preserves from their own fruit, sewing their own clothes, and insulating their own attics, not only because it pays but because it symbolizes a small triumph of quality over mediocrity and of individualism over the System.[17]

Maybe that's why we rejected terrestrial energy—because nobody understood it.

Part Four
Terrestrial Energy

Interlude

Cooper Nuclear Power Station, Brownville, Nebraksa

The Cooper Nuclear Station sits on a two-square-mile plot of farmland next to the Missouri River about 60 miles south of Omaha. Brownville has only 150 inhabitants, so the plant draws workers from all over eastern Nebraska. "There probably isn't a home anywhere within 30 miles of here that doesn't have a relative or friend working at the plant," says Eugene Toesche, a Vietnam vet and former weekly newspaper owner who guides me through the plant. "We have people driving down every day from Lincoln and Omaha." Like most nuclear plants around the country, Cooper is very popular locally as a quiet and non-polluting source of energy, jobs, and tax revenues.

For most of its history, Cooper was one of the nation's most problematic reactors. In 2002, Nebraska Public Power was prepared to close it down. "Then they looked at what it was going to cost to replace it and had second thoughts," says Toesche. "They called in Entergy, and their people said, 'You've got a good plant here, you're just not running it right.' They signed a contract to take over operation of the plant. Now we're using what we have more efficiently."

Toesche leads me through the maze of radiation checkpoints reminiscent of an airport security gate. At one point a machine scans our ID cards and checks our identity. From the machine next to mine suddenly erupts Homer Simpson's familiar, "Doh!" The employee has made a mistake in scanning his ID card. "Homer Simpson is very popular around here," chuckles Toesche. "We're all big fans."

He leads me past the control room. Through a plate-glass window, I watch a group of about eight well-dressed operators standing around in casual conversation. The control panel is smaller than some of the old photographs, with brightly colored lines forming a circuit that points from one indicator to another. "The logic chains are all mapped out right on the panel," says Toesche. "It's much more organized than it used to be." Across from the plate-glass window in the narrow hallway, a series of framed photos shows the plant during various stages of construction in the 1970s. "It's amazing that all this was once a hole in the ground, isn't it?" says Toesche.

Now he takes me over to the simulated control room on the other side of the building. As we enter the narrow hallway, I suddenly confront the same photographs on the wall. It is completely disorienting. "Wait a minute," I say. "Are we back where we were before?"

"Everything here is duplicated to a tee," laughs Toesche. "If they dropped you in here blindfold, you'd never know if it is was simulated or real."

In the simulated control room I meet James Florence, an enthusiastic young training coordinator in his late 30s. Like 70 percent of nuclear plant operators, Florence came to it through the Navy, the legacy of Admiral Hyman Rickover. When I mention Three Mile Island, everybody in the simulator is eager to talk.

"We're a completely different industry now," says Florence.

"Those early plants never took account of the human factor. The people who built them never had to run them. In an emergency, you had 500 lights blinking at you at once, but nothing telling you what to do. Now we've got everything set by priorities." He shows me the circuit-board diagramming on the control panel. "It's all if-then propositions," he says. "It's very logical."

"We're prepared for any eventuality," he continues, "fires, tornados, earthquakes. Here, want to see an earthquake?" He sits down at his computer and rifles through a few files. As I sit in my chair, the room begins to rumble. Suddenly everything is shaking, lights are flashing, horns sounding. "That's an earthquake—about .1 g of acceleration," he says. "See, the reactor has already shut down. The ECCS is operating. We've lost our main power but the back-up generators have kicked in. Everything is under control.

"The school kids love that one," he winks.

Toesche takes me outside to show me the back-up facilities. Like most other reactors, Cooper refuels every 18 months. "A truck pulls up to the gate and unloads the new fuel rods," he explains. "They're basically harmless. You can handle them with gloves." Plans for Cooper's next refueling in 2008 have been long since completed. "We're now working on 2010."

As we return to the main entrance, we pass a large tricycle. "There aren't any cars allowed inside the gate," says Toesche, "and the messengers got tired of walking long distances outside in the cold. So they asked if they could ride bicycles. The NRC considered it for eight months and decided bicycles were too dangerous, but they would allow a tricycle. Somebody got smart and installed a seatbelt on the tricycle just for a joke but it didn't go over. You can't joke about safety here."

Another round of buildings and Toesche takes me up to the storage pool for spent fuel rods on the top floor. We peer over a metal railing into the 40-foot depths glowing with faint blue of the Cherenkov effect. "You could swim in there," says Toesche. "The water shields you. We send divers down every once in a while if something gets stuck." In fact, workers at other plants have occasionally been caught taking a dip in the storage pool, earning huge fines and reprimands from the NRC.

Ironically, Cooper is now one of the grand old ladies of Nebraska Public Power's fleet. All but two of its coal plants have been built more recently. The company added five new combined cycle gas boilers in the 1990s but gas is now so expensive that they run only during summer peaks. Meanwhile, Cooper is operating so successfully that NPP is selling spare power to other states.

"They're making a pile of money on this plant," says Toesche as he leads me to the gate. "It's a great technology."

Chapter 17

The Pioneers

Up to the twentieth century "reality" was everything humans could touch, smell, see and hear. Since the initial publication of the chart of the electromagnetic spectrum humans have learned that what they can touch, smell, see and hear is less than one-millionth of reality.

—R. Buckminster Fuller

This chapter is something of a love letter to the men and women who made possible the discovery of terrestrial energy. They are among the forgotten heroes of the Twentieth Century. Albert Einstein, of course, is well known and was *Time's* "Man of the 20th Century." But Einstein's contribution was a lightening bolt—brief and dazzling. His Theory of Relativity illuminated the path ahead. But the groundwork was done by a battalion of men and women whose names are not so familiar—Ernest Rutherford, Wilhelm Roentgen, Henri Becquerel, the incomparable Marie Curie, Neils Bohr, Erwin Schrodinger, James Chadwick, Enrico Fermi, Lisa Meitner, Ernest Lawrence, Eugene Wigner, Leo Szilard, and many others. Although they do not always appear in the history books, many have been honored by having new elements named after them.

Besides being brilliant theorists, the pioneers of nuclear energy were above all *experimentalists*. Beyond formulating mathematical equations, they were eager and willing to follow their formulas into the laboratory. Enrico Fermi, the brilliant Italian scientist who was the prototype, made a career of following his intuitions. At the University of Chicago he would ask his students, using the information already in their heads, to calculate the number of piano tuners in Chicago. Working from various educated guesses—the number of households, how many households might have pianos, how many pianos a tuner could service in a year—they could almost always come within a good ballpark figure—around 100. Physicists have since called this "The Fermi Solution"—a commonsense approach that makes use of the information at hand.[1]

Here is how this worked in practice. On July 16th, 1945, Fermi stood on a hillside in Alamogorda, New Mexico along with dozens of other Manhattan Project scientists witnessing the world's first atomic explosion. As the fireball expanded ten miles away, Fermi released a handful of confetti behind his head, letting it float to the ground. From measuring how far the shock wave carried the tiny pieces of paper, he was able to calculate with his ever-present slide rule the magnitude of the 10,000-ton blast. When the most sophisticated instruments returned their own measurements a few days later, Fermi's estimate matched within one percent.

One other story is worth recounting. Back in Rome, in the 1930s, when Fermi and his colleagues first realized they could perform alchemy-like transmutations of one element into another by bombarding them with newly discovered neutrons, Fermi set himself the task of going through the entire periodic table of elements, from hydrogen to uranium, to see what could be created. Working in the converted mansion that housed the Rome Institute of Physics, Fermi and his colleagues bombarded the elements with neutrons in one room, then scurried down the hall to test for new elements in the chemistry lab. Because many radioactive isotopes last only a few minutes, Fermi and his colleagues—who prided themselves on their athletic ability—began racing each other down the corridor.

One day a distinguished Spanish scientist came to the lab looking for "His Excellency Dr. Fermi," whose fame already spread across the continent, earning him the title, "The Pope of Physics."

"Is he here?" the Spaniard inquired of a young lab assistant.

"That was him who just ran past you," the assistant replied.[2]

For Fermi, the world made perfect sense. It was physical, not theoretical. All that was necessary was to find an intuitive pathway into the invisible subatomic world.

Chapter 18

The Road to Los Alamos

The problem which was already being mooted by such scientific men as Ramsay, Rutherford, and Soddy, in the very beginning of the twentieth century, the problem of inducing radio–activity in the heavier elements and so tapping the internal energy of atoms, was solved by a wonderful combination of induction, intuition, and luck by Holsten so soon as the year 1933.

So began *The World Set Free,* H.G. Wells' prophetic novel about the coming nuclear age. Written in 1913, when the interior dimensions of the atom were just being conceptualized, Wells got the dates almost exactly right. It was in 1932 that James Chadwick discovered the neutron and 1934 when Enrico Fermi announced the transformation of elements. Only then did Wells' vision of a world set free by boundless nuclear energy begin to come into view.

Eighteen years before Wells wrote, Wilhelm Roentgen, a

professor of physics at the University of Wurzburg, had first opened the door. On November 8, 1895, while studying electron beams produced by a cathode ray, Roentgen noticed that a screen on a nearby table was giving off a strange glow. What aroused his curiosity was that the glass tube emitting the cathode ray was completely wrapped in black paper. The electron—which were already familiar to the world—should have been blocked by the paper. Instead, some other form of energy was passing right through the barrier.

A month later, Roentgen published an article describing this new phenomenon. Along with it were two photographs—one of gold coins taken through a leather pouch, the other of the bones in his wife's hand. Even though the Internet lay a century ahead, these two photographs circled the globe in a matter of days. There was something otherworldly about this new energy that could pass right through flesh and other soft substances. Roentgen called them "x-rays."

The initial explanation of x-rays was that they must be some kind of solar energy, since all energy comes from the sun. Henri Becquerel, a French scientist, pursued Roentgen's discovery by investigating potassium uranyl sulfate, a uranium salt that gives off a phosphorescent glow. Theorizing that the glow must ultimately derive from the sun, Becquerel exposed the salts to sunlight, and wrapped them in a black material next to a photographic plate with a Maltese cross in between. Sure enough, the plate showed an image of the cross. Solar energy had somehow been captured in the salts.

Then one day, as he was about to repeat the experiment, the sun went behind a cloud. Becquerel wrapped the cross, salt and plates and put them in a drawer, waiting for a brighter day. When he came back to them a few days later, however, he was struck with an intuition. Without exposing them to sunlight, he developed the plates. Sure enough, there was the same image of the Maltese cross. Sunlight had nothing to do with it. The mysterious emanations were coming from the uranium itself.

(Many people believe Becquerel's paternity in the discovery of radiation is the reason the French have so enthusiastically embraced nuclear power.)

The mysterious properties of uranium soon became the obsession of a young assistant in Becquerel's laboratory, Marie Sklodowska. The daughter of a Warsaw professor, Sklodowska had immigrated to Paris to study physics. There she met Pierre Curie, already a noted scientist, and they were married. Although Pierre had already made several important discoveries in electromagnetism, he put aside his work to pursue Marie's passion, which she named "radioactivity."

The Curies' focused on pitchblende, an ore found in Czechoslovakia in which Becquerel had unearthed his mysterious uranium. When uranium was extracted from the ore, however, the Curies discovered—to their utter astonishment—that the pitchblende became *more* radioactive. Working in extreme poverty, the couple spent four years processing tons of pitchblende in a backyard shed, desperately trying to analyze its contents. Lacking blankets, they often piled chairs on the bed for warmth. After almost ruining their health, the Curies were able to isolate two new elements, radium (88) and polonium (84), which Marie named after her native Poland. Although they did not realize it, both radium and polonium are the "daughters" of uranium decay. Their half-lives are much shorter and they are therefore more highly radioactive. This is why the pitchblende became even more radioactive when uranium was removed.

In 1901, Roentgen received the first Nobel Prize for Physics ever awarded for his discovery of x-rays. Becquerel and the Curies shared the 1903 Nobel for radium and polonium. Tragically, Pierre Curie was killed three years later when he was struck by a horse-drawn wagon while crossing a Paris street in a rainstorm. Marie won a second Nobel Prize in 1911 for her chemical isolation of radium—the first person ever to win two Nobels.

As with many mysterious new phenomena, radiation was

soon hailed as a wonder drug. Marie Curie found it could cure gangrene and toured the battlefields of World War I using radium to treat wounded soldiers. It proved miraculous, curing 90 percent of the cases, and became the standard agent until it was replaced in the 1930s by sulfa drugs, which were less effective.

Since Roman times, people across Europe had flocked to "hot springs" to improve their health. For centuries it was believed that the warm waters, or perhaps sulfur in the water, had a beneficial effect. Now it was realized that this heat was produced by radioactivity in the rocks and that this radiation might be the key. "Baths" that had operated for centuries now began advertising their radiation. Physicians also promoted radiation treatment for a variety of ailments, particularly bursitis, arthritis and rheumatism. It was even used in cosmetics. Someone once marketed uranium underwear.

In 1932, however, Eben M. Byers, a Pittsburgh steel manufacturer, former golf champion and health fanatic, died of radium poisoning after taking two bottles a day of Radithor, a popular elixir. Then a group of women were found to be suffering from extreme radiation poisoning and cancer after working in a watch factory in a Paterson, N.J. They had been painting radium glow-in-the-dark numbers on the dials and pointed the tips of brushes with their tongues. Among the 774 women who were exposed, 48 suffered death from painful bone cancers. The public became aware of the dangers of radioactivity and enthusiasm waned.[1]

In the meantime, the world of physics had been turned upside down by an obscure, 26-year-old clerk in the Swiss patent office. In the "annus mirabilis" of 1905, Albert Einstein published *five* landmark papers, any one of which could have won the Nobel Prize. He identified the well-known Brownian movement as the random movement of microscopic particles being bombarded by molecules. He identified the photoelectric effect, the basis of photovoltaic electricity. Most important, Einstein proposed his

Special Theory of Relativity, which posited that there was no absolute measure of space or time in the universe except the speed of light. Almost as an afterthought, the Special Theory included the equation $E=mc^2$.

Mass and energy were different forms of the same thing, Einstein hypothesized. Although the precise mechanism was not clear, Einstein's theory suggested that radioactivity was some form of energy resulting from the transformation of matter into energy within the atom.

The biggest impediment to understanding radioactivity at this point was that *no one knew what an atom was*. Our familiar model—tiny electrons orbiting around a large nucleus—was only in the earliest stages of formulation. The researcher who pursued the question to its fullest was Cambridge University professor who had arrived from New Zealand by way of Canada—Ernest Rutherford, the "father of nuclear physics."

In 1909, a year after he had already won the Nobel Prize for his work in chemistry, Rutherford began bombarding a thin gold foil with positively charged alpha particles from radioactive polonium. Although the vast majority passed directly through the foil, he found that one in 8,000 bounced straight back. "It was as if you fired a 15-inch naval shell at a piece of tissue paper and the shell came right back at you," said the astounded Rutherford.[2]

Faced with this stunning evidence, Rutherford posited a model for the atom. A positively charged mass, he suggested, must be concentrated in a very small space at the center—the nucleus. Particles of the opposite charge, the electrons, occupied the remaining space—7999/8000ths—even though they had very little mass. Ten years later Rutherford bombarded nitrogen (7 protons) with the same alpha particles and found a very small portion turned into oxygen (8 protons). The newspapers were filled with excited reports that Rutherford had "split the atom." He demurred, saying he was only "playing with marbles."[3] In fact, he had achieved the dream of the alchemists, transforming one element into another.

Rutherford's theory neatly explained Dmitry Mendeleyev's 1869 arrangement of the elements into the periodic table. The progression of elements from hydrogen to uranium by their "atomic number" would match the number of positively charged protons in the nucleus. An identical number of negatively charged electrons would then occupy the remaining outer region, giving the atom a neutral charge. There was only one problem. The elements progressed much more steeply by atomic *weight* than atomic number. For every element beyond hydrogen, the atomic weight is at least double the atomic number and for the largest it is 2½ times as great. In 1920, Rutherford proposed the existence of a third particle—a neutron—that had almost the same weight as a proton but no electrical charge. The neutron, he speculated, also resided in the nucleus. In 1932, James Chadwick, a colleague of Rutherford's at Cavendish, identified the elusive neutron.

The vistas being opened by these discoveries attracted some of the most brilliant minds of the century. The problem was that, according to Newtonian physics, the atom should be impossible. If the negatively charged electrons were really orbiting the nucleus, like planets whirling around the sun, then they would have long since lost momentum and been drawn into the nucleus by its positive charge, collapsing the structure. In 1913, Neils Bohr, the brilliant Danish theorist and father figure for the entire generation of physicists, formulated a new and puzzling model for the atom. The electrons, he said, were not like roaming satellites. Instead, they could occupy only discrete orbits. When light was emitted from an atom—always in a discrete spectrum line—it was because an electron had jumped from one energy level to another in a "quantum leap." Bohr's theory became the basis of quantum physics.

In 1925, Louis de Broglie, a young French physicist, showed that the electron, which was thought of as a particle, also exhibited the properties of waves. In 1927, Erwin Schrödinger formulated his famous equation showing that subatomic particles could be described as waves of *probability* representing the likelihood that one would be in a given position at any given moment. In the following year, Werner Heisenberg formulated his famous "Uncertainty Principle," which asserted that the position and velocity of a particle could *never* be measured simultaneously.[4]

Bohr accepted this strange universe, saying that anyone who thought he really understood the quantum world was probably mistaken. Subatomic particles were not Newtonian billiard balls, bouncing around according to the laws of classical mechanics. Certainty had to be abandoned. The position of a particle at any point in time could only be described in terms of probability. Bohr likened it to throwing dice. Einstein, who never really accepted quantum mechanics, dissented. "God does not throw dice," he intoned.

It was a thrilling time for the "international family of nuclear physicists." Developments came rapid-fire. Nobel Prizes abounded everywhere. Yet before another decade was out, all these brilliant minds would find themselves locked in a tragic trajectory of world events. As totalitarianism spread across the globe, all would end up playing leading roles on a stage where the shadow of Einstein's equation grew larger and larger.

Hovering over these discoveries was the ultimate question of whether this subatomic world could be tapped for the vast stores of energy that Einstein's theory predicted. Rutherford was certain it couldn't be done. Leo Szilard, an enthusiastic young Hungarian researcher, thought differently. In 1932 he took out several patents on techniques for splitting the atom. When Szilard went into Rutherford's office a year later to discuss the possibilities, the Father of Nuclear Physics unceremoniously threw him out. Rutherford didn't calm down for weeks.[4]

Einstein tended to side with Rutherford. He told the American Association for the Advancement of Science at the 1934 meeting that trying to break apart individual atoms was a dubious endeavor. "It is like trying to shoot birds in the dark in a country where there are not many birds," he said.[5]

In 1934, Frederick Joliot and Irene Joliot-Curie (she was the daughter of Pierre and Marie) bombarded an aluminum sheet with alpha particles and found it produced an artificial radioactivity. Yet the amount of energy created was miniscule compared to the amount needed to produce it. The difficulty was apparent—the positively charged nucleus was repelling any positively charged alpha particles, which are made up of two protons. Electrons would be attracted to the nucleus, but they were far too small to create enough impact.

At this point entered young Enrico Fermi, a self-taught Italian who had found a place in the laboratories of Professor Orso Mario Corbino. Fermi seized upon the 1932 discovery of the neutron and added it to the physicist's artillery. At first he had no success. Then, on a brilliant intuition, Fermi decided to screen the neutron source with paraffin, a wax made mostly of hydrogen and carbon. As if by magic, the door to the subatomic world opened. Apparently slowed by the paraffin, the neutrons sailed right into other nuclei, breaking atoms apart left and right. Every nuclear reactor now has a "moderator"—a small-molecule substance such as water, carbon or beryllium—that slows neutrons so that absorption can take place.

In 1934, Fermi announced he had transformed small quantities of uranium into different elements. Probably, he said, the neutron had been absorbed by the uranium nucleus. Then it had split into a proton and electron, moving one notch up the periodic table to become unknown Element 93. He was awarded the 1938 Nobel Prize for this effort.

Yet there was some dissent about Fermi's conclusions. Ida and Walter Noddack, another husband-and-wife team working in Vienna, expressed skepticism. They were a formidable couple, generally regarded as the world's foremost experts in rare earths after isolating the element rhenium (75) in 1925. Writing in a German chemistry magazine, Ida offered another explanation:

> It would be equally possible to assume that when a nucleus is demolished in this novel way by a neutron, `nuclear reactions' occur that may differ considerably from those hitherto observed in the effects produced on atomic nuclei by proton and alpha rays. It would be conceivable that when heavy nuclei are bombarded with neutrons, the nuclei in question might break into a number of larger pieces, which would no doubt be isotopes of known elements but not neighbors of elements subjected to radiation.[6]

Noddack's letter was ignored. Busy looking for transuranic elements, artificial elements heavier than uranium, Fermi and his colleagues failed to consider that something much more remarkable was happening. The uranium might be splitting into two much smaller elements, releasing vast amounts of energy.

In his acceptance speech in Stockholm on December 12, 1938, Fermi recalled, "[W]e, in Rome, used to call [these new] elements 93 and 94 Ausenium and Hesperium respectively. It is known that O. Hahn and L. Meitner have investigated very carefully and extensively the decay products of irradiated uranium, and were able to trace among them elements up to the atomic number 96."[7] The Italian press paid little attention, instead criticizing Fermi for not wearing the Fascist uniform or giving the Fascist salute when receiving his prize. Two days later, Fermi and his wife Laura, who was Jewish, fled to London, abandoning Mussolini's Italy. Yet before his acceptance speech was published six months later, Fermi was forced to amend it. The thesis upon which he had won the Nobel Prize had been mistaken.

Lise Meitner and Otto Hahn, mentioned in Fermi's address, had collaborated on nuclear physics at the Kaiser Wilhelm Institute in Berlin for twenty-five years. When Hitler came to power in 1933, most Jewish scientists were dismissed. Although Jewish, Meitner was protected by her Austrian citizenship.

Then on March 12, 1938 Hitler's armies marched into Austria and eight months later came *Kristallnacht*, the beginning of the Jewish pogrom. Meitner, no longer protected, fled to Denmark, where she took up residence in a lonely seaside cottage. Fritz Strassman, a younger colleague, took her place in the laboratory. The Joliot-Curies in Paris had been questioning Hahn and Meitner's identification of transuranic Element 96 in the debris of Fermi's bombardment, but Hahn had dismissed their doubts. Now Strassman decided to take a second look.

At Christmas 1938, Meitner's nephew Otto Frisch, one of Neils Bohr's students, visited Meitner at her seaside refuge. With him he brought news that Hahn and Strassman had published another paper. The element created in Fermi's experiment was not a transuranic. It was barium, they wrote, an element little more than half the size of uranium. Critics were confounded and some were saying there must be a mistake.

Meitner immediately recognized what was happening. The uranium atom had cleaved in two. Huge amounts of energy must have been released. She wrote Bohr in Copenhagen. Less than a month later, Bohr was on a ship bound for New York to discuss the matter with Albert Einstein, now a scholar in residence at Princeton. Meitner named the new phenomenon "atomic fission."

Rutherford had died in 1937, still firmly believing that the discoveries of his laboratory would have no bearing on the outside world and that the energy bound in the nucleus would never be liberated. Now every important physicist in the world knew differently. The amount of energy released would be enormous—2 million times that of a chemical explosion.

And the world was headed toward war.

By 1939 the map of Europe was unrecognizable. Germany and the Soviet Union were dividing Poland. France and Britain had declared war. Jews were being persecuted in Germany and everywhere the Nazis conquered. Many nuclear scientists—Einstein, Bohr, Eugene Wigner, Victor Weisskopf, plus the amazing Hungarians, Szilard, Edward Teller, and John von Neumann—were Jewish. They soon joined Fermi and Meitner in exile. Only Werner Heisenberg, Hans Seuss and a few others remained in the Third Reich.

Two native Russians, Fritz Houtermans and George Gamow, had returned to the Soviet Union in the 1930s, only to find themselves caught in Stalin's purges. Houtermans had all his teeth knocked out by inquisitors before "confessing" to inventing a device that could instantaneously measure the speed of incoming aircraft—an impossible feat that nevertheless got him released. Houtermans escaped back to Germany. Had the Soviet authorities not brutalized him, he might have built them an atomic bomb. Gamow made two unsuccessful attempts to escape with his family, one through Afghanistan and another across the Black Sea, before finally defecting at a conference in Brussels.

All knew what the rest of the world did not—that the energy at the nucleus of the uranium atom might be sufficient to build a bomb with an explosive power beyond anything ever imagined. In *The World Set Free*, Wells had envisioned the first nuclear bombs being dropped in 1948—missing the mark by three years. However, he saw them as exploding *perpetually*, like radioactive isotopes. Not even the greatest prophetic novelist of the twentieth century could grasp a nuclear weapon's awe-inspiring power.

The tension of preserving secrecy became almost unbearable. Szilard proposed that nuclear scientists in the West cease publishing their discoveries for fear of giving them away to the Germans. At the same time, Germany stopped exporting uranium. Scientists on both sides became convinced that the other side was developing a bomb.

The original concept had come to Szilard one afternoon in October 1933 while he was waiting for a traffic light in Vienna. It was known that a radioactive element could disintegrate when it absorbed a neutron. What would happen, Szilard wondered, if the disintegration were to release two more neutrons? Wouldn't those neutrons be able to split other atoms and on and on, initiating a chain reaction? Szilard filed patents and began trying to tell Rutherford and others, without success. Now in 1939, with the fission of the uranium atom an established fact, the idea seemed more plausible than ever.

The key question was whether enough neutrons were released when the uranium atom split in two. Szilard began a series of experiments at Columbia University to find out. Edward Teller would later recall sitting at the piano in his Washington apartment playing a Mozart violin sonata with a friend when the phone rang. He picked it up and heard Szilard's voice at the other end of the line. "I found the neutrons," he said.[8]

Szilard, Wigner and Teller were all Jewish and fearful for the fate of European Jewry. Now convinced that Hitler was working on a bomb, they began meeting with state department officials and military men, trying to convey a sense of alarm, even though the world was not yet at war. Unfortunately, to the military, rumors about ultimate weapons were old hat. "We have a regimental goat," growled one army officer at a meeting. "The other day someone brought me one of these death rays. I tried it on the goat and he's as lively as ever!"[9]

With war clouds permanently settling over Europe, Szilard and Wigner decided there was only one thing left to do. They would approach the most famous scientist in the world, Albert Einstein, and ask him to write a letter to the most powerful man in the world, President Franklin D. Roosevelt. Surely one could get the other's attention.

Wigner was a colleague of Einstein's at Princeton but the sage was on vacation. Through a third party, Wigner got a message through to Einstein and asked for a visit. Word came back that he was staying in a cabin near the beach in a small town on eastern Long Island. Wigner thought it began with a

"P." With Teller driving, Wigner and Szilard set out on a hot July afternoon in 1939 for Patchogue on the East End.

After a two-hour journey, the trio arrived in the resort town but could find no trace of Einstein. Szilard consulted the map. "Could it be Peconic?" he asked. "Yes, that's it," said Wigner, and off they went again.

In another hour they were in Peconic but still with no success. They inquired of several vacationers but no one knew anything. Szilard was ready to give up. "Perhaps fate never intended it," he said, but Wigner wanted to persist. Finally, Szilard had an idea. "Let's ask one of the local people," he said. They spotted a deeply tanned seven-year-old boy carrying a fishing pole.

"Do you know where Dr. Einstein lives?" they asked.

"Sure," said the boy, "I'll show you," and led them to a seaside cottage.

And so the age of atomic power began.

> In the course of the last four months it has been made probable—through the work of Joliot in France as well as Fermi and Szilard in America—that it may become possible to set up a nuclear chain reaction in a large mass of uranium, by which vast amounts of power and large quantities of new radium-like elements would be generated. Now it appears almost certain that this could be achieved in the immediate future.[10]

So began Einstein's letter to the President. Even with a missive from the most famous scientist in the world, however, approaching the President of the United States was not an easy task. Szilard had a friend, Alexander Sachs, a Wall Street businessman, who visited Roosevelt regularly. On October 11, 1939—five weeks after Germany and Russia had invaded Poland—Sachs arrived at the White House carrying Einstein's letter. To make sure Roosevelt didn't put it aside, he read it aloud.

The effect was not what was intended. Neither man really understood the subject and both became weary trying to comprehend. Roosevelt said it sounded interesting but he didn't know what he could do about it. Dismayed at his ineptitude,

Sachs managed to wangle another appointment for breakfast the following morning.

"That night I didn't sleep a wink," Sachs recalled later. Instead he paced the floor of the Carleton Hotel, going outside three times to sit in the park across from the White House. Finally, he had an idea.

Next morning, Sachs again addressed Roosevelt. "Mr. President," he said, "during the Napoleonic Wars, a young American inventor approached the emperor and offered to build a fleet of steamships by which he could cross the English Channel. `Ships without sails?' said Napoleon, and sent him away. That man was Robert Fulton. Historians believe if Napoleon had listened more carefully, nineteenth century European history would have taken a completely different turn."

Roosevelt pondered this for a long time. Finally he sent for a bottle of Napoleonic brandy that had been in the Roosevelt family for generations.

"What you're saying," he said, offering Sachs a glass, "is that you want to make sure the Nazis don't blow us up."

"Precisely," said Sachs.

Roosevelt summoned General "Pa" Watson, his military attaché, and handed him the letter. "Pa," he said famously, "this requires action!"[11]

The newly appointed Uranium Committee met the following week at the Naval Laboratory in Washington. At Columbia University in New York, Fermi was working to create an "atomic pile" that would produce a nuclear chain reaction. At this point, everyone believed a bomb would require tons of uranium. Then in 1940, Otto Frisch (Meitner's nephew) and Rudolf Peierls, both now refugees in London, noted that U-235, the isotope that constitutes only .7 percent of the natural ore, is much more fissile than the predominant U-238. If it could be "enriched" into almost pure U-235, a bomb could be made with only a few pounds of material. The trick would be to separate the two isotopes, an almost impossible task since they are chemically identical and would respond the same to any chemical reaction.

Mark Oliphant, head of Britain's uranium research, sent a report to the U.S conveying Frisch and Peierls' proposal. When no response came within five months, Oliphant flew to Washington and asked Lyman Briggs, the lethargic chairman of the Uranium Committee, what was going on. Briggs said the letter was locked in a safe. "The U.S. is not at war," he explained patiently. Oliphant immediately took a train to New York to tell Fermi and James Bryant Conant, chairman of the National Defense Research Committee. Britain did not have the resources to attempt isotope separation, he said, but America did. By October the National Academy of Sciences recommended an all-out effort. The Uranium Committee decided to move ahead. The crucial vote was taken on Saturday, December 6, 1941.

After Pearl Harbor, things moved swiftly. General Leslie Groves, an MIT-educated engineer who had supervised construction of the Pentagon, was chosen to direct a project that would absorb tens of thousand of people, a billion dollars, and eventually grow to the size of the auto industry. Although most scientists were still on the East and West Coasts, a more inland location was preferred for fear of German submarines. Fermi's experiment was moved to the University of Chicago. Two huge tracts of land were acquired at Oak Ridge, Tennessee and Hanford, Washington. At Oak Ridge, families that had lived on the land for generations were removed in a matter of weeks. Less populated, the Hanford site eventually grew to 1,000 square miles. Oak Ridge would begin the painstaking enrichment process, trying to raise the U-235 content from 0.7 to 90 percent. Hanford would produce the newly discovered transuranic element, plutonium, which had been found to be more fissile than U-235. Plutonium was emerging as a dark horse candidate, even though little was known about it. According to

Army Corps of Engineers tradition, the effort was named after its original headquarters city—"The Manhattan Project."

To oversee the scientific effort, Groves chose Robert J. Oppenheimer, a much-loved professor of physics at Berkeley, even though Oppenheimer had left-wing ties. (A long-time girlfriend had been a devout Communist.) As the two men searched the country for a remote headquarters, Oppenheimer suggested the Ranch School, an outdoor-oriented academy for boys just outside Los Alamos, New Mexico, near where Oppenheimer had once vacationed. Within a month the school was closed and nuclear scientists from all over the world began to congregate in the desert.

Probably not since fifth century Athens had there been such an assembly of minds—and may never be again. Nobel Prizes were almost the price of admission. Richard Feynman, only twenty-four, was the bad boy of the group, picking locks (he once cracked the safe containing the only plans for the bomb) and taunting the censors (he instructed his wife to cut up a harmless letter into tiny pieces so the bureaucrats could spend days piecing it back together). Yet when Enrico Fermi arrived, it was Feynman he pulled aside to test his latest ideas. "Those people out there will all nod their heads and say I'm brilliant," he told a colleague. "This young man isn't afraid to tell me if I'm wrong."[12] When Neils Bohr arrived, traveling under the pseudonym of "Nicholas Baker," he was overwhelmed. "I told people you couldn't build a bomb unless you turned the whole country into a factory," he said. "Now you've gone and done just that!"[13]

On the other side of the Atlantic, things were not moving as swiftly. Hitler knew of the atomic bomb but only vaguely grasped its dimensions. At one point he had the post office working on the project. But Werner Heisenberg, who had remained in Germany, was leading the task and seemed to be making progress.

In 1941, Heisenberg visited Denmark to see his old friend

and professor Niels Bohr, an historic encounter dramatized in the 2000 play, *Copenhagen*. His mission is suspected to have involved two things: 1) to find out if the Allies were building a bomb, and 2) to see if there was a better method than amassing huge amounts of uranium. The Frisch-Peierls idea of using the rare U-235 had not reached him. After the war Heisenberg claimed he only stayed in Germany to sabotage the effort, but there is still enormous doubt about what transpired at the meeting.

Then on December 2, 1942, theory became reality. In a squash court beneath the grandstands at Alonzo Stagg Field at the University of Chicago, Fermi oversaw the world's first atomic chain reaction. The "pile" consisted of layers of graphite bricks containing small spheres of natural uranium. The entire structure was supported by a wooden framework and laced with cadmium rods that could be drawn in and out to control the reaction. One particularly large rod entered vertically and was raised and lowered by a rope. Beside it was stationed a young assistant, one Normal Hilberry, who held an axe, ready to cut the rope at a moment's notice if the chain reaction accelerated out of control. His moniker—"safety control rod axe man," abbreviated as "scram"—has ever after been used to describe a rapid, unplanned reactor shutdown.[14]

All through November, lab workers piled graphite blocks to a height of almost three stories. If positioned just right, with the graphite moderating the neutrons, a small chain reaction should occur. Sawing the bricks by hand left the technicians as blackened as coal miners. Finally, Fermi calculated the pile had reached a critical mass. Slowly the control rods were withdrawn while a needle measuring neutron activity climbed the gauge. Three times the needle approached the critical point, then leveled off—as Fermi said it would. On the fourth try an alarm sounded. The warning for excess neutrons had been set too low. Fermi called a break for lunch.

By mid-afternoon the group was ready to begin again. This

time the neutron alarm was set higher and the needle kept rising. Everyone except Fermi became nervous. As he wrote later:

> At this moment we knew that the self-sustaining reaction was under way. The event was not spectacular, no fuses burned, no lights flashed. But to us it meant that release of atomic energy on a large scale would be only a matter of time.[15]

The click of Geiger counters became deafening. Fermi let the self-sustaining reaction continue for twenty-eight minutes before giving the order to lower the fuel rods. Everyone breathed easier. The chain reaction had produced sixty watts, enough to illuminate a single light bulb.

Achieving a nuclear chain reaction was one thing, detonating a bomb was something entirely different. As the scientists assembled in Los Alamos, they knew such a device would require bringing together a much greater amount of fissionable material in a single instant. If it happened too slowly, the material would fissile prematurely and "fizzle" into a dud. At the same time, once the explosion began, its shock wave would push neutrons away from the center, which would also cause a fizzle. The bomb's shell had to somehow be "tampered" so that neutrons reflected back toward the center. The interval in which all this concentration of material had to take place was 0.00001 seconds.

Under Hans Bethe, the great German physicist in charge of the theoretical aspects, the teams developed several potential strategies. Two small quantities of U-235 might be fired into each other from a small cannon, producing a critical mass. Or the bomb could be cut in two like halves of a grapefruit and rapidly slid together. As a third option, the bomb might be surrounded by chemical explosives and imploded upon itself. The subject occupied some of the world's best minds for nearly three years. The last option seemed least practical.

There is 28-page paper, now available on the Internet, entitled "Los Alamos Primer." It begins with this introduction:

> The following notes are based on a set of five lectures given by R. Serber during the first two weeks of April 1943, as an "indoctrination course" in connection with the starting of the Los Alamos Project. The notes were written up by E. U. Condon.[16]

The original pages are typewritten single-space and covered with hand-scrawled equations, sketched graphs, and crude drawings of the bomb's possible configurations. At the time, it must have been the most valuable document in the world. A single copy delivered to the Germans or Russians or Japanese could have changed history. (An annotated version of these notes was published in 1992 as *The Los Alamos Primer: The First Lectures on How To Build An Atomic Bomb* by the University of California Press.)

Meanwhile, at Oak Ridge, three techniques were developed to separate the U-235 isotope. Gas diffusion involved filtering uranium fluoride gas through microscopic pinholes, with the lighter isotope passing through more readily. Electromagnetic separation involved shooting the gas through a magnetic field, with the lighter isotopes once again taking a slightly different path. Thermal diffusion involved heating a container so that the lighter isotopes rose toward the top. The plant set up to carry out these processes was a half-mile long, six stories high and covered 43 acres. After two years of laborious effort, Oak Ridge was able to separate enough U-235 to produce one bomb—the one dropped at Hiroshima.[17]

Across the continent, plutonium was being manufactured

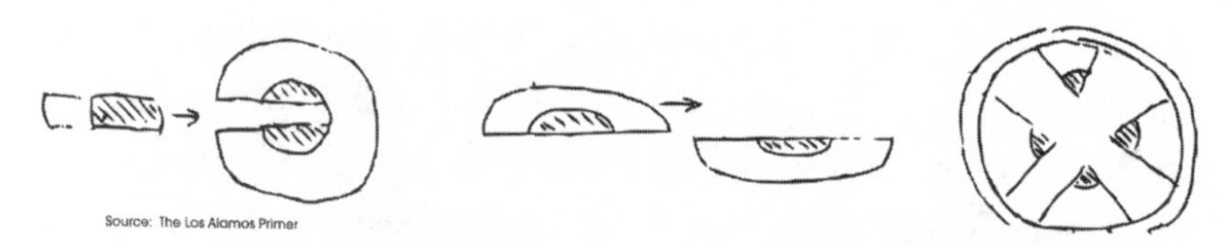

Source: The Los Alamos Primer

Original Drawings for Bomb Designs at Los Alamos

at Hanford through a process called "breeding." When U-238 absorbs a neutron it decays quickly into Pu-239, which is even more fissionable than U-235. A graphite moderator is used to accelerate the process. (This is what the Soviets were doing at Chernobyl.) When the first reactor-bred plutonium was delivered to Los Alamos, however, it was contaminated with Pu-240 and Pu-241, Pu-240 being an isotope that is far less fissionable. The gun design that shot the two halves together was deemed *too slow* to create a critical mass. Implosion—the dark horse—became the front-runner. (It was this change in plans, along with crude drawings copied from the Primer, that Klaus Fuchs, the British spy at Los Alamos, passed on to the Soviets, helping them build their own bomb by 1949.)

On July 16, 1945, a plutonium bomb dubbed "Trinity" was mounted atop a metal tower near Alamagordo, 210 miles southwest of Los Alamos. Detonation was scheduled for sunrise. Months earlier, someone had conjectured that the intense heat from the bomb might be enough to fuse oxygen and nitrogen atoms in the atmosphere, setting off a thermonuclear chain reaction that would ignite the whole atmosphere, obliterating the planet. Edward Teller had been assigned to investigate this possibility. After determining the temperatures required to fuse nitrogen and hydrogen, however, he decided it couldn't happen.

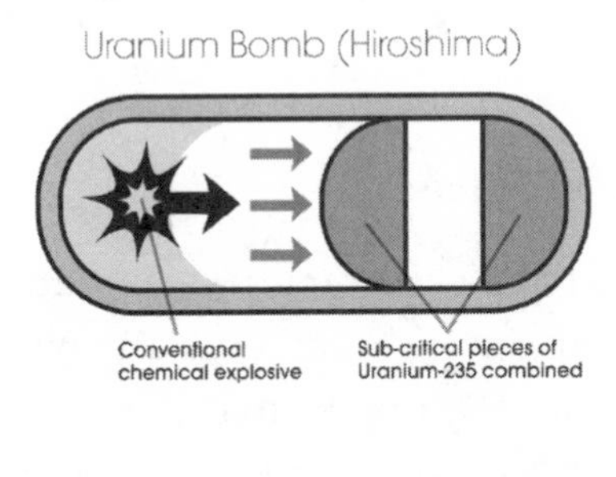

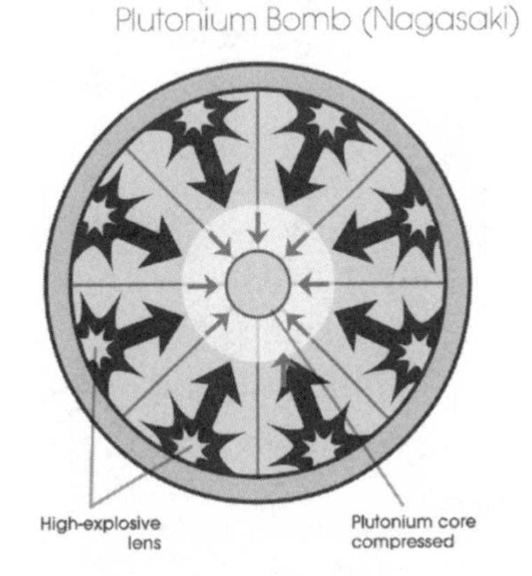

Source: Wikipedia, created by User:Fastfission

Two Types of Nuclear Weapons

Now, as the scientists assembled, Teller ran into Serber, the author of the primer. The generator at Los Alamos had failed once again and the two were groping their way through the dark. Teller mentioned that Oppenheimer had warned of rattlesnakes at the observation site. "What are you going to do?" he asked Serber.

"I'll take a bottle of whiskey," Serber replied.

Then Teller began recounting his investigations on whether the whole world might blow up when the bomb was detonated. He didn't think it could possibly happen, he said, although you could never be a hundred percent certain.

"What do you think of that?" asked Teller.

Serber pondered this for a moment, and replied, "Maybe I'll take two bottles."[18]

People a hundred miles away saw the flash. But most took it for a distant thunderstorm. Oppenheimer, a scholar of the classics who had read much of world literature in the original, recalled a line from the *Bhagavad Gita*: "I am become Death, the destroyer of worlds." In *Brighter Than a Thousand Suns*, a 1958 history of the Manhattan Project, Robert Jungk quoted the entire passage:

> If the radiance of a thousand suns
> were to burst into the sky,
> that would be like
> the splendor of the Mighty One—
> I am become Death, the shatterer of Worlds.

"We all hoped that with the end of the war emphasis would be shifted decidedly from the weapon to the peaceful aspects of atomic energy," wrote Fermi later. "We hoped that perhaps the building of power plants, production of radioactive elements for science and medicine would become the paramount objectives."

Chapter 19

Atoms for Peace

The first atomic bomb used in warfare was dropped on Hiroshima on August 6, 1945. As mentioned, the bomb was equal to 15,000 tons of TNT. The largest bomb ever dropped had been one ton. When the bomb was at its takeoff point on the tiny Pacific island of Tinian, a general knowledgeable of the situation refused to believe "Little Boy" was the "secret weapon." It was far too small. Nonetheless, the results were devastating. Precise numbers have never been established because large numbers of soldiers were stationed in the city, but estimates of the fatalities range from 80,000 to 120,000.

Before the bomb was dropped, the six members of the Japanese war cabinet had been split 3–3 on whether to continue the war. Amazingly, after Hiroshima, not a single vote changed. It was only after the second nuclear bomb was dropped on Nagasaki a week later that the Emperor personally intervened, announcing directly over the radio, "We must endure the unendurable." It was the first time the Japanese people had ever heard his voice. Even then, there was a revolt against the decision among the war faction of his cabinet. War Minister Korechika Anami put down the rebellion, believing his first duty was to the emperor. Then he went home and committed *Hara Kiri* over the shame of losing the war.[1]

Whether the bomb should have been dropped on a populated city or demonstrated in some remote area will be endlessly debated. It is worth remembering, however, that only six months before Hiroshima, 7,000 marines died taking Iwo Jima, an island of eight square miles. On the Japanese mainland, the size of California, the entire population was dug in and ready to fight to the death. The U.S. military had already manufactured 500,000 purple hearts and was anticipating 100,000 to 500,000 new deaths. Casualities among the Japanese could have been as high as 10 million. The firebombing of Tokyo on May 26, 1945 had killed more people than either Hiroshima or Nagasaki, yet the Japanese hadn't flinched. On the island of Saipan in 1944, several thousand Japanese civilians—many holding children in their arms—threw themselves off high cliffs rather than surrender to U.S. Marines.[2]

Another factor to consider is the Soviet Union, which entered the war for only a week before Hiroshima and received North Korea as the prize. Stalin invaded North Korea the day Nagasaki was bombed. Communications flew back and forth until an American general, glancing at a map, suggested dividing the country at the 38th parallel. The Soviets miraculously accepted. Had they continued south, the U.S. had no troops to prevent it. Had the Russians participated in an invasion of Japan, we would have likely had a North and South Japan instead of North and South Korea. At the same time, historians note, Stalin was already making moves in West Berlin. Had the war in the Pacific tied us up for another year or two, Western Europe might have fallen as well.

There was one more thing. After the second bomb was

dropped on Nagasaki, we had no more nuclear weapons in our arsenal. It would have taken at least a month to build another. By that time the shock value would have worn off. More than anything, it was the surprise and the fear of the unknown that caused the Japanese capitulation.

The use of the bomb against the Japanese stunned many at Los Alamos. They had seen the Manhattan Project as an effort to defeat Hitler. Robert Serber and two others physicists wrote a letter to a Japanese former colleague and attached it to a radiation-measuring device that followed the bomb down at Hiroshima, urging him to persuade the Japanese General Staff to end the war. The letter reached the physicist several months after the war ended.

The rapid capitulation of the Japanese set off huge rejoicing. Los Alamos celebrated with an all-night party and fireworks displays. There are photographs of the entire populations of Oak Ridge and Hanford, children included, mobbing the streets holding up newspapers of the Japanese surrender. Up to that point, most inhabitants of these cities had no idea what they were working on.

With the bomb came the public's first encounter with radiation. Although there had been the brief vogue in the 1920s and x-rays were common, the levels of exposure from the bomb were unprecedented. Much coverage of Hiroshima and Nagasaki concentrated on the radiation sickness suffered among people who survived the fireball. There was concern that survivors would suffer long-term cancers or birth defects.

At home the story was the same. It had been easy enough to gloss past the dangers of radiation during the wartime emergency, but now the hazards came into sharp focus. In Los Alamos only a week after Nagasaki, Harry Dagnian, a 26-year-old physicist, working with two small quantities of fissile material, accidentally set off a chain reaction for a fraction of a second. His right hand received a huge dose of radiation. Within a week, he was suffering severe radiation sickness—delirium, hair loss—and within a month he was dead.

Eight months later, Louis Slotin, a young daredevil who was a favorite at Los Alamos, suffered a similar accident. Slotin had fought in the Spanish Civil War and tried to join the Royal Canadian Air Force in the early years of World War II but his scientific genius routed him to Los Alamos. His specialty had become testing the criticality of nuclear material. One model for the bomb, you will recall, involved two halves of a grapefruit. Slotin's job was to bring them together, inch-by-inch, to determine when they became critical. On May 21, 1946, while preparing for the second major bomb test at Bikini, Slotin's screwdriver slipped and the material flashed a dazzling blue light, the beginning of a chain reaction. He tore the two halves apart with his hands, averting a small nuclear explosion. Workers standing a few yards away suffered only minor doses but Slotin was fatally exposed. He died nine days later.

After these incidents, a group of physicists at Los Alamos formed a committee they called the "Association of Atomic Scientists." It launched a campaign to warn the public of the dangers of excessive radiation.

Much of the early news was that the "Atomic Age" would be an epoch of wonders. In *Atomic Energy in the Coming Era,* published less than three months after Hiroshima, David Dietz, a Scribbs-Howard science reporter promised:

> A nonstop flight around the earth's equator in 24 hours.... A 240,000-mile rocket trip to the moon.... These are but two of the miracles just head of mankind in the coming Era of Atomic Energy.[3]

Testifying before Congress, Alvin Weinberg, then an outstanding young nuclear scientist at Oak Ridge, tried to steer attention toward the positive aspects of the new technology. "Atomic power can cure as well as kill. It can fertilize and enrich a region as well as devastate it. It can widen man's horizons as well as force him back into the cave."[4] There was serious talk of atomic powered ocean liners as well as home heating plants and even small generators that could be wired into clothing for warmth.

The first great question was whether atomic energy should be under military or civilian control. The military moved swiftly and within six months Congress was considering the May-Johnson Bill, which would give entire responsibility to the army. Leo Szilard rushed to Washington and testified in opposition. His performance roused others and May-Johnson was soon replaced by the Atomic Energy Act of 1946, creating the Atomic Energy Commission, a civilian agency that would control the technology for military purposes.

For a while there was considerable sentiment among Los Alamos scientists for sharing the secrets of atomic energy with the Soviet Union. These were the dawning days of the United Nations and there was much enthusiasm for world government. In December 1945, James Bryant Conant, the president of Harvard who had played a large role in the Manhattan Project, went to the Four Power Conference in Moscow ready to present an agenda for an exchange of scientists and international controls. To his chagrin, he was not even allowed to speak. Any faint hope of continuing the Soviet-American wartime alliance was dashed in July 1946 when Russian foreign minister Andre Gromyko publicly rejected Conant's offer at the United Nations.

In early 1948, the Communists took over Czechoslovakia. Next Stalin blockaded Berlin, leading to the Berlin Airlift of 1948–1949. Still, America's sole possession of the atomic bomb seemed to give us enormous advantage. The most pessimistic predictions said the Soviets could produce a bomb of their own in five to ten years. Some thought it would take longer, although veterans of the Manhattan Project were less sanguine. Then in August 1949, a B-29 "flying laboratory" aloft in the Far East detected traces of radioactive material in the atmosphere.

Their origin soon became obvious—the Russians had exploded an atomic device somewhere in the vast reaches of Siberia. The nuclear arms race had begun.

Recriminations were immediate. A few months later, Klaus Fuchs, who had been at Los Alamos and now directed the British bomb project, was unmasked as a Russian spy. Shortly afterward, the Rosenbergs were found to have passed information from David Greenglass, who was also at Los Alamos. The result was a fanatical effort to root out spies and impose loyalty oaths, eventually leading to the McCarthy Era. In the long run, though, this espionage probably played only a minor role. As the Los Alamos veterans knew, the technology was not that complex and the Russians had good scientists.

The question now became whether to develop a hydrogen bomb. Around the time Edward Teller was calculating whether Trinity would blow up the world, he theorized that the blast from an atomic bomb could set off a *fusion* reaction among hydrogen atoms, duplicating the process of the sun. Teller suggested wrapping a fission bomb in a blanket of deuterium or tritium, the two heavy isotopes of hydrogen. A Hiroshima-type bomb would become a trigger for a bomb 1,000 times its size. The "H-Bomb" was tested successfully at Enewetak Atoll in the Pacific in November 1952. Within a year, the Soviets had detonated their own thermonuclear weapon as well. It was designed by Andrei Sakharov, who later became a prominent Soviet dissident.

The hydrogen bomb fractured whatever unity was left among Los Alamos veterans. Most protested, forming committees, signing petitions and testifying before Congress. A few supported Teller and the bomb. During the war, the military had tolerated the "eggheads," many of whom had left-wing sympathies. Now opposition to the H-bomb became a symbol of disloyalty. In 1954, after a lengthy inquisition, Congress revoked Robert Oppenheimer's security clearance for his dissent. Andre Malraux, the great French writer-diplomat, remarked, "He should have stood up and said, `Gentlemen, I *am* the atomic bomb!'"

With the world moving rapidly toward Armageddon, President Eisenhower tried to reverse the momentum in December 1953 by introducing the "Atoms for Peace" program. "I feel impelled to speak today in a language that in a sense is new—one which I, who have spent so much of my life in the military profession, would have preferred never to use. That new language is the language of atomic warfare," began his dramatic speech before the United Nations:

> The United States pledges...to devote its entire heart and mind to find the way by which the miraculous inventiveness of man shall not be dedicated to his death, but consecrated to his life.[5]

To jumpstart the process, Congress revised the Atomic Energy Act, ending the government monopoly on nuclear technology and finally making civilian use possible. The utility industry expressed extreme satisfaction. In a highly symbolic gesture, Lewis Strauss, a businessman, was appointed head of the AEC. Speaking before a conference of the National Association of Science Writers in New York on September 16, 1954, Strauss made a famous claim that would echo through the decades: "It is not too much to expect that our children will enjoy in their homes electrical energy too cheap to meter."[6]

A much more realistic assessment had been expressed only a few months earlier in a *National Geographic* article, "Man's New Servant, The Friendly Atom:"

> Today's big question about peacetime atomic energy is when it will be used for producing industrial power. One authority has predicted economical atomic power "in a very few years, certainly less than ten." Other estimates run up to 30 years.... Electric power produced by atomic energy cannot be much cheaper than present-day power, however. Fuel represents only about 20 percent of the cost of electric power to the consumer today. Even if atomic energy eventually proves cheaper than coal, the remainder of the process of making electricity will

> be the same as now, so that the other 80 percent of the cost will not be reduced.[7]

Strauss's words have lived much longer.

Admiral Hyman Rickover was born the son of Jewish immigrants who had fled Poland for Chicago in 1905. A bright and determined young man, he won an appointment to the Naval Academy—an astonishing accomplishment for a poor Jewish boy from the slums. Falling in love with the Navy, Rickover became a lifer, gaining a master's degree in electrical engineering from Columbia University while in uniform. During the 1930s he commanded two submarines and a minesweeper. When the war came, he was appointed head of the Electrical Section of the Bureau of Ships. His service won him the Legion of Merit.

After the war, Rickover became obsessed with the vision that submarines should run on nuclear power. At the time, submarines operated on electrical batteries that lasted only a few hours at top speed. Subs had to surface continually to recharge, a process that left them vulnerable to attack. In theory, nuclear submarines could remain submerged for months. Rickover went to Oak Ridge to learn the science, then garnered an assignment at the Division of Reactor Development in the AEC. Later he returned to the Bureau of Ships as director of the Naval Reactors Branch. Leveraging this experience, he was able to launch the *Nautilus*, the Navy's first nuclear powered submarine, in 1954. Hitting the water in New London with the message, "Underway on nuclear power," the *Nautilus* immediately set all-time records for speed and endurance in the first week, racing 1,300 miles underwater to Puerto Rico in less than ninety hours. The nuclear submarine redefined warfare, becoming part of the nuclear-armed triad—submarines, land-based silos, and constantly aloft bombers—that promised instant dispatch of nuclear weapons for "mutually assured destruction."

Rickover knew every square inch of his submarines and was relentless in pushing his projects through the bureaucracy.

Ted Rockwell, a veteran of Oak Ridge who became Rickover's biographer, tells the story of a submarine captain who called Rickover in desperation as he was about to launch. Someone had dropped a screw into the reactor vessel and he was terrified it would clog the reactor, causing a meltdown at sea.

"Go down to the bottom deck where the reactor vessel emerges," said Rickover. "At the very bottom there's a small plate. Remove that plate and you'll find the screw right there." Sure enough, Rickover was right. He had anticipated everything.

Another time, Rickover was trying to replace a navigational gyroscope that he complained could not stand the shock of depth charges. After six months of work, a manufacturer came to his office and presented a gyroscope. "This can withstand any kind of impact," he said.

Rickover took the gyroscope and threw it against a radiator, smashing it to pieces. Then he walked out without saying a word. On such principles was the Nuclear Navy built.[8]

By 1955, five utilities had announced plans to build nuclear electric plants but none had acted. The Joint Congressional Committee on Atomic Energy, which oversaw the AEC, was growing restless. Democrats on the committee wanted the AEC to build reactors itself. Hoping to keep the government out of utility business, President Eisenhower instructed Rickover to "beach" one of his submarine reactors on a riverfront site owned by the Duquesne Light Company in Shippingport, Pennsylvania, 25 miles west of Pittsburgh. It became America's first civilian plant. (The Soviets built one in 1954.) Shippingport was barely bigger than a submarine motor, producing only 60 MW, but after more than a century of the country's worst air pollution, the coal capital was eager to embrace the new technology.

The reactor was scheduled to open in September 1957 when a problem developed. Progressive Era reforms dictated that insurance companies set their rates according to actuarial records. But nuclear had no record. Traditionally, new technologies were underwritten by European insurers but in 1957 European companies had not yet recovered from their war losses and Washington was reluctant to involve them anyway. General Electric told Congress it would not "proceed with a cloud of bankruptcy hanging over its head." Westinghouse agreed it "could not proceed without some kind of government backing."[9]

So Congress responded by passing the Price-Anderson Reactor Safety Act. The bill organized a consortium of private insurers to provide 60 million dollars in coverage. Then the federal government pledged another 500 million dollars. The measure passed in September 1957. Two weeks later, on October 10, the graphite moderator at the Windscale reactor in England caught fire, releasing 20,000 curies of Iodine-131, contaminating large quantities of milk and farm produce. (By comparison, Chernobyl released 7 million curies, Three Mile Island only 15 curies.) Nuclear accidents were indeed a possibility. Nevertheless, Shippingport opened on schedule.

Critics have argued ever since that Price-Anderson proves nuclear is too dangerous to operate without government protection. Yet the federal government insures all kinds of activities. The Federal Deposit Loan Insurance Corporation insures the banking system. The Employee Retirement Income Security Act insures corporate pensions. The *Catalogue of Federal Insurance Activities*, the General Accounting Office's listing of all actions insured by the federal government, is 90 pages long.[10]

The hundreds of Army Corps of Engineers hydroelectric dams that dot the countryside carry *no* insurance. As government-owned facilities, they are exempted from liability by sovereign immunity. When the federally owned Teton Dam burst in 1976, private insurance companies invoked their flood exemption and the 500 million dollars in damages to victims of the flood went completely uncompensated. The federal government responded only by offering low-interest loans. If large-scale carbon sequestration is ever practiced, it will dwarf the risks covered by Price-Anderson. A catastrophic release of carbon dioxide would be much more serious than a nuclear accident.

In 1967, Price-Anderson was changed to no-fault. The utility companies waived their right to defend against tort claims and anyone who proved damages could collect. In exchange, the utilities' overall liability was capped at $560 million. Worker's Compensation works the same way. So does the Black Lung Disability Trust Fund. In 1999, President Clinton and Congress proffered the same liability cap to the computer industry for anticipated Y2K damages claims *without* offering any compensating coverage to anyone who might be injured.

Today the federal government is completely off the hook with Price-Anderson. The primary coverage provided by private insurance companies now stands at $300 million. In addition, each of the nation's 104 reactors can be assessed *$100 million a piece* for accidents at *other* reactors. That puts total coverage above $10 billion. No other industry carries anywhere near as much insurance. It is no wonder reactor owners have become so fanatically concerned with *each other's* safety. As the saying goes in the industry, "We are all hostages to each other."[11]

Shippingport kicked off a golden age of nuclear construction. In 1959, Dresden I outside Chicago became the first reactor built entirely without government financing. In the next two years, Yankee Rowe in Connecticut and New York's Indian Point were built privately. None was over 150 MW but in 1963 General Public Utilities (which eventually built Three Mile Island) announced a 515 MW reactor at Oyster Creek in southern New Jersey. Although larger than any power plant ever built, Oyster Creek was soon dwarfed by seventeen more 800 to 1,000 MW projects licensed over the next five years.

General Electric and Westinghouse competed fiercely for the business, offering "turnkey" projects at preset prices, regardless of construction costs. Losses mounted into the hundreds of millions for both companies, yet they felt they were fighting for the future. "It's going to take a long time to restore to the treasury the demands we put on it to establish ourselves in the nuclear business," commented one GE executive.[12]

With concern over air pollution mounting, nuclear power seemed ready to sweep all before it. Glenn Seaborg, winner of the 1951 Nobel Prize for his discovery of plutonium, was now chairman of the AEC. A native Californian, he joined the board of the Sierra Club and persuaded it to endorse nuclear as an alternative to the Glen Canyon Dam. When Con Edison proposed a pumped storage plant at Storm King Mountain on the Hudson in 1962, its environmental opponents proposed a second reactor at Indian Point in its place. When a group of left-wing college students met in Port Huron, Michigan in 1962 to form the Students for a Democratic Society, their opening manifesto, the "Port Huron Statement," read as follows:

> Our monster cities, based historically on the need for mass labor, might now be humanized, broken into smaller communities, powered by nuclear energy, arranged according to community decision. These are but a fraction of the opportunities of the new era: serious study and deliberate experimentation, rooted in a desire for human fraternity, may now result in blueprints of civic paradise.[13]

To be "anti-nuclear" in the early 1960s did not mean being against nuclear power but nuclear weapons. Both the Russians and the U.S. had tested hydrogen bombs for over a decade. After throwing up the Berlin Wall in 1961, Nikita Khrushchev exploded a 100-megaton bomb in northern Siberia, the largest ever detonated. A few months later, Kennedy and Khrushchev squared off in the Cuban Missile Crisis—the closest we ever came to nuclear war.

Eight months after this heart-stopping confrontation, events reversed with stunning swiftness. On June 10, 1963, Kennedy gave a surprisingly conciliatory speech at American University inviting Russia to join a nuclear test ban treaty. Khrushchev was favorably impressed and stopped jamming Voice of America, allowing the Russian people to hear the offer. On July 15, British, American and Soviet delegates sat down in Moscow for ten days and hammered out a ban on all above ground testing. None of the three countries has exploded a weapon in the atmosphere since.

One of the things that drove Kennedy forward was the growing recognition of "radioactive fallout." Strontium-90, a long-lived uranium by-product, is taken into the body, where it mimics calcium. It had been found in children's bones and mother's milk. Rachel Carson's *Silent Spring* had just been published and the realization was dawning that pesticides and radioactive isotopes could concentrate in the food chain.

Once testing was banned, however, many anti-nuclear organizations were bereft of an issue. Without missing a beat, they began concentrating on nuclear *power*, without acknowledging there was any difference. This soon created rifts in the environmental movement, with David Brower quitting the Sierra Club in 1969 to found Friends of the Earth. Paul Ehrlich, author of *The Population Bomb*, turned his attention to nuclear, arguing, "Nuclear power represents the single greatest threat to the health and safety of humanity." In the introduction to *The Silent Bomb*, published by Friends of the Earth, Ehrlich wrote:

> These dangers are immediate, awesome and unprecedented: risk of catastrophic accident, release of poisonous wastes, sabotage, the building of atomic bombs by terrorists—these are only a few possibilities hanging over our heads. Any one of these can have devastating short-run consequences for the thousands of people who will die slowly and for the tens of thousands who will suffer.... [H]ow did our government and industry ever get away with erecting a nuclear network under out noses? Why has the debate on these questions been postponed until now?[14]

Reactors were being constructed near population centers and scientists knew there was legitimate concern for safety. In 1964, the AEC commissioned a study with the Brookhaven Laboratory to evaluate the worst-case scenario. A previous study had estimated 3,400 fatalities. With technological improvements, the AEC was hoping the danger had been alleviated. To its dismay, the study found risks had *increased*. With larger reactors, fatalities could rise as high as 45,000. The AEC went to quite a bit of trouble to keep the study under wraps.[15]

The major concern was a meltdown, particularly one that expanded into William K. Ergen's "China Syndrome." In smaller reactors, a loss of cooling water was not critical because temperatures were not that high. With the larger new reactors, however, the decay heat would be enough to melt the fuel rods. As a safety measure, the AEC added a new requirement—an Emergency Core Cooling System (ECCS) that would pump high-pressure water into a damaged reactor vessel.

Although some computer models seemed to show the ECCS would work, there was no way to test it in a full-scale reactor. Critics remained skeptical. The Union of Concerned Scientists, founded in 1969 to oppose pesticide use, now started to challenge the AEC. Led by Henry Kendall, a physics professor at MIT, and staffed with numerous outstanding scientists, UCS presented a formidable opponent to the AEC.[16]

To prove the ECCS would work, the AEC built a miniature test reactor at the Idaho National Laboratory. To its dismay, the test showed the ECCS failing. Steam around the fuel bundles resisted intrusion and the cooling water blew right past the core and out the breach in the pipes. Tests at Oak Ridge also showed the zirconium-alloy cladding of the fuel rods would swell, clogging the core. Alarmed by these findings, the AEC once again suppressed the report.

In January 1971, the AEC opened up what it expected to be a "brief" hearing on the ECCS in Washington. The proceedings lasted two years. UCS lawyers hammered away at the testimony until it became clear the AEC was sitting on mountains of disconcerting information. Fearing that reactor construction might stall, the AEC released an "interim report" in July 1971 endorsing the ECCS, based once again on the computer models. Just to be sure, however, the commission recommended that new reactors limit their operations to 80 percent of full power. The UCS called for a moratorium on all new licenses and demanded that existing reactors be shut down until the issues could be resolved.[17]

Thermal pollution was also attracting public attention. At first, the AEC refused to deal with the problem, saying it only had jurisdiction over radiation. Congress changed the law but by that time the utilities had started building cooling towers anyway, fearful of lawsuits and public opposition. As Samuel Walker, the official historian of the AEC, writes: "[Thermal pollution] played a vital role in transforming the ambivalence that environmentalists had demonstrated toward the technology into strong and vocal opposition."[18]

With the ECCS under intense scrutiny, Glenn Seaborg—the last of the Los Alamos generation—stepped aside in August 1971 to be replaced by James Schlesinger, a rising star in the Nixon Administration. Schlesinger immediately reversed Seaborg and said the AEC would honor the new National Environmental Policy Act, which required an environmental impact statement for every federal project. "There's a *Bird Watcher* Running the Atomic Energy Commission," crowed *National Wildlife* magazine. Schlesinger's efforts to placate environmentalists soon faltered, however, and in 1974 Congress finally ended the AEC's dual role. The agency was split into the Nuclear Regulatory Commission, which would regulate the industry, and the Energy Research and Development Administration, which would do research. Both were eventually subsumed into the new Department of Energy.

A few months after the break-up of the AEC, on March 22, 1975, the TVA's Brown's Ferry reactor in Decatur, Alabama suffered the nation's worst nuclear accident to date. This is the one where a technician using a lighted candle to check for air leaks in a tightly enclosed compartment filled with electrical circuitry set fire to the insulation. The fire disabled several key safety systems. For five hours, AEC officials insisted on trying to smother it with CO_2 and dry chemicals used for electrical fires. All the while, the fire chief from nearby Athens insisted it was not an electrical fire but an insulation fire that could be doused with water. AEC official finally accepted his recommendation and the blaze was put out in twenty minutes.[19]

Browns Ferry threw a huge scare into the industry. The AEC tried to dampen fears by issuing a "Reactor Safety Study," prepared by MIT professor Norman Rasmussen. When the Rasmussen Report appeared six months after Brown's Ferry, it created its own firestorm. The report's most memorable conclusion was that the chances of being killed in a nuclear accident were about the same as being hit by a meteorite. This raised howls of derision from anti-nuclear activists. Critics said the fault-tree approach ignored exactly the sort of "common mode failures" that had occurred at Brown's Ferry. The UCS calculated that Rasmussen had underestimated the chances of a serious nuclear accident by a factor of ten.

However persuasive their concerns, though, nuclear opponents did not have much to offer in its place. Environmental groups had been opposing dam construction for ten years. The Arab Oil Boycott had dashed hopes that oil could replace coal or hydro. The "Natural Gas Shortage" of early 1977 had abruptly halted a switch away from oil. Other than a full-scale return to coal, how were we going to run our electrical system?

It was at this point that Lovins published his famous *Foreign Affairs* article, claiming that both coal and nuclear could eventually be phased out and replaced by solar energy. The anti-nuclear movement was suddenly galvanized. They were no longer just *against* things but *for* something—conservation and solar energy. Nuclear was no longer simply a bundle of unknown dangers but a positive impediment to progress.

Alvin Weinberg became a major advocate of nuclear, debating Lovins in several public forums. Always willing to be objective, he had published a 1971 article in *Nuclear News* called "The Moral Imperatives of Nuclear Energy" in which he admitted, "We nuclear people have made...a Faustian bargain...with society." The price of inexhaustible source of energy would be "both a vigilance and a longevity of our social institutions that we are quite unaccustomed to."[20] In a phrase that would echo through the decades, he said it might require a "nuclear priesthood" to keep watch over the technology. The phrase certainly didn't resonate with the industry, but environmentalists saw it as a blueprint for a remote technological elite.

With the Vietnam War and nuclear testing now resolved, student radicals began to focus on nuclear energy. Ralph Nader weighed in with *The Menace of Atomic Energy* (1977) and founded Critical Mass, which became a fountainhead of anti-nuclear tracts.[21] In April 1977, students from all over New England occupied the Seabrook Nuclear Station in New Hampshire, bringing construction to a halt. The "Clamshell Alliance" had taken its name from concerns about thermal pollution, but the opposition to nuclear was now general.

At about the same time, the Atomic Energy Board of Canada published an obscure report called "Risk of Energy Production," by Canadian scientist Herbert Inhaber.[22] The study attempted a broad assessment of the risks of every kind of energy production, from the chances of a nuclear accident to the risks of falling off a roof while installing a solar collector. Inhaber claimed that natural gas was the safest technology with nuclear second. Solar technologies had a high risk factor and coal and oil were the worst.

"The Inhaber Report" circulated among professional circles for a while before being published in the British journal *New Scientist* in 1978 and then in the prestigious American journal *Science* in 1979.[23] Once again the report drew thunderous denunciations. Critics charged that Inhaber had absurdly understated the risks of a nuclear holocaust while adding logging accidents in the risks of solar energy. One of Inhaber's principal assumptions was that technologies such as wind and solar collectors would require constant backup from coal. Writing in *Mother Earth News*, Paul Ehrlich responded:

> The facts are as follows: First, any power plant used for base load electricity production (that is, one which will be "online" as much as possible—whether conventional or unconventional) needs some backup capacity for those occasions when the base load plant is out of service. The nonconventional plants treated by Inhaber have built-in energy storage, and require no more backup than conventional coal or nuclear plants (whose "stand in" requirements Inhaber ignores).[24]

Just what the "built-in energy storage" was for the non-conventional plants mentioned, Ehrlich did not specify.

The battle of the reports continued through early 1979. In January, after years of effort, Congressman Morris Udall, chairman of the House Subcommittee on Energy and the Environment, finally persuaded the NRC to withdraw the Rasmussen report. After several unfavorable reviews, the Atomic Energy Board of Canada pulled back on Inhaber as well.

Then suddenly all the studies became incidental. At 4 AM on March 28, something occurred that had not figured in either report. Because maintenance was being conducted in another part of the plant, several indicator lights on the control panel of a reactor near Harrisburg, Pennsylvania were temporarily out of order. In compliance with NRC regulations, the operators had hung large cardboard tags on their switches. When an unexpected pressure surge arose in the reactor, the operators checked to see if auxiliary feedwater valves were open. In fact they were closed, but an indicator light conveying this information was partially obscured by a maintenance tag. Failing to notice their mistake, the operators set in motion a series of events that would shake the industry to its roots and make the world familiar with a sandbar in Susquehanna River known as Three Mile Island.

Telephoning the head of the Nuclear Regulatory Commission from the control room hours later, a shaken NRC official wiped away a decade of predictions with a single sentence. "It's a failure mode that's never been studied," he said.[25]

Chapter 20

Three Mile Island and Its Aftermath

At 4 AM on Wednesday, March 28, as *The China Syndrome* was playing its second week in theaters around the country, operators at the six-month-old Unit 2 of the Three Mile Island Nuclear Station near Harrisburg experienced a "scram." With the reactor operating at 97 percent capacity, a minor malfunction occurred in the secondary cooling circuit, causing the temperature in the reactor vessel to rise slightly. Sensing the change, the system shut down within one second. Reactors, after all, were designed to respond automatically to almost any irregularity.

In a pressurized water reactor (PWR), like the one Westinghouse had designed at Three Mile Island, water in the primary cooling loop draws heat from the core but is prevented from evaporating by being kept under high pressure. This pressure is regulated in an auxiliary relief tank, where a "pilot-operated relief valve" keeps it from mounting too high.

When Unit 2 shut down, the pilot-operated relief valve

opened for ten seconds to relieve pressure as it was supposed to do. At that point it should have closed automatically. Instead, it stuck open. Water began streaming out of the relief valve, draining the primary loop that cools the core.

The reactor was programmed to deal with this. Responding to loss of coolant, high-pressure pumps began injecting additional water into the core. However, the operators didn't immediately recognize what was happening. When the pilot relief valve stuck open, they sent an instruction for it to close. Although a light on the control panel indicated the instruction had been *sent*, nothing on the room-sized, space-age assembly of gauges, lights and indicators told them whether the valve had actually *closed*.

Pressurized water rushed into the relief tank, seeking to escape out of the open valve. This temporarily raised pressure in the tank. Reading this from a gauge, the operators thought there must be *too much* water in the primary loop. They became concerned the system would "go solid," meaning completely fill with water, a dangerous situation that makes it difficult to maintain circulation.

Thinking they were relieving excess pressure in the primary loop, the operators overrode the automatic safety system and turned off the water pumps. In less than fifteen minutes, the primary loop had lost enough water so that the fuel rods were exposed. Their decay heat raised the temperature to 1000°. What water remained in the cooling system began to evaporate. Some of this radioactive steam leaked through the open relief valve and into the containment building. Not until radiation monitors started going off in the control room did the crew realize it had made a serious mistake.

Once the error was recognized, the crew turned the pumps back on. The steam clogged the pipes, however, and much of the water blew right past the core—just as the 1970 tests in Idaho had predicted. Through a long series of pressure adjustments, the steam was eventually bled out the relief valve and pumped into waste tanks in an adjoining building. A seal on one of the tanks leaked, however, and a small amount of radioactive steam escaped into the atmosphere. The rooftop reading was 1200 millirems—about four times annual background in most parts of the country. This radiation quickly dissipated and readings in adjoining neighborhoods remained normal.

Word of the accident quickly spread through the company and to the Nuclear Regulatory Commission in Washington. The press got word within hours and soon hundreds of reporters were converging on the scene. Meanwhile, Metropolitan Edison and NRC sat down to try to figure out what was happening. They soon became concerned about a hydrogen bubble that had formed in the primary loop. Heat from the white-hot core was splitting water molecules. The oxygen was combining with the cadmium coating of the fuel rods, leaving some of the hydrogen free. If this came in contact with free oxygen, there would be an explosion. Twelve hours into the accident, a sudden pulse had occurred in the containment. At first the crew thought it was a ventilation damp slamming shut, but later they realized it as a small hydrogen explosion. Now a 1000-cubic-foot bubble had accumulated at the top of the reactor vessel and was hindering the cooling. If a second explosion occurred, it could rip the roof off the containment and scatter radioactive debris across the countryside.

Before the end of the first day, officials from General Public Utilities had already alienated the press, offering only vague answers and stumbling through press conferences until reporters became openly hostile. Governor Richard Thornburg of Pennsylvania, on the other hand, was a commanding presence, maintaining his composure and giving straight answers. Still it was obvious that that he was just as mystified as anyone by the technology. NRC officials had been late getting to the scene and quickly made things worse by being unable to give a clear explanation about the hydrogen bubble.

The NRC did not hit its stride until North Carolina-born Harold Denton, director of the Office of Nuclear Reactor Regulation, emerged as its spokesman. Denton's down-home folksy ways reassured the public. When Bob Abernathy of the *Today* show marveled that such an accident could happen, he responded, "We're still on a learning curve, but our hindsight is getting better all the time." When asked if *The China Syndrome* had been accurate, he confessed it "took some liberties" but added, "I really enjoyed Jane Fonda's acting." Meanwhile, theaters around Harrisburg added special showings of *The China Syndrome* to meet the surging demand.[1]

For three days, public attention was riveted on central Pennsylvania. Heeding concerns about the possible hydrogen explosion, Governor Thornburg announced on Friday that pregnant women and children within a five-mile radius should leave the area. A full-scale evacuation would cause enormous disruption, emptying several hospitals and prisons, but plans were made anyway. Meanwhile, workers at the Mallinckrodt Chemical factory in St. Louis were roused out of bed at 3 AM Saturday and spent the next twenty-four hours manufacturing potassium iodine pills that were flown to Harrisburg by private jet. The pills would block uptake of the radioactive iodine that causes thyroid cancer. When they arrived, however, Governor Thornburg decided not to distribute them for fear of causing further panic—even after a local radio station threatened to broadcast where they were stored. In the end, the pills were never distributed.

Across the rest of the country, Three Mile Island produced one hair-raising report after another. Conrad, the *Los Angeles Times* cartoonist, drew a picture of a mushroom cloud emerging from a cooling tower. "Radiation!" screamed a one-word headline in the *New York Post*. A national correspondent swore he saw radiation dripping down the side of the cooling tower. (It was probably water vapor, since radiation is invisible.) Over 1,000 reporters and newscasters covered the story and people around the country thought of little else for an entire week.

Finally, on Sunday April 1, 1979, four days after the accident began, Jimmy Carter and his wife Rosalynn brought a sense of relief by arriving at Three Mile Island. Donning yellow smocks and shoe covers, the President and his wife strolled through the control room, chatting amiably with operators and NRC officials. Public reassurance was enormous. Carter was, after all, a former nuclear engineer under Admiral Hyman Rickover.

After several more days, the NRC determined that the threat of the hydrogen bubble had been exaggerated. Only after a week or more was there any possibility it could self-ignite, as oxygen seeped into the system. The bubble was bled out and temperatures in the reactor were gradually brought under control. Although the crisis was passing, it had been a terrifying experience.

Vowing to get to the bottom of things, Carter appointed a special commission to study the accident. John Kemeny, a renowned computer expert and president of Dartmouth University, was named chairman. The commission would have no representatives from either the nuclear industry or the anti-nuclear movement, only neutral experts. As Carter spoke, many of the 22,000 people who had evacuated the area around Three Mile Island began to filter back to their homes.

Even as NRC officials began to feel more confident, environmental groups charged that they were covering up the true scope of the accident. Within a week, the Environmental Coalition on Nuclear Power, a Pennsylvania group that had opposed the construction of Three Mile Island, sent Chauncy Kepford, a Ph.D. in chemistry, to tell a committee of the House of Representatives that "hundreds, maybe thousands" were likely to die as a result of the accident. Meanwhile, a quickly conducted Harris Poll showed 66 percent of the public believed a reactor accident could cause a "massive nuclear explosion." (Because the U-235 content of reactor-grade fuel is only 3 percent, it cannot explode under any circumstances.)

The Kemeny Commission convened immediately, vowing

to get its report out before the end of the year. If a prompt explanation was not forthcoming, there might be a grass-roots movement to close down all nuclear reactors, leaving the country woefully short of electricity.

As the commission took shape, however, it turned out to be long on lawyers, public relations specialists, and NASA engineers but extremely short on people who understood nuclear power. Almost no one had any idea of how a reactor worked. Desperate to avoid relying on the industry, Kemeny turned to the Navy for help. The brass sent back the name of Ronald Eytchison, a captain who was just completing a term on the Atlantic Fleet Nuclear Propulsion Examining Board.

"When I joined the investigation, there seemed to be a general perception that because the accident had been initiated by a stuck-open power-operated valve, if the valve were redesigned, then the cause of the accident would be eliminated," wrote Eytchison years later. "There was little apparent concern for or understanding of the human factor in the accident." Eytchison knew differently. "I suspected the accident more likely had been a result of human error than simple equipment failure. And in my Navy experience, training deficiencies frequently had underlain human performance problems."[2]

Eytchison was assigned a squad of Dartmouth Law School students and began exercising the commission's subpoena power, summoning crate loads of records from the industry. Then the team requisitioned a Chevy station wagon and began touring the country, taking depositions from company executives.

> On one trip, as we rolled down U.S. 29 toward B&W [Babcock and Wilcox], an Ivy Leaguer handed me a document from the backseat. He thought it looked significant. My response when I read it was, "Dynamite!" I was looking at a desk memo from a B&W engineer commenting on an incident that had occurred at a plant called Davis-Besse.... [I]t turned out an event quite similar to that at TMI had taken place in September 1977.[3]

Indeed, the exact same valve—the pilot-operated relief

valve—had stuck open at Davis-Besse in Toledo in September 1977 while the reactor was operating at only 9 percent of full power. In fact, the malfunction had occurred *nine times* before in other reactors. As at TMI, operators at Davis-Besse had misread the situation and overridden the automatic emergency cooling system. Fortunately, after twenty minutes, one operator had guessed—*guessed!*—what was wrong and closed the valve, stemming the loss of coolant.

After studying the event, a Davis-Besse engineer concluded that, had the reactor been operating at full power (as TMI would be 18 months later) "it is quite possible, perhaps probable, that core uncovery and possible fuel damaged would have occurred." However the top brass at Babcock & Wilcox were unimpressed. They elected to sit on the information, failing to pass it along to the NRC or other owners of B&W reactors.[4]

The nuclear industry did not communicate very well. In fact, it did not communicate at all.

As the Kemeny Commission gathered information, a bizarre technological wonderland began to emerge. The premise had been that the people who designed reactors were geniuses while the people who operated them understood nothing—and indeed many operators were no more than high school graduates. The resulting strategy was to make reactors "idiot-proof"—capable of running themselves with almost no interference from the crew. Even the utilities seemed to know little about how their reactors worked. In 1972, a utility vice president had confessed to Daniel Ford, "The average utility knew as much about the nuclear plant it was buying as the average car-buyer knows about cars."[5]

In control rooms around the country, everything seemed crazily mismatched and inappropriate. Those gleaming instrument panels with their rows of space-age buttons and lights were actually more like cartoon replications of futuristic technology borrowed from an old Flash Gordon set. No sensible human being could possibly comprehend them. As the Kemeny Report concluded:

> There seems to be a persistent assumption that plant safety is assured by engineered equipment, and a concomitant neglect of the human beings who could defeat it if they do not have adequate training, operating procedures, information about plant conditions, and manageable monitors and controls.... Overall, little attention had been paid to the interaction between human beings and machines."[6]

Every reactor, it turned out, had been built to a customized design. As utility executives and engineers had been drawn into the new technology, they wanted to make their own contributions. The Atomic Energy Commission—anxious to move forward—was always willing to oblige them. State utility regulators had often added their own suggestions. The result was a hodgepodge of unique designs that made every reactor an island unto itself.

Nor had there ever been any clear delineation of who was in charge. Manufacturers would build a plant on a "turnkey" basis, then leave the utility to figure out how to run it. Matt Groening, the creator of *The Simpsons*, who grew up with the ill-fated Trojan reactor in Ranier, Oregon (demolished in 2006), probably had it right when he imagined reactor builders leaving an envelope on the utility company's doorstep saying, "Congratulations, you are the new owner of a nuclear plant." At Three Mile Island, no one had ever decided whether Metropolitan Edison (the Pennsylvania company) or General Public Utilities (the New Jersey parent) had final authority.

The result was a checkerboard of isolated utilities trying to run reactors they really didn't understand. At TMI, the training department had spent only one two-hour session *per year* discussing the operating experience at other reactors.[7] Because plants were so different, employees rarely transferred from one facility to another. There was no cross-fertilization of experience or expertise.

Within individual plants the eccentricities bordered on the absurd. Instrument panels were designed with no awareness of what information was needed or even whether they were going to be operated by human beings. As Samuel Walker, the official historian of the NRC, would later write:

> Within a few seconds after the accident began, the plant's alarm systems, including a loud horn and more than a hundred flashing lights on the control panels, announced the loss of feed-water in the secondary loop, the turbine trip, the reactor trip, and other abnormal events. But they offered little guidance about the cause of those occurrences and did not differentiate between trivial and vital problems. One of the operators, Craig Faust, later commented, "I would have liked to have thrown away the alarm panel. It wasn't giving us any useful information."[8]

Incredibly, although the water level in the core was the most important safety condition in the reactor, *not a single gauge told what it was*. The operators could only infer the status from other readings. The gauge that gave the best indication was tucked far behind the control panel, requiring operators to run awkwardly back and forth. Absent any established system of flagging problems, the operators had regularly attached "trouble tags" to the various levers and buttons that needed attention. It was one of those tags that had obscured the warning light at Three Mile Island.

Practices at other reactors around the country were found to be on the same level. In one instance, an important gauge was so high up on the console that operators had to climb a ladder to read it. In another plant, two switches that had to be thrown simultaneously were so far apart that it took two operators to coordinate the effort. In an instance that came to symbolize the mismatch between humans and technology, the control panel in one reactor featured an identical pair of levers that manipulated the control rods. One lever moved them up, the other moved them down. Yet there was no indication as to which was which. To keep them straight, operators had mounted two different brands of beer cans on the knobs.[9]

Evaluating the situation a year later in a brilliant article in *Reason* magazine, industrial psychologist Adam Reed argued that the culture of secrecy in the old Atomic Energy Commission had prevented crucial post-war advances in industrial safety from penetrating the nuclear industry.

> In the early 1950s, research established that about 80 percent of all industrial accidents were due, not to defects or malfunctions in industrial equipment, but to error and confusion on the part of human beings operating it.... Before the end of the '50s, some principles of good human factors design had been established, and insurance company safety consultants were bringing the new discipline of engineering psychology to bear on the design of industrial plant equipment. By 1965, hardly any new equipment could be put into operation unless it conformed to the standards for safe human factors established by Underwriters Laboratory. The exception was equipment for the new and burgeoning nuclear power industry.[10]

Even in a conglomerate such as General Electric, the nuclear branch had been completely isolated from other parts of the company.

> [M]anufacturers of nuclear power reactors had highly competent engineering psychologists working for their other divisions, but the AEC insisted on keeping nuclear reactor work secret and isolated. By 1970, no new design for a toaster or blender at General Electric could get off the drawing board without being examined by an expert in human factors. Yet the same company was designing, manufacturing, and delivering nuclear reactors that had never been seen, much less examined by an engineering psychologists...
>
> One of the basic principles of human factors design is compatibility of control movements: a control lever should move up to move things up, down to move things down.... Gauges for different variables should be as distinctive as possible, so

> that the operator will know at a glance which gauge he or she is looking at. In most nuclear power plants, however, the operator is confronted by a sea of identical controls, distinguished from one another only by discreet uniformly colored labels that are invisible except at close proximity.
>
> It was only after the loss of the Three Mile Island plant in 1979 that engineering psychologists asked what the hell was going on in nuclear power plant control rooms. What they saw made them shiver.[11]

When the Kemeny Commission published its report in November—a remarkable six months after the accident—it reached an identical conclusion:

> The control panel is huge, with hundreds of alarms, and there are some key indicators placed in locations where the operators cannot see them. There is little evidence of the impact of modern information technology.... During the first few minutes of the accident, more than 100 alarms went off, and there was no system for suppressing the unimportant signals so that operators could concentrate on the significant alarms.... [G]iven all the above deficiencies, we are convinced that an accident like Three Mile Island was eventually inevitable.[12]

Reporters at the scene had marveled at the stoicism of Dauphin and Lancaster County residents during the accident. Now the backlash began to emerge. Rumors of sick animals and cancer epidemics filled the air. Stories circulated of pets acting strangely and cows that wouldn't give birth. One farmer claimed to have lost a whole year's herd of calves. (A subsequent investigation showed it had been a bacterial infection.) A local couple conducted an informal survey and discovered a virtual epidemic of cancer and birth defects in the surrounding population. Not until 1985 did a more scientific survey by the Pennsylvania Department of Health show there was no unusual incidence.

In was not until August 1980 that the first clean-up workers were allowed to make brief forays into the containment building. Cameras lowered into the reactor vessel showed a pile of rubble. It was not until seven years later that it was determined 70 percent of the core had been damaged and 33 percent had melted. The debris was shipped to Hanford, Washington for disposal. Remarkably, the Japanese government paid $18 million to haul away a small portion for its own research.

In March 1981, when GPU announced it wanted to reopen Unit One, neighborhood stoicism gave way to outrage. Protestors charged that GPU had not yet changed its management or even taken responsibility for the catastrophe at Unit 2. The utility "lacks the requisite character to safely operate a nuclear reactor," wrote one petitioner to the NRC. GPU responded that it was being forced to buy 2000 MW a day on the expensive spot market. If Unit One didn't reopen soon, the company would bankrupt.[13]

The argument focused on the psychological impact on local residents. One opposition group, People Against Nuclear Energy, argued that the National Environmental Policy Act of 1969 should extend to the psychological impact on surrounding neighborhoods. A federal court held in their favor and Unit One was put on hold. The U.S. Supreme Court, however, reversed the decision in April 1983. Meanwhile, GPU was indicted for providing the NRC with false information about leaks shortly before the accident. Not until 1985 did Unit One finally go back online.

In Washington, the Nuclear Regulatory Commission responded to the accident by stepping up oversight. The target of criticism on all sides, the commission became absolutely draconian in its demands. Fines that had been up to this point $1,000 a day were now raised to $100,000. Plants were shut down for the smallest infractions. When cracks appeared in the pipes of one plant, every similar plant was closed for weeks until they had been inspected. Instead of being integrated into the rest of the economy, the nuclear industry was becoming ever more isolated.[14]

As the NRC tightened the screws, construction of new reactors ground to a halt. NRC licensing was a two-part process. First, the commission issued a construction license that allowed a utility to put shovels in the ground. Then when the reactor was built, the utility had to apply for an *operating* license to open the plant. At best construction took five to seven years and by that time many new safety features had been invented. The NRC began requiring that these new developments be incorporated. Compliance became a moving target. If the NRC did not demand the updates, anti-nuclear groups would intervene, asking the courts to impose them. By 1990, the passage from construction license to operating license was averaging *fourteen years* and several projects took more than twenty.

In addition, a new 1982 federal law required the utility to develop government-led evacuation plans in case of an accident. Nearby states were required to join. As critics in Congress warned, many governors decided to block the opening of plants simply by refusing to participate. When New Hampshire's Seabrook Plant was finally completed in 1987, Governor Michael Dukakis of neighboring Massachusetts delayed the opening for three years by refusing to submit an evacuation plan. The NRC finally issued a license without Massachusetts' approval. In New York, Governor Mario Cuomo successfully scuttled the $7 billion Shoreham plant on Long Island by refusing to participate in an evacuation plan. The loss ruined the Long Island Lighting Company's finances and led to its takeover by the New York State Power Authority in 1998.

In Washington state, the largest public default in history occurred when the Washington Public Power Supply System (WPPSS, pronounced "Whoops"), a consortium of 88 public utilities, tried to undertake the most ambitious construction program in the country. With few hydro sites left, the consortium planned five 1000 MW reactors. Electrical demand slumped after the 1970s Oil Crisis, however, and WPPSS was left with huge surpluses. As the NRC began demanding design changes, construction costs skyrocketed. In 1983, WPPSS reneged on $2.25 billion in construction bonds. Only one reactor was ever completed.

By 1984 the retreat had turned into a rout. In the space of a few weeks, Public Service Co. of Indiana announced it was canceling further work on the 2,260 MW Marble Hill Reactor, already half-completed at a cost of $2.5 billion. Philadelphia Electric cancelled two reactors at Limerick after spending $3 billion. In Illinois, the NRC temporarily denied Commonwealth Edison an operating license after completion of its $3.7 billion Byron plant. And Cincinnati Gas & Electric announced it would convert the Zimmer Reactor in Moscow to coal after it was 97 percent complete.

Most utilities were ready to quit. "The first lesson we've learned is `don't build nuclear plants in America,'" said Don Beeth, director of nuclear information at Houston Lighting & Power. "You only subject yourself to financial risk and public abuse."

"No utility executive would consider ordering a nuclear plant today," echoed Robert Scherer, chairman of Georgia Power and head of the U.S. Committee for Energy Awareness. "He would be either certified or committed."[15]

The golden age of nuclear energy was over almost before it had begun.

Chapter 21

The Revival

If there was one slender ray of hope for nuclear power, it was the Institute for Nuclear Power Operation (INPO). Organized by a group of utility executives right after Three Mile Island, INPO was a voluntary association of reactor owners that was given wide authority to enforce safety standards and supervise operator-training programs. Headquartered in Atlanta, it soon became very independent. "INPO gave structure to an industry when it was in total disarray," says Joseph V. Rees, a professor of public administration at the Virginia Polytechnic Institute, who did an extensive study of the post-TMI nuclear industry. "Until then the industry had never understood itself."[1]

It was INPO that first put out the word of how nuclear-owning utilities were all "hostages to each other." As Walter J. McCarthy, CEO of Detroit Edison, described it in a 1990 speech to fellow executives: "It took the shock of TMI to make us realize for the first time that an event at a nuclear plant anywhere in the country could affect every other nuclear plant. Each licensee is a hostage to every other licensee. We truly are all in this together."[2]

It didn't take long for INPO to diagnose what was wrong. Too many utility executives had too little understanding of the technology. "There had been a mentality that nuclear was just an extension of coal," says Rees. "INPO concluded that utility executives weren't treating the technology with enough respect."

As one utility executive admitted later:

> In the fossil fuel business the general philosophy is run it till it breaks. Then you shut it down, fix it, and run it again. Every minute you don't use that capital for production purposes you are running costs.... That's no good for a nuclear plant. The potential consequences of a breakdown are too great. And if something breaks, you can't just walk in and fix it.[3]

For inspiration, INPO drew on Admiral Hyman Rickover, the father of the nuclear navy, and still a hero to the industry. In November 1983, INPO persuaded Rickover and his staff to visit Three Mile Island. The admiral gave General Public Utilities a favorable review and recommended that Unit 1 be restarted. INPO also decided to use Rickover's approach as its model for monitoring reactor safety. The Navy, after all, had never had a serious accident. "Admiral Rickover had a philosophy that nuclear is a very unforgiving technology," writes Rees. "You have to treat it differently."

After Three Mile Island there had been strong sentiment in Washington for turning operator training over to the federal government. Senator Daniel Patrick Moynihan campaigned for a nuclear academy in the nation's capital, but the industry feared once again getting lost in the bureaucracy. It persuaded Congress to settle for an industry-run effort in Atlanta. The result was a complete upgrade of nuclear training. No longer were applicants taken out of high school. In fact, training to operate a nuclear power plant has become one of the most sophisticated regimens in the entire economy.

By the early 1990s there had been no repeat of serious accidents and minor incidents were declining. While half the plants in the pipeline after Three Mile Island were cancelled, 40 went on to completion. The last of the 1970s cohort was the Watts Bar, opened by the TVA in 1996, twenty-three years after receiving its construction license. In 1979 there had been 72 reactors operating in the country. By 1992 there were 112.

Still the supposed benefits of nuclear power did not materialize. Capacity factor—the time a plant is up and running—was stuck at 60 percent, the utility industry average. Nuclear was still more expensive than coal and was competing poorly against the new natural gas plants. The industry seemed at a dead end. "Suppliers of nuclear plant components already are dropping out of the business and the pool of university-trained nuclear engineers is shrinking," reported the Associated Press.[4]

The administration of George H. W. Bush tried to keep the industry on life support. The 1992 Energy Policy Act made licensing a one-step procedure. Utilities could now obtain both construction and operating licenses *before* putting shovels in the ground. The new chairman of the NRC, Landon W. Zech, Jr., promised to cut the construction process from fourteen years to six. It didn't make much difference. No one was applying for licenses anyway.

Then came what seemed a deathblow—electrical deregulation. "Nukes Face a New Crisis-Competitive Markets," said a 1992 headline in *Public Utilities Fortnightly*. "Rather than extending the life of nukes, more people are preoccupied with the impending shutdown of an entire generation of operating plants," reported Stephen Maloney, an industry consultant.[5]

Nuclear reactors had become the industry's white elephants. Many were saddled with huge construction debt. Combined-cycle plants burning natural gas were being built at a furious pace. While nuclear's costs could remain hidden under the old regulatory regime that guaranteed utilities a profit, deregulation would leave these aging reactors glaringly exposed. As Amory Lovins gleefully proclaimed: "Nuclear has gone from too cheap to meter to too expensive to matter."

In 1998, the Department of Energy reduced its nuclear research budget to zero for the first time since World War II. When a utility official questioned the move, a DOE official told him, "What's the difference? Deregulation is going to kill nuclear anyway." The Washington International Energy Group, an industry think-tank, predicted half the nation's nuclear reactors would drop out of the deregulated market. Looking into its crystal ball, DOE's *1999 Annual Energy Outlook* predicted that nuclear's share of electricity would decline from 18 percent to 7 percent by 2020. "Electricity generation from nuclear power declines significantly over the projection period," said the forecast. "Of the 99 gigawatts of nuclear capacity available in 1997, 50 gigawatts are retired, and no new plants are constructed by 2020."

It looked like the end of the line.

In 1991, Don Hintz, CEO of Entergy, a merchant energy company spun out of Southeast Utilities, had a problem. Actually it was two problems—Arkansas Units 1 and 2. Built in the 1970s, these twin lumbering giants were continually plagued by outages and safety alarms. Ralph Nader had just named them the second and tenth worst run reactors in the country in his annual "Nuclear Lemons" award. "Both were running less than 50 percent of the time," recalls Hintz. "The industry average was 60 percent. My sense was that someone had to go down there and either run them right or shut them down."

The person he chose was Jerry Yelverton, a vice president of Southern. Yelverton had always been a fan of nuclear power. When he arrived in rural Russellville, he decided to pursue a different strategy. "The problem was that nuclear had been seen as an extension of coal," he says. "With a fossil fuel plant you run the boiler until something breaks down, then you stop and fix it. It's not a bad strategy. You need to give the boiler a rest anyway. So you end up running it about 60 percent of the time.

"The same philosophy had carried over to nuclear. But I knew in the Navy they did things much differently. They run those reactors on submarines and aircraft carriers for two years without stopping. I started asking people, `Why can't we do the same thing here?'"

Yelverton discovered that most outages had nothing to do with the reactor itself. Instead, it was mechanical and electrical equipment that kept breaking down. "There was always a turbine tripping or a wire shorting out somewhere or a voltage regulator on the blink. People had accepted this as routine because that's what happened in a coal plant. I thought we could do much better."

Concentrating on upgrading equipment and tightening ordinary procedures, Yelverton slowly began to improve the reactors' performance. By 1995 both Arkansas units had climbed toward a capacity factor of 75 percent. This was not only an improvement, it was unknown territory. No generating station of any kind had ever reached a capacity of more than 70 percent. Executives at Entergy headquarters in Jackson, Mississippi began to take note.

"We had tried just about everything in energy in the mid-90s," says Gary Taylor, trim and lithe at 52 with salt-and-pepper hair and mustache, who became CEO of Entergy in 2000. "We were running coal plants. We were building gas plants in the United Kingdom. Then we got hit with a windfall profits tax in England and decided to pull back. We weren't sure what to do next. One day we sat around in an all-day skull session and asked, `What is our core business?' Don Hintz said, `I think the thing we do best is run nuclear reactors.' Then somebody said, 'Let's go out and buy some.'"

Entergy decided to buck the national trend. While the rest

of the industry was trying to unload its reactors, Entergy began buying them up. Some were picked up for a song. In 1998, the company bought Maine Yankee, a small New England plant, and the Pilgrim reactor near Boston. Next came Vermont Yankee, which supplies 70 percent of the Green Mountain State's electricity. Venturing into New York, Entergy bought Con Edison's Indian Point and the FitzPatrick plant on Lake Ontario. Northern utility executives quietly chuckled at this naïve group of Mississippians who were taking their biggest liability off their hands. Soon Entergy had the second largest nuclear fleet in the country.

At about the same time, Oliver Kingsley, a former Navy submarine commander, had taken over operations at Commonwealth Edison in Chicago. Like Yelverton, Kingsley found Commonwealth's fleet–then the largest in the country–performing far below expectations. "Of the twelve reactors, six were shut down and the other six were operating at 47 percent," recalls Kingsley, still tall and trim at 67, sitting in his retirement home in Alabama. "I told people, `If you're going to play for Coach Kingsley you're going to have to change your offense, your defense, and your kicking.'"

Kingsley took the same approach as Yelverton, concentrating on equipment improvements and avoiding mechanical and electrical shutdowns that the business regarded as routine. "A lot of people wanted to blame the NRC for nuclear's problems, but I always thought it was our own fault," says Kingsley. "I remember sitting down one day in a casual conversation with Zack Pate, who was chairman of INPO. We counted up and there were maybe only five or ten reactors around the country that had never been in trouble at some point."

Like Yelverton, Kingsley concentrated on improving even the smallest operating details. "I have spent months personally chairing meetings with employees from all levels of the company," Kingsley recalls. "You have to get everyone involved. The Finns are refueling in 15 days. Why is it taking us three months? Remember, the worker knows much more about what's going on than management. You have to empower people."

The biggest improvement came in refueling. In the old days, the operation was performed by plant employees who took three months and often regarded it as a vacation. Now Entergy and Exelon set up roving teams of specialists that moved from plant to plant, choreographing refueling operations to the minute. Soon the fuel rods were being replaced in less than a month. Repairs were carefully scheduled so they could be performed during shutdowns. Both companies also developed special safety teams that could be dispatched anywhere in the country at a moment's notice to meet emergency repairs. "If anything goes wrong at a reactor, the whole industry knows about it within a few hours," says Taylor. Scrams and unplanned outages, which had consumed more than 100 days a year, were soon reduced to less than 40. These improved practices were emulated throughout the industry and by the turn of the century, the nation's entire nuclear fleet of 104 reactors was being upgraded.

The first reactor had been sold in 1998. By 2000, Duke Power, Progress Energy, Constellation Energy, and Florida Power & Light had entered the bidding. Since then almost half the nation's fleet of reactors has been transferred into the hands of merchant energy companies specializing in operating reactors. It was the obverse of Alvin's Weinberg's "nuclear priesthood." Instead of an isolated corps of nuclear acolytes, the industry was now being run by proud Navy veterans and open-handed American businessmen who communicated on a constant basis.

Realizing their reactors had more potential than previously achieved, companies began applying to the NRC for "upratings," which would allow them to run at anywhere from 2 to 20 percent higher levels. Reactor upratings have now added 4,500 MW of capacity to existing reactors. Meanwhile, capacity factor for the nation's entire fleet reached *90 percent* in 2001 for the first time. (Coal, responding to the competition, has pushed up to 70 percent.) Together, these upratings and capacity factor improvements have provided the equivalent of *23 new 1,000-MW reactors* to the national grid. Nearly one-third of electrical growth since 1990 has been met merely by upgraded performance from nuclear plants.

The percentage of electricity provided by the nuclear fleet has climbed from 17.8 percent in 1989 to 19.8 percent in 2008, even though *five older reactors had retired and only one new one had added.* (Browns Ferry 1, which experienced the candle fire in 1976, was reopened in 2006.) Natural gas now has a *nameplate* capacity of 443,000 MW–30 percent more than coal and more than four times that of nuclear. Yet natural gas only provides 20 percent of our electricity, as opposed to 49 percent from coal and 19.8 percent from nuclear. Reactors run almost year-round while gas plants operate only at 30 percent capacity—about the same as the best windmills. This is because, most of the time, gas is too expensive.[6]

Even critics were impressed. "When private companies started buying reactors, people said, `they'll run them until they melt down, collect the decommission money, and move on to something else,'" said David Lochbaum, nuclear safety engineer with the Union of Concerned Scientists, which has been critical of nuclear power since the 1960s. "That hasn't happened. These companies know their future is riding on safety. If one nuclear reactor melts down, they'll lose their whole fleet. With proper management you can serve both masters."

Nevertheless, sticking to the old way, UCS published a 2006 study, "Walking a Nuclear Tightrope," claiming that the safety conditions in the industry had hardly improved since Three Mile Island:

> Is nuclear power in the United States safe enough today just because a reactor has not experienced a meltdown since 1979? The answer is a resounding no. In the 27 years since the TMI meltdown, 38 U.S. nuclear power reactors had to be shut down for at least one year while safety margins were

restored to minimally acceptable levels. Seven of these reactors experienced two year-plus outages.

Though these reactors were shut down before they experienced a major accident, we cannot assume we will continue to be so lucky. The number and length of these shutdowns testifies to how serious and widespread the problem is.[7]

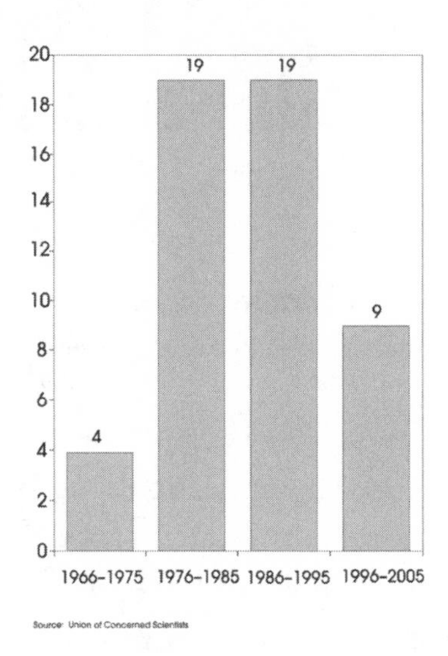

Reactors Experiencing Year Plus Outages

In fact, only *one* reactor—good old reliable Davis-Besse—has experienced a one-year shutdown since 1997. In order to reach this conclusion, UCS lumped eight shutdowns from 1996–1997 into the following decade to make it seem like things had not improved. *The New York Times* swallowed the story whole, running it under the headline, "Reactors Prone to Long Closings, Study Finds." The headline could have just as easily been, "New Ownership Brings Vast Improvement in Nuclear Performance."[8] (See graph on following page.)

In 1999, Constellation Energy, the major utility in Maryland, took a new tack. It applied to the NRC for a 20-year renewal for its Calvert Cliffs plant. To everyone's astonishment, more than a hundred citizens, including the entire town government,

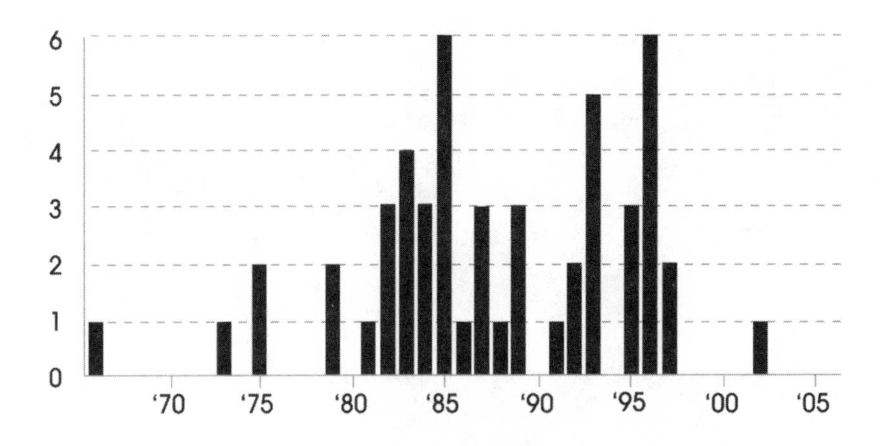

Source: *New York Times*

Reactors on Ice

Number of nuclear reactors shut down for more than a year, plotted by the year the shutdown began

showed up to support the application. "In 1960, we were the poorest county in Maryland," said Patrick Buehler, a Calvert County commissioner. "Now we have one of the highest per-capita incomes." The reactor not only provides high-paying jobs but pays enough taxes to make the schools, parks and hospitals some of the best in Maryland. In its first year of operation, the plant paid more in taxes than the entire county budget from the previous year.

This is one astonishing thing about nuclear. Most people who live next to reactors love them. Instead, opposition tends to come from national and regional groups that oppose nuclear on principle. Bisconti Research, which does public opinion polls for the Nuclear Energy Institute, has found that 70 percent of the public now views nuclear favorably—an all-time high. Meanwhile, support among people with a reactor in their town is an incredible *90 percent*.[9] "There is no `not-in-my-backyard' syndrome with nuclear," says Ann Bisconti, director of the firm. "People who live with reactors have no problem."

Connecticut paid nuclear an even higher complement in

2006 by proposing a *windfall profits tax* on reactors. Writing in the Hartford Courant, attorney general Richard Blumenthal expounded:

> Rising oil and natural gas prices have caused rates to rise nationwide, but nowhere near Connecticut's stratospheric heights.... Only sweeping change—now—will prevent this crisis from becoming a catastrophe.
>
> First, we need a windfall profits tax aimed at generators who have reaped outlandish and undeserved profits through irrational market rules. In 2006, the nuclear Millstone II and III generators in Waterford will have profits of $274 million and $419 million respectively, while the coal-fired Bridgeport Harbor plant will enjoy profits of $113 million.... The generators using low-cost coal and nuclear fuel reap these same high prices and enjoy ever-ballooning profit margins.[10]

Only a few years before, nuclear opponents had been gloating the technology was "too expensive to matter." Now nuclear had become too profitable to be allowed to keep its rewards.

Nuclear reactors now operate close to two years without interruption, twenty-four hours a day, seven days a week, twelve months of the year. They are the ultimate "base load" electricity. In 2007, the industry set all-time highs in capacity factor (91.8) and kilowatt-hours produced (807 billion), while achieving an all-time low in production cost at 1.68 cents per kwh, beating the old mark of 1.72 cents in 2005.

The record for uninterrupted service by a single reactor, however, remained unbroken. It was established in 2003 at 688 days, just six weeks short of two years. The record is held by Unit 1, Three Mile Island, Harrisburg, Pennsylvania.

Interlude

Brattleboro, Vermont

Mike Gosekamp is a volunteer fire captain, helps with Boy Scouts, and roots for his son's soccer and basketball teams. He's also a certified nuclear power plant operator and training supervisor at the Vermont Yankee Nuclear Station. This makes him part of an elite corps that produces 80 percent of Vermont's electricity, making it the greenest state in the nation. Wyoming, with its coal operations, produces more greenhouse gases in one day than Vermont produces in a *year.*

"Signing on for this job is like entering the medical profession," says Gosekamp, who oversees the simulated control room at the plant's $10 million training center. "You have to commit yourself to years and years of education. You're on call 24 hours a day, whether it's 8 o'clock Saturday night or 2:30 AM Wednesday morning. There's a whole ethic of responsibility and professional dedication."

Tucked into an inconspicuous cinderblock building just north of Brattleboro, Vermont Yankee's simulator duplicates bolt-for-bolt, lever-for-lever, the actual control room at the 540-MW reactor ten miles south. Eight feet tall and fifty feet wide from end-to-end, the control panel still has a space-age look with rows of lights and gauges, but all are connected by logic trees of different color lines. "Everything is an if–then proposition," says Gosekamp.

Mandated since 1979 by the Nuclear Regulatory Commission, the simulator is one of 65 at nuclear sites around the country. Each year Vermont Yankee schools ten new initiates while keeping the plant's 50 licensed operators on their toes through constant retraining. "Even our most experienced operators spend four days out of every six weeks in here," says Gosekamp.

As with about half the nation's 4,000 licensed nuclear operators, Gosekamp first encountered the technology in the U.S. Navy. After serving four years as a nuclear machinist's mate, he signed on with Vermont Yankee in 1988. About half his pupils are college graduates. Those who aren't Navy veterans usually come from engineering schools or any of the nation's maritime academies.

Even so, six years of experience aboard nuclear submarines only earns you a ticket through the front door. "We start everybody at square one," says Gosekamp, still trim and athletic at 47. "They do ten months training before they ever touch a valve. At that point they can enter the plant as unlicensed auxiliary operators." This apprenticeship lasts another two to five years, at which point the candidate can *begin* official licensing courses. That's 14 more months on the job and in the simulator. If they pass, candidates can then take the NRC licensing exam.

"All this training enables us to find the employees we want," says Gosekamp. "We're looking for operators who can think for themselves. We want them to assume responsibility. Nobody around here is a slacker. The quality of people you end up working with is unparalleled."

INPO's National Academy of Nuclear Training in Atlanta now accredits and supervises training schools at each of the nation's 104 reactors. "Our mission is to promote the highest standards of safety and reliability," says Jim Caulk, director of communications at INPO. The payoff for the industry has come with lower insurance rates. This gives reactor owners an enormous incentive to cooperate on safety matters and share information with the rest of the industry.

The training operates under a give-and-take system developed by the military during the Cold War. "There's constant updating and feedback," says Gosekamp. "We'll have a class come to us and say, `Wouldn't it be better if you did it this way?' We've made equipment changes in the plant based on things that happened right in this simulator."

Maintaining performance is always a challenge in a job where—ideally—almost nothing ever happens. Week after week, month after month, nuclear plants hum along, only occasionally adjusting their output. At the actual plant on the Connecticut River, the control room is manned by ten-member crews working 12-hour shifts. "We've found people can stay alert for that long and prefer to have whole days off," says Gosekamp. "We try to balance work and family." Licensed operators make around $100,000 and live well. Still, the job carries huge responsibilities. "You have to make sacrifices," says Gosekamp. "You can't just shut the plant down and go home."

Although generally non-political, nuclear operators often feel called upon to defend the technology. "When people ask me about nuclear bombs, I tell them I don't even know how a bomb works," says Gosekamp. "It's a completely different technology. Nuclear fuel rods are only 3–5 percent fissionable material while bombs are more than 90 percent. I often wonder how anyone could ever use our technology to make a weapon."

Like firemen or crewmembers on elite navy ships, nuclear operators have developed an esprit de corps built on professional performance and pride in their unique responsibilities. "We're a group that's likes to step up," says Gosekamp. "We all want to be the guy on the line taking the foul shot at the end of the game."

Chapter 22

Radiation

In the early 1980s, a Taiwan steel company accidentally mixed a quantity of highly radioactive cobalt-60 into a commercial batch. The steel was then used in the construction of 1,700 apartments. As a result, people living in these buildings were subject to background radiation some 7,000 times the amount that would come from living next door to a nuclear reactor.

When dismayed officials discovered this enormous error fifteen years later, they surveyed past and present apartment dwellers, expecting to find an epidemic of cancer. Normal incidence would have predicted 160 cancers among the 10,000 residents. To their astonishment, the researchers discovered only *five* cases of cancer—a 97 percent reduction from the anticipated amount. These findings were published in the *Journal of American Physicians and Surgeons* in 2004. As one researcher phrased it, exposure to high levels of background radiation had apparently bestowed upon residents "an effective immunity from cancer."[1]

Radiation is very simple to understand. It is energy. According to Einstein's formula, $E=mc^2$, you will recall, the world is divided into matter and energy. Anything traveling *at* less than the speed of light will have "kinetic energy" but is considered "matter." Anything traveling at the speed of light is pure "energy."

In fact, the term "speed of light" is a little misleading. We call it the speed of light for the same reason the British call a flashlight a "torch"—it is a term a carried over from a previous era. It should really be call the "speed of radiation" or the "speed of energy." Light is only thin sliver of a much broader phenomenon in nature, the electromagnetic spectrum.

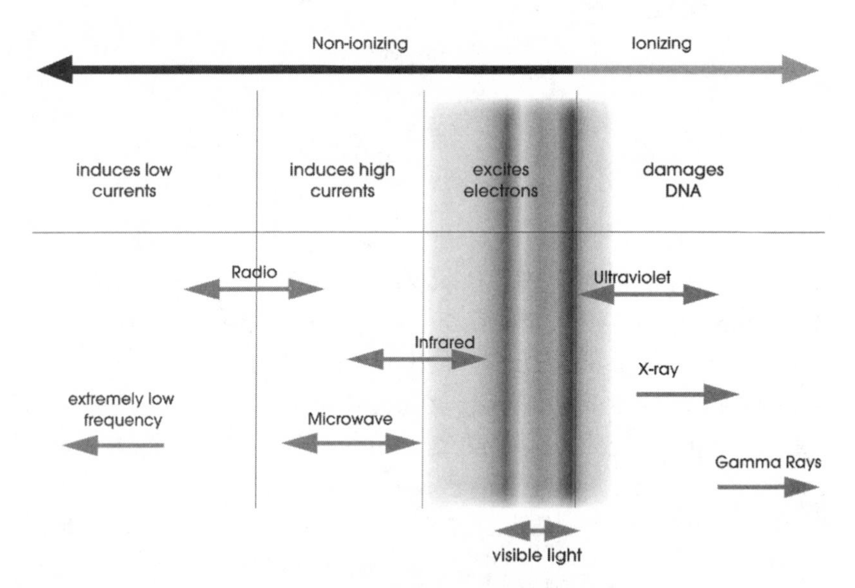

The Electromagnetic Spectrum

For most of human history, the only portions of the electromagnetic spectrum perceptible to humans was visible light and infrared heat. All other forms of radiation are undetectable to our senses. Then in the nineteenth century, James Clerk Maxwell demonstrated that both light and radio waves, which had just been discovered, were part of a family of wave phenomena, all traveling at the speed of radiation and differing only by the frequency of their vibrations and the length of the wave. We call them "radiation" because they radiate outward from a single source, like the sun. But there is a plethora of sources. Radiation is everywhere. It is the way energy travels through the universe.

The electromagnetic spectrum is a great rainbow of energy. At the low end are waves of very low frequency and an extraordinarily long wavelength. The radiation given off by a household electric current, for example, may oscillate at three cycles per second and have a wavelength of half the continental United States. At the opposite end is gamma radiation, which can have a wavelength of 0.0000000000000001 (10^{16}) meters and a frequency of 10,000,000,000,000,000,000,000,000 (10^{24}) cycles per second. In between, in order, are audio frequencies from electric generators; radio waves (also in the range of nuclear magnetic resonance—MRI); VHF (very high frequency) and UHF (ultra high frequency), used to transmit television signals; radar; microwaves; infrared heat waves; visible light; ultraviolet light; x-rays, which resonate from inner electrons; and gamma rays, which emanate from the nucleus of the atom. There are an almost unlimited number of divisions, as we prove when we auction off individual bandwidths in the small sector of the spectrum for commercial radio broadcasts.

The speed of radiation—187,000 miles per second—is the grand cosmic speed limit of the universe. This was Einstein's great insight—neither matter nor energy can exceed it. The Theory of Relativity says that the other dimensions—time and distance—adjust themselves to accommodate this one absolute, the speed or radiation.

Just as we are constantly confronted with matter, so we are constantly bombarded by energy. We encounter it from outer space, from the sun, from the air, from the earth (terrestrial energy), and from radioactive elements within our own bodies. Radioactive potassium and radioactive carbon are two isotopes built into our flesh and bones. Radioactivity is a product of nature, as natural as the sun or the grass that grows. Although we cannot see or feel most of its forms, our bodies have long since adjusted to it.

What we call "nuclear radiation" actually comes in three different forms, only one of which is pure energy. Ernest Rutherford was the first person to identify these, giving them the names of the first three letters of the Greek—alpha, beta, and gamma. "Alpha" radiation is actually a helium atom stripped of its electrons—two protons and two neutrons traveling at about one-twentieth the speed of light. "Beta" radiation is an electron traveling at 94 percent of the speed of light. Gamma radiation is pure energy and very powerful, traveling at the speed of radiation. "Cosmic rays" are actually naked protons that rain down from outer space at half to 99 percent the speed of light. They are believed to be the nuclear fallout from exploding stars.

Traveling at such high speeds may make these subatomic particles seem like they could penetrate anything, but their small size diminishes their momentum. Alpha particles can be blocked by a piece of paper or the tough outer layer of human skin. Beta particles can be stopped by tin foil, but will penetrate the outer layer of skin, causing reddening. Gamma rays, which have the greatest momentum, can penetrate skin and are only stopped by four inches of lead or two feet of concrete. Gamma radiation is the most hazardous for human exposure, although alpha and beta particles can be equally dangerous if they are inhaled or ingested and get past the skin's outer layer.

Outer space is filled with cosmic rays and other radiation. That's one reason astronauts must wear spacesuits. On earth we are largely shielded by the earth's atmosphere, which does a wonderful job of filtering most of the electromagnetic spectrum-with the happy exception of long radio waves and visible light. Otherwise the earth would be almost completely dark.

Since the 1970s, opponents of nuclear power have tended

to play on the public's unfamiliarity with radiation by positing that: (1) all radiation is dangerous, and (2) nuclear power is our major source of exposure. Typical book titles have been *Killing Our Own: The Disaster of America's Experience with Atomic Radiation, The Enemy Within: The High Cost of Living Near Nuclear Reactors, Radiation: The Silent Killer*, and *No Immediate Danger? Prospects for a Radioactive Earth.*[2] (The earth is already radioactive.) "The closer you get to the plants, the more babies are dying," proclaimed Dr. Ernest Sternglass, the famous anti-nuclear crusader of the 1970s. In *The Careless Atom*, one of the earliest anti-nuclear tracts, Sheldon Novick presented this grim picture of the biological effects of radiation:

> When one of these particles or rays goes crashing through some material, it collides violently with atoms or molecules along the way.... In the delicately balanced economy of the cell, this sudden disruption can be disastrous. The individual cell may die; it may recover. But it if does recover, after the passage of weeks, months or years, it may begin to proliferate wildly in the uncontrolled growth we call cancer.[3]

Bernard Cohen, the tireless University of Pittsburgh physics professor who has spent thirty years trying to calm public fears of radiation, offers this rejoinder:

> Before we shed too many tears for the poor fellow who was struck by one of these particles of radiation, it should be pointed out that every person in the world is struck by about 15,000 of these particles of radiation every second of his or her life.[4]

Radiation, like anything else, can be harmful in excessive doses. Sound waves (which are vibrations of air, not part of the electromagnetic spectrum) can damage the eardrum. Excess exposure to microwaves can cause biological damage. (That is why our microwave ovens are shielded.) Infrared radiation can roast people alive. Visible light can cause blindness—as Galileo discovered when he turned his telescope to the sun. Ultraviolet light can burn the skin or cause skin cancer.

Above the range of visible light, radiation becomes so powerful that it can knock the electrons off individual molecules and even break molecular bonds. Therefore it is called "ionizing radiation." It is more dangerous than radiation in the lower frequencies. Ionizing radiation can cause radiation sickness or various cancers at very high doses. Just what the effect is at much *lower* doses has been a subject of long and bitter debate.

Ionizing radiation is measured in a bewildering variety of ways. There are three methods for calculating its intensity and—just as you might expect—also American and International standards, just as with meters and inches. The actual numbers of atomic disintegrations releasing radiation particles or photons is measured in "becquerels" (Standard International), and "curies" (non-SI), with one becquerel equal to 37 billion disintegrations per second and one curie equal to 3.7×10^{10} becquerels. "Roentgens" are another measure, usually used for calibrating x-rays machines.

While curies and becquerels measure radiation activity, "rads" and "grays" measure the dose *absorbed* by human beings. A gray is the System International (SI) unit for one joule per kilogram of body weight—a fairly large dose. The American "rad" (radiation absorbed dose) is 1/100th of a gray. As usual, the European units are more scientifically pleasing while the American units correspond better with everyday life. Most common exposures are at the level of rads and millirads.

Perhaps the best unit is the "biologically effective dose." This is a measure of absorbed radiation that gives weight to the relatively mild effects alpha and beta particles as opposed to the more serious effect of gamma rays. Europeans use "Sieverts," named after a Swedish doctor who researched the biological effects of radiation. The American unit is the "rem" (radiation equivalent for man), with 100 rems equal 1 Sievert. Most everyday exposures are at the level of "millirems," 1/1000th of a rem (or .001 millisieverts). The average American absorbs about 350 millirems of radiation per year from various sources, or .35 millisieverts.

The Radiation Effects Research Foundation, a Japanese-U.S. organization, has studied the effects of Hiroshima and Nagasaki on 36,000 survivors. The incidence of leukemia was 2.7 percent for those standing within a kilometer of the bomb, descending to 0.06 percent for people more than 1.5 km (about a mile). The incidence of all other cancers was 6.6 percent for the closest victims, descending to 0.3 percent in those more than a mile away. Although there was a noticeable incidence of mental retardation among infants exposed in the womb—particularly those who were between 8 and 15 weeks—no pattern of long-range birth defects has ever showed up among the succeeding generations of the survivors of Hiroshima or Nagasaki.[5] In fact, some studies are showing that certain low-dose survivors of Hiroshima and Nagasaki are living longer than their peers.[6]

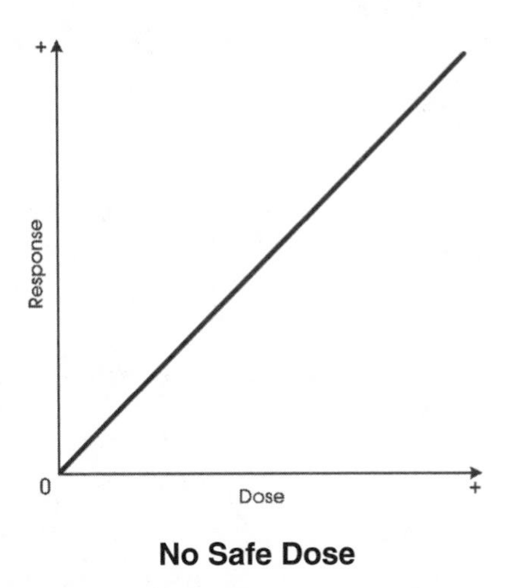

No Safe Dose

The pattern of greater incidence with greater exposure to a damaging factor is called the "dose-response" relationship. Such a relationship is generally assumed to be linear, meaning that incidence and exposure rise and fall in a fairly straight line.

For radiation, this would mean the more exposure the more cancer. The highest exposures in Hiroshima were around 600 rems. Cancer incidence fell off below 100 rems and below 10 rems there was no detectable effect at all. If 10 rems were about the middle of the graph, then the rest of the line extending down to zero is only *projection* of the upper right-hand proportion. If any cancers were caused by these low levels of exposure, they would be impossible to find in a large population.

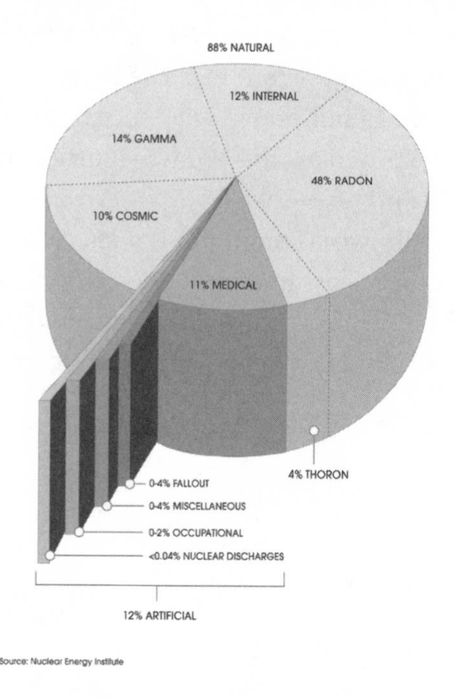

Source: Nuclear Energy Institute

Annual Radiation Exposure in the U.S.

Of the 350 millirems to which the average American is exposed every year, about four-fifths come from natural sources and half are man-made. About 35 millirems come from cosmic rays. Another 45 millirems come from terrestrial energy—uranium and thorium in the earth. Radon gas, a daughter of uranium decay, is ubiquitous in the air and provides about half our natural exposure. The remainder comes from *within our own bodies*. Potassium-40, which has been around since the dawn of the earth (half-life 1.3 billion years), is taken up by plants and deposited in our organs. Carbon-14 is constantly being created by cosmic rays striking nitrogen atoms in the atmosphere—although atmospheric atomic testing doubled its incidence for a brief while. It too is taken up by vegetation and migrates to our bones and organs. Its half-life of 5,730 years allows archaeologists to date organic material back to 60,000 years. Together these two elements provide about 38 millirems.

Most non-natural exposure comes from medical sources. X-rays and the various forms of radioactive tracers provide about 35 millirems. Fallout from nuclear testing—mostly strontium-90—adds 2 millirems. The routine exposures from nuclear power plants contribute less than one-tenth of a millirem. Living on the property line of a nuclear plant all year round would expose a person to 1 millirem. Traveling in an airplane also adds about 1 millirem, since the thinner atmosphere gives less protection from cosmic rays.

Throughout the world, background radiation differs widely. People who live in the mountains are more exposed to more cosmic rays because of the thinner atmosphere. Granite has much larger concentrations of uranium and thorium than sedimentary rocks. Thus, living in New England or the Rocky Mountains exposes people to more background radiation than living in the Midwest or the Mississippi Delta. In other parts of the world, high concentrations of uranium and thorium create even higher background levels. Kerala, a state in India, has eight times the normal radiation in the rocks. The Beach of Guarapari in Brazil has 175 times normal background levels and the Ramsar region of Iran radiates 18 rems—400 times the world average.

People living in these regions do not seem to suffer any ill effects. If anything, they often seem *healthier*. Cancer rates among people living in the Rocky Mountain States, which has the highest exposure in the country, are one-third lower than the rest of the nation, while people living in the Louisiana Delta, which has the lowest exposure, have the *highest* rates of cancer in the country.

Why don't all levels of radiations cause cancer? The answer comes from the sixteenth century German scientist Paracelsus, commonly known as the "Father of Toxicology," who famously wrote, "All things are poison, only the dose matters."

Our bodies are continually fending off environmental insults. We have an airtight skin that keeps out bacteria. Our immune system fights off infection. When harmful ultraviolet rays strike our skin, most people's skin darkens for protection. People who suffer very high exposure to nuclear radiation also develop a "radiation tan" as did many of the unfortunate emergency workers at Chernobyl. The melanin pigment actually works to deflect some radiation, just as it does for the sun's ultraviolet rays.

Nevertheless, these defenses can be overwhelmed. The skin can be lacerated, letting in germs. The immune system can encounter a particularly virulent new invader. Even people with the darkest pigmentation can suffer ultraviolet damage. And exposure to very high levels of ionizing radiation can be equally damaging.

The situation is complicated in that many environmental agents that are harmful at high levels can be beneficial or even essential at lower doses. Sunlight may cause skin damage but it also stimulates vitamin D. People deprived of sunlight suffer from rickets, as did the inhabitants of industrial cities of Nineteenth Century England. High concentrations of selenium in grass cause "blind staggers" in farm animals and humans, yet selenium deficiencies can lead to high blood pressure, cardiac dysfunction, and certain forms of arthritis and anemia. Even vitamins can be toxic at very high doses and multi-vitamin regimens have recently been linked to elevated prostate cancer.[7]

How can we express this paradox? The answer is that there are different kinds or dose-response curves. If a substance is toxic right down to the smallest exposures, then the dose-response curve will be a straight line that runs through zero. If there are levels below which it is not toxic, however, there will be a "threshold" below which damage does not occur. Such a dose-response curve would look like this:

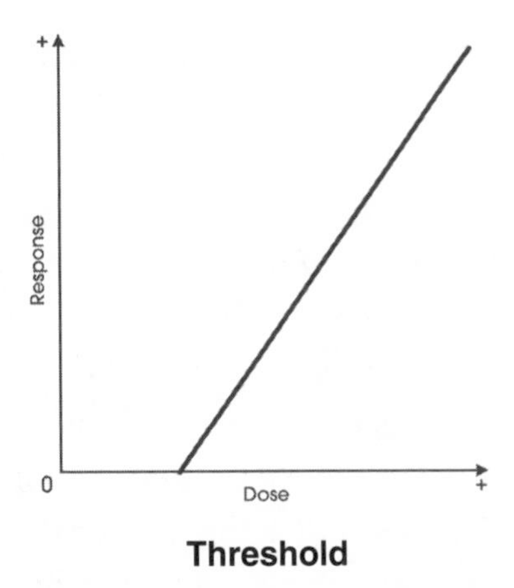

Threshold

The largest controversy that has dogged nuclear power over the past fifty years is whether there is a *threshold* to the damage caused by low exposures to ionizing radiation—in other words, whether there is "no safe dose." We know high levels of radiation can cause sickness and cancer. There is no evidence that very low doses do the same. In order to be safe, however, most public authorities have *assumed* the "linear, no threshold (LNT)" hypothesis in setting safety standards. But the LNT hypothesis has never been anything more than a presumption. There is ample information suggesting that low levels of exposure may not be harmful—and may even be beneficial.

With radiation *sickness*, there is a clear threshold. A dose of 1,000 rems kills 100 percent of the victims outright. A dose of 500 rems—the worst that came from the first atom bomb—kills 50 percent of the victims. Below 100 rems mortality is around 10 percent and below 10 rems there is no visible effect. Thus, the relationship is nearly a straight line with a threshold around 10 rems—about 300 times normal background levels.

Cancer is another matter because the effects may not show up for 20 years and 45 percent of any given population will contract cancer anyway. Among the 800 Hiroshima and Nagasaki survivors who absorbed 600 rems, about .9 percent died of cancer (9 in 1000). Among 14,000 arthritis patients in England treated with 300 rems of x-rays in the early 1950s, .7 percent got cancer. When 900 German patients were treated with 900 rems of radium injections for spinal tuberculosis during the same period, .6 percent got cancer.[8] Among the "radium girls" who painted the watch dials, radium deposits in their bones reached an appalling level of 1,700 rems. Many of the women were disfigured or died very painful deaths. The rate of bone cancer among 775 women was 7 percent (48 cases).[9] Outside the watch-painters, which were probably the worst exposures ever, the most extreme exposures have produced an incidence of about 1 percent. This compares to the 8 percent rate of smokers who get lung cancer. Radiation is a relatively mild carcinogen.

Directly after World War II, the Atomic Energy Commission tended to discount any effect from very low levels of radiation and thresholds were set fairly high. As concern mounted in the 1960s, however, the National Academy of Sciences was asked to intervene. The NAS set up a Committee on the Biological Effects of Ionizing Radiation (BEIR), which has issued periodic, comprehensive studies on the latest findings, the latest—BEIR VII—published in 2005.

As a precaution and to allay public fears, the BEIR committee *assumed* the linear, no-threshold hypothesis. This continues today, even though the hypothesis has come under more criticism. BEIR-VII states, "At doses of 100 mSv [i.e., 10 rems] or less, statistical limitations make it difficult to evaluate cancer risk in humans."[10] Nevertheless, the report estimates that at a dose of 100 millirems—50 cross-country flights from New York to Los Angeles and back per annum—one person in 1,000 will contract cancer. Since 450 of 1,000 people contract cancer anyway, one additional cancer is impossible to detect.

Other federal agencies also set thresholds at around this level. The EPA permits voluntary maximum acute doses of 75 rems for rescue workers in life-saving situations. For non-life-saving efforts, the limit is 25 rems. It estimates a dose of 10 rems raises cancer levels by 0.8 percent and considers 1 to 5 rems an "action level." People living in the Ural Mountains are annually exposed to 10 rems of natural background and have lower-than-normal cancer rates. The residents of the Taiwan apartments were receiving annual doses of 7.4 rems.[11]

The real problem begins when we try to extrapolate these numbers down to the level of incidental exposures. The most intense x-rays deliver about 200 millirems and the average American receives 110 millirems in x-rays annually. The highest level recorded outside Three Mile Island during the accident was 46 millirems. The exposure to the public from the country's entire fleet of nuclear reactors is estimated to be .5 millirems per year.[12]

These numbers may seem small, but when applied to large populations they can add up. Using EPA's standards of one new cancer in 10,000 for each millirem of exposure would mean there should be 2,990 additional cancers among the 650,000 residents of Lancaster and Dauphin Counties who were exposed at Three Mile Island. In the New York Metropolitan Area, the routine operation of Indian Point would be causing 500 cancers annually among the region's 10 million people.

The 2,990 cancers never did show up around Three Mile Island. The 500 New York cancers would be hard to find in a population where 45,000 people get cancer every year. Yet it is this chain of logic that allows Greenpeace to claim that 70,000 people will die from Chernobyl or Dr. John Gofman, a leading anti-nuclear advocate of the 1980s, to proclaim: "Licensing a nuclear power plant is, in my view, licensing random premeditated murder."[13]

The logic of no threshold certainly does not apply to all hazards. If 100 out of 100 people will die from falling from a 100-foot-tall building, that does not mean 1 in 100 will die from stepping off a one-foot sidewalk. If swallowing 100 aspirin will kill 100 out of 100 people, that does not mean taking 2 aspirin will kill 2 people. Clearly there are thresholds below which the body's defenses can deal with an environmental insult.

As the certainty has taken hold that *any* level of radiation is dangerous, however, the efforts at trying to prevent exposure have reached absurd proportions. After Three Mile Island, the Nuclear Regulatory Commission forced the industry to spend $2 billion *per reactor* for safety features to reduce emissions to 10 millirems per year. The NRC's own Reactor Safety Study estimated this amounted to $2.5 billion per life saved—assuming any lives were saved at all. Bernard Cohen compares this to the amount of investment that could save lives in other situations—$24,000 for equipping ambulance corps with mobile intensive care units, $90,000 for providing every woman in America with a free Pap test, $12,000 for equipping every home in America with a smoke alarm, or even $50 per life saved for providing measles immunizations in the Cameroon.[14]

Yet when the NRC decided in 1992 to classify emissions below 5 millirems as "below regulatory concern," anti-nuclear crusaders stormed Congress to crush the attempt. Ralph Nader's Public Citizen proclaimed the change would cause 12,412 cancer deaths and called it "a trade-off of people's lives in favor of the financial interest of the nuclear industry."[15] When the NRC tried again in 2002, David Ritter of Public Citizen proclaimed: "The American public has spoken loudly and clearly on this issue before. That's why Congress banned the `Below Regulatory Concern' policy in 1992, conceding that radiation is always a concern."[16] The measure failed once again.

These fears are certainly not peculiar to America. After Chernobyl, *100,000 European women had abortions,* fearing their babies had been damaged by radiation exposure. The highest dose experienced anywhere on the continent was 100 millirems—100 times lower than the threshold below which no effects were found at Hiroshima and Nagasaki. These levels returned to normal after three weeks.[17] Yet Bruno Comby, a French nuclear engineer and founder of Environmentalists for Nuclear Energy, reports that eight years after Chernobyl, a woman told him she was afraid to go windsurfing in the Baltic Sea for fear it was still contaminated by radiation.[18]

This wild hyperbole reaches a climax in the portrayal of plutonium as "the most dangerous substance ever known to man." Plutonium is the man-made element formed when U-238 absorbs one neutron. It is fissionable and can be used to make bombs. Plutonium is also an alpha-emitter, the least dangerous form of radioactivity, although it can be very harmful in the lungs. As for toxicity, plutonium is about as dangerous as caffeine. Arsenic, mercury, and lead are much worse. The botulin toxin, "the most dangerous substance known to man," is deadly at 0.0000003 parts per million—10,000 times more toxic than plutonium. Anthrax spores—employed in that strange, only recently solved episode of homegrown terrorism three weeks after September 11th—are only slightly less toxic than botulin.

During the 1980s debates, Ralph Nader regularly stated, "A pound of plutonium could kill every person on earth." In order to do this, however, it would have to be ground into a fine dust and inhaled by every individual. Then cancer of the lungs might emerge after twenty years. But plutonium does not grind easily into dust and quickly falls out of the air from its own weight. The task of inserting it into each person's lungs would have to be performed manually. As the late Peter Beckman wrote, "Even tomorrow's production of hatpins could kill a million people if placed in each individual heart."[19] In the 1980s, Bernard Cohen offered to go on television and eat as much plutonium as Ralph Nader would eat caffeine. Nader never took up the offer.

The question of whether exposure from nuclear plants is causing cancer has been addressed in numerous studies. In 1990, the U.S. Department of Health and Human Services, the Public Health Service, and the National Institute of Health undertook an exhaustive survey of cancer deaths between 1950 and 1984 in 107 counties with nuclear installations. The three-volume study found there was "no evidence to suggest that cancer mortality in counties with nuclear facilities was higher than, or was increasing in time faster than, the mortality experience of similar counties in the United States."[20] In 1991, the authors published a synopsis in the *Journal of the American Medical Association* that concluded:

> Although public concerns have been raised with respect to Fernald, Rocky Flats, Hanford, Three Mile Island and others, this survey has not detected excess mortality due to leukemia or other cancers that might have been caused by radioactive emissions.[21]

The U.S. Navy has now logged 5,500 reactor-*years* aboard nuclear submarines without any indication of increased cancer in crewmembers, who work within a few feet of the reactor (with lead shielding, of course). The nuclear industry itself has logged 3,100 reactor-years without any indications of increased cancer among the thousands and thousands of employees. Instead, most studies find *lower* cancer rates among nuclear workers.

In fact, the evidence that small doses of radiation are not dangerous and may even be beneficial has eventually spawned a science of its own. It is called "hormesis."

Although Dr. Edward Calabrese just turned sixty, he still has all his hair, even if it has turned gray. "My father lived to be 95 and he had all his hair right to the end," says Dr. Calabrese, sitting in his small but comfortable office at the University of Massachusetts. "I'm not sure, but there may be some hormetic effect somewhere."

When Dr. Calabrese first came to UMass in 1976, he had never heard the word "hormesis." It had only been invented in 1943 and still doesn't appear in most spell checks or dictionaries. Yet in May 2007 Calabrese hosted more than a hundred scientists at the 6th Annual Conference of the International Hormesis Society in Amherst. "Almost half the papers were about radiation," he says.

Dr. Calabrese's journey to discovery began in 1966 while he was taking an undergraduate class in plant physiology. "We were assigned to put a growth inhibitor in the soil and measure the dose-response curve in peppermint plants," he says. "I did the experiment and found it was actually encouraging growth. We went back to check and found we had accidentally diluted the inhibiter by a factor of 100. We repeated our mistake several different times and always got the same effect."

Calabrese was puzzled but didn't pursue the idea for twenty years until one day he came across a notice for a conference on the stimulating effects of low-level electric currents, sponsored by the Electric Power Research Institute. The brochure mentioned the term "hormesis." "I wondered if what I had seen with the peppermint plants might not be the same thing," he says. Calabrese attended the conference and discovered there is a body of literature describing the phenomenon he had observed in many fields.

In 1990, he founded BELLE—the Biological Effects of Low Level Exposure (www.belleonline.com)—to investigate the hormetic effect. "It's an extension of the immune response," says Calabrese. "If you expose a person to a small amount of a virus, the immune system will develop a resistance to larger exposures. Hormesis works the same way. We're postulating that the body has mechanisms for dealing with environmental insults—chemicals, toxins, radiation—and that small exposures are beneficial because they stimulate the defense system into action. Hormesis is the rule, not the exception." A hormetic dose-response curve would look like this:

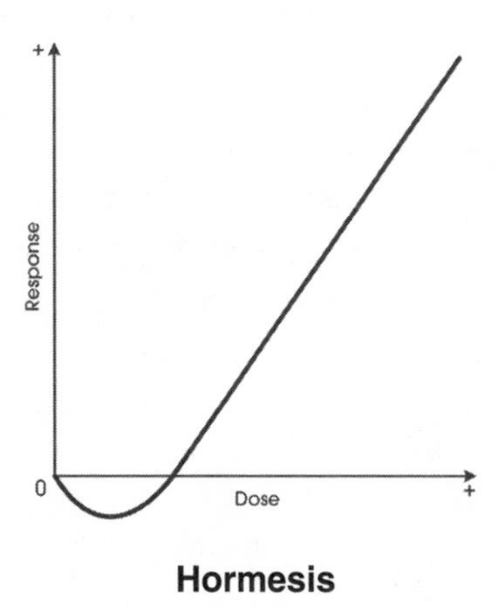

Hormesis

Dr. Calabrese has assembled a database of thousands of synthetic and industrial chemicals acting on dozens of organisms, from test animals to bacteria and fungi. "We find evidence of hormesis everywhere," he says. "There's not an organism that doesn't display it and not a chemical where you don't find a response.

In the 1970s, T.D. Luckey, a former chairman of the department of biochemistry at the University of Missouri (Columbia) School of Medicine, began collecting data on low-level exposure to radiation in animal experiments. "Extensive literature indicates that minute doses of ionizing radiation benefit animal growth and development, fecundity, health and longevity," he concluded. Among other things, researches have found that when paramecia and other single-celled organisms are *shielded* from cosmic radiation, they grow sluggish and don't reproduce. When the shielding is removed, their behavior returns to normal. Far from being dangerous, he posited, steady doses of background radiation may actually be necessary to living organisms.

In 1980 Luckey published *Hormesis With Ionizing Radiation,* citing over 1,000 studies that described hormetic effects of radiation in both plants and animals, while suppressing background radiation is detrimental. "The hormetic effect is well established," he said. "On the other hand, there has never been the slightest confirmation of the linear-no-threshold hypothesis."

Summarizing twenty years of exhaustive review of the literature in 1997, Luckey concluded:

> Exposed nuclear workers and military observers of atmospheric atomic bomb tests with carefully selected control populations provide 13 million person-years of experience with low-dose radiation. These carefully monitored studies show conclusively that low doses of ionizing radiation reduce premature cancer mortality in humans. When person-years were used to obtain a weighted average, the cancer mortality rate of exposed persons was only 65.9 percent that of unexposed controls.[22]

In 1980, the US Department of Energy (DOE) awarded a contract to the School of Public Health of Johns Hopkins University to study the effects of radiation on nuclear shipyard employees. The workers had been continually exposed to gamma radiation from cobalt-60—the same steel-reinforcing isotope used in the Taiwan apartments. Running from 1980 to 1988, the $10 million effort compared 28,000 shipyard workers who had worked close to nuclear reactors to 32,500 who had not. Cancer rates were significantly *lower* among the nuclear workers and death rates from all causes were 24 percent below the non-exposed cohort. Although the final report was approved by its Technical Advisory Panel and submitted to DOE in 1991, it has never been published.[23]

"Short bursts of radiation appear to stimulate the immune system," wrote one of the authors of the shipyard study in a subsequent article. "If the study aim had been to look for health benefits of ionizing radiation, it would have been a huge success. As a study to find radiation risks, it was an abysmal failure. This may explain the reason the study has yet to be published."[24]

Natural occurrences certainly don't confirm the no-threshold hypothesis and often confirm the hormetic effect. People in the Guangdon province of China are exposed to three times the world average through high levels of radioactive thorium in nearby mountains. A study of 70,000 residents found no significant health effects compared to a similar population in nearby low-radiation areas. Almost identical results are found in the high-background regions of India and Brazil. Residents of Ramsar, Iran, on the Caspian Sea, live with the highest doses in the world—10 rems—almost 28 times the world average and twice the amount permitted to workers in power stations. Yet tests have found these people to be healthier than surrounding populations. When their blood samples were subject to a "challenge dose" of gamma radiation, only half as many of their cells suffered chromosome damage.[25]

Despite these findings, regulatory agencies continue to cling to the no-safe-dose hypothesis. When improved home insulation began restricting the circulation of air in American homes, the Environmental Protection Agency became concerned about increasing concentrations of radon gas. In 1992, the EPA published a *Citizen's Guide to Radon*, warning that radon was causing 14,000 lung cancer deaths per year—one out of every eight deaths normally attributed to smoking. The study reached this conclusion by taking the death rates of early uranium miners who were exposed to extremely high levels of radon and extrapolated them downward to residential exposures, using the no-safe-dose hypothesis. On this basis, EPA launched a $100 million initiative to reduce household radon exposure. Many homes now cannot be sold without undergoing radon abatement.[26]

In response to the EPA alarm, Bernard Cohen undertook an exhaustive comparison of radon levels and lung cancer rates in 1,729 counties encompassing 90 percent of the population of the United States. Cohen found lung cancer varies *inversely* with radon exposure. Counties with the lowest radon levels have lung cancer rates 30 percent above those counties with the highest radon levels. "Our study has tremendous statistical power, including effectively nearly a million lung cancer deaths and 20 million deaths from other causes," wrote Cohen.[27]

Yet in 1998, the National Research Council's Committee on the Biological Effects of Ionizing Radiation devoted its entire *BEIR-VI Report* to the presumed dangers of radon. Sticking with the no-safe-dose hypothesis, BEIR-VI, "estimated 15,400 to 21,800 deaths attributed to radon [that] constitute an important public health problem...There is no safe level of radon—any exposure poses some risk of cancer." In considering the results of Dr. Cohen's study, a BEIR committee member commented: "Most of us would not be willing to discard a useful theory on the basis of such a test."[28] Yet there are no tests that confirm the EPA's alarm about radon.

Prodded by New Mexico Senator Pete Domenici, DOE has initiated a $20 million program to study the health effects of low-level radiation. As an indication that there is some dissent going on, the EPA has also co-sponsored BELLE conferences at the University of Massachusetts. At the policy level, however, the no-safe-dose hypothesis reigns supreme. The EPA is currently proposing a multi-billion-dollar mandate to lower groundwater standards around nuclear plants from 25 to 15 millirems. World governments may spend $1 trillion over the next twenty years trying to shield the public from low-level exposure. "The great irony of all this expenditure," says Bernard Cohen, "is that it is probably reducing health."

Not everyone in the world is willing to go along with this. The Taiwan apartment experience, plus the numerous other studies showing a hormetic effect, are having an impact. The French Academy of Medicine has issued a statement rejecting the no-safe-dose hypothesis. Invoking the U.S. National Council for Radiation Protection, the communiqué concluded:

> [T]here are detected no global increase of cancer risk, increase of congenital malformations, or abnormalities induced by cytological radiation effects with newborns, in the well-studied population of the particular highly-exposed region of Kerala India to external radiation and internal contamination. Identical conclusions are obtained in exposed Chinese populations. And last, as stated by the NCRP in the United States: "It is important to notice that the evidence of cancers in most of the exposed populations to low-dose radiation has not been found to be increased, and that in most of the cases this incidence seems to have been reduced."[29]

Says Edward Calabrese, perhaps the world's foremost advocate of hormesis theory, "On the scientific side, I think we're winning. On the political side, though, I'm not sure we'll ever win."

One of the most fateful outcomes of radiation phobia has been the food poisonings that regularly occur because the Food and Drug Administration refuses to approve the sterilization of most foods by radiation. It was Marie Curie who first cured bacterial gangrene on the battlefields of Europe. Today Cobalt-60, a fission product, is used to kill microbes in food and medicine. In hospitals, gamma radiation sterilizes syringes, gloves, medical instruments and blood for transfusion, all of which would be damaged by heat sterilization.

Food irradiation was initiated in the 1960s but soon ran aground from political opposition. Critics spread the rumor that it makes food radioactive or denatures vitamins. Forty-one countries now use radiation to treat 200 different types of meats, fruits and vegetables, but the FDA—bowing to pressure from Naderesque groups—has approved it only for red meat and a few other items. Nader's Public Citizen claims irradiation "gambles with the health of the American people."[30]

As a result, *E. coli* outbreaks in vegetables and non-red meats continue to plague the nation. In 2006, an outbreak in tainted spinach sickened hundreds of people. Two months later, a similar poisoning occurred in lettuce served at Taco Bell on the East Coast. Seventy people were infected and eight suffered kidney failure. A salmonella outbreak in 2008 infected tomatoes. Said Michael Osterholm, director of the Center for Infectious Disease Research at the University of Minnesota: "If even 50 percent of meat and poultry consumed in the United States were irradiated, the potential impact on food-borne disease would be a reduction in 900,000 cases, and 350 deaths."

After the Taco Bell incident, the *Wall Street Journal* puzzled as to why people accept pasteurization of milk with no problem but suffer anxiety of the irradiation of food. It suggested the term "Curieization" be adopted to personalize the technology and make it appear less threatening.[31]

For decades we have been counting all the hypothetical deaths of people killed by nuclear technology—3,000 at Three Mile Island, 70,000 at Chernobyl, countless numbers from the low-level exposure around nuclear plants. Yet no one keeps count of the real number of lives that could be saved through greater implementation of radiation technology.

Whether the hormesis theory will prove right or wrong, it raises one certainty—public fears about small exposures to radiation have been highly exaggerated. Radiation is a normal part of life. The solar system was far, far more radioactive at its birth than it is today and life evolved in a highly radioactive environment. Cockroaches, one of the oldest life forms, are almost immune. It would not be at all surprising to find we have also evolved defenses against it.

One of the great ironies of the coal-nuclear debate debate is that, in addition to being the world's largest source of greenhouse gases, coal plants are a far greater source of radioactive emission—sometimes up to 100 times that of nuclear reactors.[32] There are specks of uranium in coal—about 1 percent of the content. As a source of energy, these uranium specks actually have greater potential than the coal itself. It is too difficult to extract, however, and the uranium goes up the smokestack with the rest of the particulate matter and scatters across the landscape. Yet with all the hysteria about nuclear, it is the reactors that must spend $2 billion apiece to make small reductions in their emissions.

At the Cooper Nuclear Plant, the NRC took eight months to approve a tricycle. At the Zimmer Coal Plant, a bicycle sits casually leaned against the wall. Nuclear will always be held to a much higher standard. Still, like Zimmer plant supervisor Ron Frey, people in the industry are proud of it.

Interlude

Boulder, Montana

Boulder, Montana is the home to a small industry of former uranium mines that have all been transformed into health spas. The monikers are enticing—Earth Angel, the Sunshine Health Mine, Free Enterprise, the Merry Widow. All have attracted victims of arthritis, asthma, psoriasis, and rheumatism plus other aches and pains for more than a half-century.

It all began back in the 1950s when Wade Lewis, a Boulder geologist, discovered radioactivity in an abandoned silver mine just on the edge of town. The shaft turned out to contain uranium, much more valuable than silver, and Lewis began mining it for the growing nuclear industry.

Then the wife of one engineer spent a few days down in the mine and found her bursitis had been cured. She told a friend who had the same condition and got the same results. The news spread by word-of-mouth and soon people were coming to relieve their aches and pains.

Lewis did some research and found there was reason to believe that low doses of radiation might alleviate a variety of illnesses. People had been exposing themselves to radiation since Roman times. Hot springs and other geothermal sites have always been renowned for their health effects. Europeans still frequent these spas. Bad Gastein in Austria advertises its high radon count and has just been remodeled for $20 million. The Radium Palace in the Czech Republic, founded in 1906 by Marie Curie, treats 14,000 patients a year and has to turn people away. Lewis soon decided to switch his business and reopened it as the Free Enterprise Health Mine, charging people to sit in the mineshaft and absorb radiation.

In the 1950s, *Life* magazine did a spread on Montana health mines and soon 100 people a day were crowding into the 400-foot tunnel. Stories of cures abounded. Even today, the visitors include a woman who in her 70s who says she was in a wheelchair with arthritis before coming to Boulder twenty years ago. Today she is spry and healthy.

Nonetheless, traffic at the mines has slowed to a crawl since the EPA began its campaign against radon in the 1990s. The visitors today are mostly Canadians—who don't pay any attention to the EPA—and oddball Americans like a party of Amish who have taken the train from Pennsylvania. "We don't travel in airplanes," says an elderly patriarch with a full white beard and glistening teeth, "but we do take the radiation cure."

In September 2006 I spent four days sitting in the damp tunnel absorbing about 400 times the EPA's "action-level dose" of radon. There were comfortable chairs and bright lights so I could catch up on my reading. One Canadian couple down the hall played cribbage all day while others chatted and napped. "Last week a couple brought a dog that was all crippled with arthritis," remarked one Alberta wheat farmer. "After a few days that dog was running around like a pup. People say this cure is all in your head but you can't tell us that dog was just pretending he felt better."

Patricia Lewis and her husband Burdette Anderson bought the mine from her grandfather Wade in 1994. After a decade of declining traffic, however, they are stoic about the future. "Our heyday has pretty much come and gone," said Lewis glumly. "Lone Tree Mine just down the road closed up this year. The EPA's campaign against radon has definitely had a negative impact." Nonetheless, she continues to communicate with groups like the Hormesis Society, hoping for a breakthrough.

After four eight-hour days in the Free Enterprise Mine, I certainly didn't feel any ill effects. (Some people get a little nauseous in the first few days.) If nothing else, I felt refreshed. The EPA might argue that I'm going to contract get lung cancer in twenty years, but most of the people I met during the week had been coming for more than 20 years and had no complaints. Free Enterprise has never had a lawsuit.

"People keep telling us one day the scientists will eventually prove us right," says Pat Lewis. "But we've been selling radiation here for 50 years with no problems. I think it's us that's going to prove them right."

Chapter 23

Waste and Proliferation

When all the other concerns about terrestrial energy have been allayed, one last question inevitably arises, "What are you going to do with the waste?"

This is puzzling. The concept of "waste" is really a hangover from the fossil fuel era. In strict physical terms, *there is no such thing as nuclear waste.*

Recall the Second Law of Thermodynamics, around which the concept of "waste" is built. Extracting energy from nature is a matter of taking something in a highly ordered state and leaving it in a state of greater disorder. No living organism, you will remember, can survive in an environment made up of its own wastes. Once we have used energy, we cannot recycle it. We will always be looking for more sources.

Extracting energy from fossil fuels means taking highly ordered organic molecules and breaking them down into two very simple molecules, carbon dioxide and water. When that is done, both are completely exhausted of their energy potential. That is why we call them "exhaust."

Exhaust does not necessarily mean waste. Exhausted material can be used for other purposes. Water is a valuable natural resource and can be readily returned to the environment. Carbon dioxide is more problematic. It is not inherently dangerous, but threatens to return us to an age when there was much more CO_2 in the atmosphere and the earth was much hotter. Carbon dioxide can be injected into oil wells for oil recovery or used to fill soda bottles. At the very worst, it can be sequestered in huge quantities within the earth. But all this is expensive and requires huge amounts of energy.

So what we do instead is the cheapest and easiest thing–we scatter it to the four winds and let it dissipate in the atmosphere. Now it is in a state of *physical* entropy as well. It would take unimaginable amounts of energy to reassemble this material. All this was emphasized in 2008 when the Oak Ridge National Laboratory announced it had developed a way to remove carbon dioxide form the atmosphere, claiming the $25 million Branson Prize. The condition was that it would require vast amounts of nuclear energy.[1]

Because a resource has been used does not automatically make it waste. As environmentalists have long taught us, pollution is really a "resource that is out of place." Rivers and lakes become polluted from sewage, fertilizers or industrial chemicals, but all these could be recovered and recycled for other purposes. Mercury from coal is regarded as a serious health threat, yet mercury is also a valuable metal. Persistent pesticides foul the environment but they could be recovered and used to kill more insects. The problem is the Second Law of Thermodynamics—it takes far, far too much energy to recover them in a useful form. It is not worth the effort.

James Clerk Maxwell, the discoverer of the electromagnetic spectrum and generally considered the best physicist between Newton and Einstein, once tried to argue that the Second Law could be overcome and energy recycled by a hypothetical "sorting demon," a miniscule creature sitting in a container of lukewarm air sorting the hot from cold molecules. If all the rapidly moving molecules were routed one way and all the slowly moving molecules another, you could create a heat differential that could be used to run a heat engine.

Physicists have spent more than a century trying to explain why "Maxwell's Sorting Demon" can't possibly exist. Their answers have filled the textbooks and scientific journals, but boiled down, they come to this: In order to perform his task, the Sorting Demon would have to expend more energy than the sorting would create. Therefore, the Second Law still stands.[2] In the Twentieth Century, the debate veered into the matter of the *knowledge* required to make the Demon's decisions. As twentieth century Information Theory has taught us, acquiring knowledge is not free but requires energy. In one effort, a mathematician finally determined that if the Demon had an infinitely long piece of paper, he could compute all the information needed to separate the molecules. Then someone pointed out if you had an infinitely long piece of paper, you could just burn the paper and not worry about a heat engine. The same would hold true for twenty-first century computer models. Those computers, remember, run on electricity.

The sorting problem is what inhibits garbage recycling. Every municipal dump is filled with potential resources. The paper could be reused, the minerals recycled, the organic material composted. The difficulty is sorting them all out. Homeowners play the role of Maxwell's Sorting Demon when they separate their garbage at curbside, but they can only contribute so much. For most garbage, the recovery value is so small that the best thing to do it throw it all in a landfill as "waste."

Now, here's the point. *There is no entropy in the nuclear fuel cycle.* Because the by-products of fission are so incredibly compact and potentially useful, none of them need to be scattered or "thrown away." They are all sitting there, waiting to be reprocessed. *Almost 100 percent of the material in a spent nuclear fuel rod can be recycled as useful material and is being done so in other parts of the world today.* The very small amounts of material that cannot be economically reprocessed today can be stored safely until it does become financially feasible in the future.

There is no such thing as nuclear waste.

Let's take a look at the material that makes up a spent fuel rod. *More than 95 percent of a spent fuel rod is plain old uranium-238,* the same kind that comes right out of the ground. It could easily go back from whence it came. Yet because we treat it as "radioactive," to coin a phrase, we create a huge, intractable and essentially unnecessary problem.

Take a shovel full of dirt out of your backyard and there is probably some uranium in it. If so, according to federal regulations, you have a "nuclear waste problem." You cannot put it back in the ground. You must take it to a nuclear waste repository. Needless to say, this is likely to add some extra expense to tilling your garden. The U-238 is indeed radioactive, but as we learned from the previous chapter, that does not make it poisonous or deadly. *Radiation is energy*. The U-238 can be re-utilized for energy—as is being done in France, Canada, Russia and most other parts of the world.

The remaining 5 percent of a spent fuel rod is even more highly radioactive material whose energy is actually too intense to be used in a nuclear reactor. But that does not mean it can't be used for other things. Radioactive isotopes have found broad implementation in medicine and industry. *Forty percent of all medical procedures now involve some kind of radioactive isotopes.* Radioactive tracers are now a primary tool in diagnostics. Radioactive energy sources are used to treat cancer. None of these materials are "waste." They do not have to be "dumped" or "thrown away."

So why are we making such a fuss over Yucca Mountain? The main reason we are stymied over a nuclear repository is because we stopped reprocessing nuclear fuel in the 1970s. We did this because we were afraid that recycling would lead to the proliferation of nuclear weapons in other countries. This concern has since proved to be seriously mistaken. There has been proliferation of nuclear weapons around the world and there may be more, but none of it has anything to do with whether we reprocess fuel in the United States.

While the United States gave up reprocessing, France, Russia, England, Canada and Japan did not. All now have a recycling industry and very little storage problem. The only result of our abandonment of reprocessing has been to create a huge storage problem in *this* country. If we were to revive recycling, only 2 to 3 percent of the material now scheduled for Yucca Mountain would have to be stored there and the whole problem would disappear. Once again, because the power of terrestrial energy is so highly concentrated—2 million times more than chemical energy—the amount of materials that must be handled at each stage of the process is almost infinitely smaller. The "exhausts" of three years of nuclear electricity can be stored indefinitely in a dry cask about four times the size of a telephone booth.

Fissionable U-235, you will recall, makes up only 0.7 percent of natural uranium. To run a reactor, it must be raised to 3.3

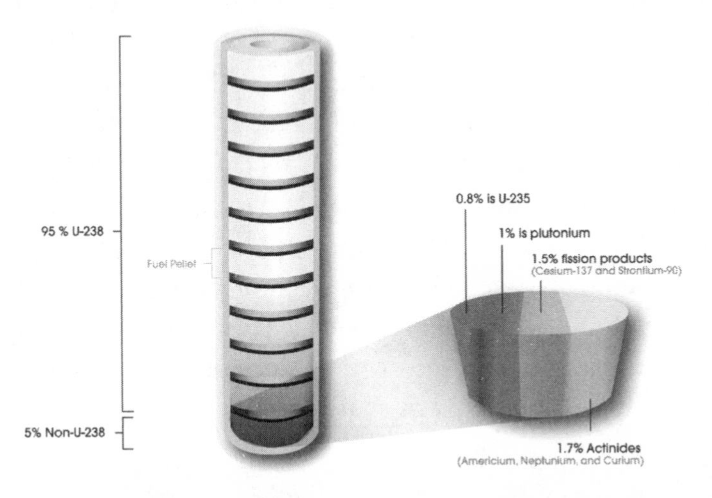

Spent Fuel Rod **Breakdown of Recyclables**

percent. After three years of producing electricity, however, most of the U-235 atoms have split into smaller elements and the U-235 content is almost back to its natural state at 0.8 percent.

Fission products make up another 1.5 percent. These are the atoms far down the periodic table—the ones that Lise Meitner first recognized in her seaside cottage. Barium, strontium, cesium and zirconium are the most common, but there are more than two-dozen in all. Many have short half-lives but some can last for decades or millions of years. Radioactivity results from the breakdown of atoms and so, the faster the breakdown, the more intense the energy release. The *intensity* of an isotope's radiation varies inversely with its *half-life*.

While U-235 atoms are breaking down into smaller isotopes, some of the U-238 will absorb neutrons and moves *up* the periodic table, forming the "transuranic elements," also know as the "actinides." Plutonium is the most common, making up 1 percent of the spent rod. The others are less abundant. They are generally intensely radioactive, and shorter-lived.

Let us look at what can be done with all of these by-products:

Fissionable Uranium. Like the glowing embers of a fire or the coal residues in a coal bin, the U-235 remaining in a fuel rod still has energy potential. Pulling it up to reactor grade would once more mean enriching it to 3.3 percent. But traces of U-235 can also work well in "mixed oxide fuels," (MOX), which are a combination of several radioactive isotopes by-products that can be mixed together to function in a reactor.

Plutonium. Plutonium is the only element besides uranium that has isotopes that undergo complete fission, splitting in two rather than merely releasing an alpha or beta particle. The bomb at Hiroshima was made from enriched uranium but Trinity and Nagasaki, plus the majority of weapons manufactured since then, were all made from plutonium.

What is not commonly recognized is that the plutonium in spent fuel rods is not "bomb-grade material." There are four plutonium isotopes—Pu-239, Pu-240, Pu-241, and Pu-242. *Only*

P-239 has the right fissionable qualities to make nuclear weapons. Fortunately for the world, fission in commercial reactors creates all four isotopes. The other three "pollute" the P-239, making it "fizzle" instead of "fissile." Purifying reactor plutonium requires another round of isotope separation that is even more difficult than uranium enrichment. The British, the Russians, and the French have all tried and failed.

So how did we build bombs with plutonium? We obtained it from an entirely different type of reactor, one that burns fuel that has been "de-enriched" of its U-235 and contains only U-238. This produces pure P-239. The other alternative is to burn the uranium for only a short time with a graphite or heavy water moderator so that the other isotopes do not have time to form. That is what the Soviets were doing at Chernobyl. It is also how North Korea built its crude homemade weapon. If all this had been clearly spelled out in 1976, we probably would have understood that plutonium from American commercial reactors cannot be made into bombs. Our noble abstinence from recycling could have been forestalled and we would have no problem of "nuclear waste."

Fortunately, the mixture of P-239 thru P-242 that comes out of a uranium reactor is good for one thing—running other reactors. When combined at a ratio of 7 percent reactor plutonium and 93 percent U-238, this creates "mixed oxide" (MOX), which duplicates enriched uranium and can be plugged into existing reactors. Thirty reactors in France, Belgium, Switzerland and Germany are already using MOX and twenty more have been licensed to do so. Most can handle only about one-third MOX with their enriched uranium, but some will take up to 50 percent. Japan aims to have one-third of its reactors using some MOX by 2010 and has licensed one reactor to run on pure MOX fuel.[3]

After the collapse of the Soviet Union, Senators Pete Domenici and Sam Nunn negotiated a remarkable deal with the Russian government whereby we are purchasing enriched uranium from their stockpile of disassembled weapons and recycling it through American power plants as fuel. *One out of every ten light bulbs in America is now lit by a former Soviet weapon.* (Twenty percent of our electricity comes from nuclear and half our fuel now comes from recycled missiles.) If ever the world has beaten swords into ploughshares, this is it![4] Recycled Soviet weaponry depressed the uranium market for years but we will probably run out in the next decade. Anticipation of this shortfall—plus the pending nuclear revival—send uranium prices skyrocketing from $10 a pound in 2004 to $140 in 2007, before dropping back to $80 in 2008.

Fission products. When uranium-235 splits in half, it divides into pairs of smaller atoms, such as strontium and xenon, barium and krypton, cesium and rubidium, with the number of protons always equal to uranium's 92. These by-products are all radioactive and will continue to shed beta particles and gamma rays until they reach a stable state, usually as lead. The lead atom, which sits at one of the highest points on the "curve of binding energy," is the ultimate form of "nuclear waste." Lead's great stability is what makes it so valuable in *shielding* against radioactivity.

Cesium-137 and strontium-90 are the most troublesome fission by-products. Both have half-lives of about 30 years, making them both long-lived and fairly radioactive. They will not disintegrate to background levels for about 1,000 years. Strontium and cesium are doubly dangerous because they mimic calcium and potassium and are taken up biologically. It was strontium-90 in mother's milk that persuaded President Kennedy to propose the nuclear test-ban treaty. Even so, both strontium and cesium isotopes also have medical and industrial uses and can be sold commercially.

Actinides. The final 1.7 percent of a spent rod are the "minor actinides"—americium, neptunium, californium, curium, einsteinium, and fermium—all of which are above uranium on the periodic table. They may have been present at the birth of the solar system but have long since disappeared due to their short half-lives. The minor actinides also take about a thousand years to reach background levels.

The main use for both the minor actinides and fission products is in medicine. Nuclear medicine is a *quarter-trillion-dollar* industry that treats a whole host of illnesses and saves tens of thousands of lives each year. Aside from the discovery of antibiotics, it is probably the greatest medical advance of the twentieth century. Tracing substances through the body with radioactive isotopes is one of medicine's most valuable diagnostic tools.

Radiological tracing is said to have begun in 1911 when George de Hevesy, a young Hungarian studying under Ernest Rutherford, became suspicious that his Cambridge landlady was recycling leftovers at the boarding table. He slipped a small sampling of radioactive material into his meal one night and then detected it several days later. De Hevesy went on to win the 1943 Nobel Prize for Medicine.[5]

Because they emit gamma rays, radioactive isotopes can serve as tiny transmitters, sending signals from within the body. Iodine-131—the most worrisome by-product of a reactor accident—is used to diagnose malfunctions of the thyroid, kidney, and liver. Iron-59 tracks metabolism in the spleen. Chromium-51 tags red blood cells. Radioactive isotopes are also used to *treat* cancer—iodine-131 for thyroid, palladium-103 for prostate, samarium-153 for the breast.

Technetium, along with promethium, is the only element below lead that does not have a stable isotope. In nature it is found only in traces, from the breakdown of natural uranium. Created in commercial quantities in nuclear reactors, however, it has proved to be invaluable in diagnostics. Technetium-99 is now the preferred isotope in 80 percent of radioisotope tests—used around the country more than 40,000 times a day. Its half-life of six hours is long enough to perform tests but short enough to minimize patient exposure. Glenn Seaborg, who won the Nobel

Prize for discovering plutonium, always considered his discovery of technetium-99 to be his greatest accomplishment.[6]

Boron Neutron Capture Therapy is yet another radiological technique used to destroy brain tumors. Non-radioactive Boron-10 is injected into the patient, where it concentrates in the tumor. Then the patient is *taken to a nuclear reactor* to be exposed to neutrons. The boron absorbs neutrons and emits alpha particles, killing the tumor cells. When Australian anti-nuclear activists recently tried to close down one of the country's few remaining research reactors, the nation's doctors protested that it would ruin nuclear medicine.[7]

The portion of the electromagnetic spectrum that includes radio waves also includes a disturbance of atoms called "nuclear magnetic resonance." These relatively low-energy waves are now used for whole-body scans and imaging. When the technology was introduced, it was called "nuclear magnetic resonance imaging." Patients and nurses proved so fearful of the word "nuclear," however, that many refused to enter the scanning rooms. It was not until the name was shortened to "magnetic resonance imaging"—"MRI"—that the technology won wide acceptance.

Industry is equally reliant on the radioactive isotopes produced both in research and commercial reactors. Reducing the ash content in coal is a key to cutting particulate pollution. Coal with more ash must be washed before burning. The ash content is measured by beaming gamma rays through the coal while it travels the conveyor belt. Paper manufacturers monitor the thickness of their product with radioactive gauges. Californium-252 and americium-241 and the fission products strontium-90 and krypton-85 are all employed in these techniques. When slipped into inaccessible places, gamma-emitting tracers become tiny x-ray machines beaming out images. Wear and tear on bearings deep within the machinery can be photographed. Pipe welds are checked by covering them with photographic paper and dropping a radioactive isotope into the pipe.

If Jack Lemmon had access to radioactive tracers, in *The China Syndrome*, his reactor might not have fallen apart at the end of the movie.

In 1973, John McPhee published a series of articles in the *New Yorker* based on interviews with Theodore A. Taylor, a Princeton nuclear expert who had begun to worry about plutonium. The articles were later reprinted in the book, *The Curve of Binding Energy*. Taylor was a dreamy, somewhat eccentric chemist and physicist who had designed some of the smallest nuclear weapons ever made. He was not altogether grounded in reality. Wrote McPhee, "I have watched him half a dozen times try to start his car with a hotel key."

Nonetheless, Taylor came to regret his work and became obsessed with the idea that other people—almost anyone in fact—would be able to assemble nuclear weapons in their garages and basements much the same as he had done. "I think we have to live with the expectation that once every four or five years a nuclear explosion will take place that will kill a lot of people.... I can imagine—in the worst situation—hundreds of explosions a year.... What we are taking with the nuclear industry is a calculated risk," he told McPhee.[8]

At that moment, Allied General Services, a consortium of Allied Chemical Corp., Gulf Oil Corp. and Royal Dutch Shell, was investing $250 million on a privately built nuclear reprocessing plant in Barnwell, South Carolina. Its purpose was to separate the radioactive materials in spent fuel rods and recycle them into useful products. The plutonium would be sold to the Clinch River Breeder Reactor, then planned for Oak Ridge.

Breeder reactors are an over-the-horizon technology that have long promised virtually unlimited amounts of fuel if and when our current uranium supplies are ever exhausted. In a breeder reactor, U-238 is wrapped in a blanket around the enriched uranium or plutonium. Neutrons from the core transform the U-238 into more plutonium so that the reactor ends up producing more fuel than it consumes. This would seem to violate the

Second Law—until we realize that ordinary fission taps only a small portion of the energy stored in U-235. Breeding taps into the vastly larger amounts of energy stored in the more common U-238. The promise of this "bottomless well" has always enthralled nuclear scientists—even though it has proved very difficult to implement in practice.

In 1976, President Ford—coming under increasing pressure during the Presidential elections—halted nuclear reprocessing. Then on April 7, 1977, only weeks before his famous fireside chat, President Carter announced, "We will defer indefinitely the commercial reprocessing and recycling of the plutonium produced in the U.S. nuclear power programs…The plant at Barnwell, South Carolina, will receive neither federal encouragement nor funding for its completion as a reprocessing facility." Private investors tried to keep the project afloat for several years and the Reagan Administration actually repealed the ban on fuel recycling, but the facility was abandoned in 1983. Its empty shell still stands at the end of a long rural road near the Savannah River, a ghostly reminder of a once promising future.[9]

Carter's logic was that recycling spent fuel would lead to nuclear proliferation. "It's an assault on our attempts to control the spread of dangerous nuclear materials," he told Congress two years later in once, again vetoing funds for Clinch River. "It marches our nuclear policy in exactly the wrong direction."[10] India had just exploded a bomb in 1974 after purloining plutonium from a research reactor given to them by the Canadians. McPhee and Taylor hypothesized that if terrorists were to build a bomb, they would do it by pilfering from the American fuel cycle. By canceling recycling and leaving plutonium in spent fuel rods, we would prevent, nuclear arms from spreading around the world.

This analysis—although it seemed logical at the time—has proved to be extremely faulty. France, Britain, Russia and Canada all went on to recycle and capture world markets. *We now import 90 percent of our medical isotopes from Canada because we do not recycle.*[11] The subsequent history of the spread of nuclear weapons also shows that our renunciation of recycling had virtually no impact.

During World War II we shared nuclear technology with Britain and the British exploded their first bomb in 1952. In 1960, France followed suit, with only minor technical assistance from the U.S. and Britain. Next came the Chinese Communists in 1964, with some reluctant help from the Soviets. Alarmed that nuclear weapons would become commonplace, Russia, England and the United States signed a Nuclear Non-Proliferation Treaty in 1968. France and China did not join until 1992. While forswearing nuclear arms, the signatories allowed for developing nuclear power for peaceful purposes. This included the right to enrich uranium.

Israel did not sign the treaty and was long rumored to be working on a bomb. In 1986 a former Israeli technician produced photographs he said were a bomb factory beneath the Dimona reactor. Today, Israel is believed to possess as many as 150–200 nuclear weapons. In 2006, while telling a German television audience why Iran shouldn't be allowed to have nuclear weapons, former Prime Minister Ehud Olmert remarked, "Iran openly, explicitly and publicly threatens to wipe Israel off the map. Can you say that this is the same level, when they are aspiring to have nuclear weapons, as France, America, Russia and Israel?" This was widely regarded as an inadvertent admission of Israel's arsenal.[12] Israel also apparently passed the technology on to South Africa in the 1980s. When the apartheid government ended in 1993, the white minority government dismantled its small stockpile of nuclear weapons, not wanting them to fall into the hands of the new regime. Nelson Mandela signed the NPT in 1996.[13]

After India tested its bomb in 1974, Pakistan embarked on a clandestine program, apparently smuggling in an entire uranium enrichment facility from Europe. It exploded five test weapons in 1998 and then announced construction of a reactor that would produce weapons-grade plutonium. Pakistan's announcement set off another round of tests by India, to the condemnation of most of the world. In 2006, President Bush signed an agreement to share nuclear technology with India, even though India has not yet signed the NPT. Bush said that India had pledged not to continue testing but the pact drew criticism at home and abroad.[14]

North Korea began a nuclear program in the 1980s with a Soviet-supplied graphite reactor—the Chernobyl kind. In 1989 the Koreans closed down the reactor for 70 days, enough time, American intelligence calculated, to extract 12 kilograms of plutonium, which would build two bombs. With concerns mounting, President Clinton sent Jimmy Carter to North Korea in 1994 to try to halt the effort. Carter came away with a pledge that North Korea would give up building a bomb if the U.S. would lift trade sanctions and provide two light-water reactors. We held up our end of the bargain but it soon became clear that North Korea had not. In 2005 Premier Kim Jong-Il announced his country had a nuclear weapon. In 2006 the North Koreans exploded a small devise underground. The country has also developed long-range missiles possibly capable of reaching the West Coast. The Bush Administration, China, and the rest of the world have engaged in negotiation to dissuade North Korea from its effort and the Koreans have recently made gestures toward cooperating.

Iraq established a nuclear program in the late 1960s and tried to purchase a plutonium reprocessing plant from France but the French refused. With French assistance, however, the country was near completion of the 40 MW Osiraq research reactor near Baghdad in 1981 when Israeli jets bombed it. In 1998, Saddam Hussein again made inquiries in Niger about purchasing uranium but apparently was unable to start another weapons program. Iraq had not signed the Non-Proliferation Treaty when we invaded the country and deposed Saddam in

2003. Despite concerns about "weapons of mass destruction," no nuclear facilities were found.[15]

In 2004, shortly after the Iraq invasion, Dr. Abdul Qadeer Khan, the "father of Pakistan's nuclear weapon," confessed to running a clandestine international ring that had provided centrifuges for uranium enrichment to North Korea, Libya, and Iran. Dr. Khan exonerated the government of Pakistan of any participation. A few hours later, General Pervez Musharraf, Pakistan's military ruler, pardoned him. It was subsequently reported that two of Khan's assistants also had discussions with Osama bin Laden. Khan has since remained under house arrest but is widely regarded as a national hero.[16]

With the help of Dr. Khan, Iran now claims to have enriched uranium, although it is never clear whether this means reactor-grade or bomb-grade material. There is a big difference. In 2006 the country held a nationally televised parade in which dancers in traditional costumes held up vials of what was said to be enriched uranium while tethered doves flew at their shoulders. This was almost certainly not bomb grade. President Mahmoud Ahmadinejad vacillates back and forth between saying he wants to build a domestic power program and saying he wants to blow Israel off the map. Probably he wants to do both. Dr. Mohamed ElBaradei, director of the International Atomic Energy Association made numerous efforts to hold back the Iranian bomb and was awarded the Nobel Peace Prize for his efforts. Russia and China supported Iran at first but eventually condemned the effort. In 2008, Iran announced it had installed another 6,000 centrifuges at its nuclear installation in Natanz. Intelligence experts now estimate Iran could probably build a bomb by 2010.

Dr. ElBaradei claims that no state has yet built a bomb while subject to international inspections under the Non-Proliferation Treaty and that without IAEA efforts 40 nations might have nuclear weapons by now. This is probably true. The problem is that several key states have refused to sign the treaty and gone and developed nuclear weapons anyway.

Now the point is this: *Our 1976 decision to end nuclear recycling has had absolutely no bearing on the spread of nuclear weapons.* The idea that terrorists or foreign nationals might break into an American nuclear facility, steal the fuel rods, load them on their backs, lug them across the Mexican border and then spirit them to some clandestine facility in Tora Bora where Al Qaeda would forge it into a bomb has always been an anti-nuclear fantasy. If terrorists were seeking plutonium today, they would head for North Korea, not Clinch River. For a rogue nation, the easiest thing to do is build a reactor or enrichment facility. The technology is not that complex.

So is there anything we *can* do to stem the tide of nuclear proliferation? President Bush put forth an excellent proposal in 2006—the Global Nuclear Energy Partnership (GNEP). GNEP is built around the principle that the nations with nuclear technology—the United States, Britain, Russia, Canada, Japan, and France—should provide civilian installations to developing nations while keeping control of the fuel. The advanced countries would build reactors and provide the fuel rods, then reclaim them three years later when they are spent. Enrichment and recycling facilities would be kept in the advanced nations. Incidentally, the program would revive our recycling industry. Outside of Iran, most countries would probably be willing to take it. GNEP offers a practical plan for providing the world with terrestrial energy while heading off nuclear proliferation. It is remarkable it has not earned more praise. Whoever succeeds President Bush will do well to continue it.

It would be wrong to underestimate the costs of nuclear reprocessing. Nuclear fuel is not completely free of entropic disorder. Although the spent fuel is all in one place, the radioactive isotopes are mixed together. Separating them must be performed by highly sophisticated equipment. When spent fuel is removed from the core, its radioactivity is so intense that it would kill everyone within 50 yards if not properly shielded. Once submerged in a storage pool, however, the radiation is blocked by five feet of water. The water gives off an eerie blue light caused by beta particles that are moving faster than light is able to travel in water—the "Cerenkov Effect," named for the Nobel-Prize-winning Russian scientist who identified it.

During recycling, extracting the uranium will actually make the remaining material *more* radioactive–just as Marie and Pierre Curie found when they extracted radium from pitchblende. But the job can be handled by remote machinery. The French have recycled spent fuel for thirty years and are now helping Japan and India do the same thing. Whatever isotopes cannot find a market are "reposited" for the future.

The important thing is this: *None of this produces waste.* No dangerous or harmful materials are indiscriminately scattered into the environment. It is all sitting in one place waiting to be recycled. And lest we forget, the 2,000,000 to 1 ratio over fossil fuels, makes the amount of material stunningly small. All the high-level by-products from fifty years of nuclear fission in this country could be assembled ten feet high on a single football field. The French store all the high-level wastes from 30 years of providing 75 percent of their electricity in *one room* in La Hague.

So what is going on at Yucca Mountain? Typically, the project is bogged down in its environmental impact statement. The Environmental Protection Agency first required that annual radiation from the site should not exceed 15 millirems for 10,000 years. Environmentalists and anti-nuclear groups said that wasn't good enough. They wanted the standard set for a *million* years. The District of Columbia Court of Appeals agreed. So the EPA set a standard of 350 millirems for the next million years (about two-thirds of what people in Denver get from natural sources). Opponents are now saying that isn't good enough either. "Isn't it a magical coincidence that the standard just happens to match the level of exposure the industry has been predicting," argued Arjun Makhijani, president of the Institute for Energy and Environmental Research, at the EPA hearings, implying a government-industry collusion.[17]

After 100 years, the radioactivity of the actinides and fission products will decline to one-tenth its original level. After 1,000 years, it will have fallen below the original uranium ore. Why do the court and the nuclear opponents want the material monitored for a million years? Because, except for the uranium, *that's how long it will take every radioactive isotope in the fuel rods to disintegrate.*[18] This will happen somewhere towards the end of time. Yet nuclear opponents will not rest until every last trace of terrestrial energy is wiped from the face of the earth.

Even as Yucca struggles, alternatives are emerging. Some utilities are using "dry cask storage," upright story concrete structures about the size of four telephone booths, fenced in by barbed wire. "Dry storage is safe on the order of 50 to 100 years," says Allison Macfarlane, co-editor of *Uncertainty Underground*, about the Yucca effort. "Geological repositories are the ultimate solution but there's no need to rush into one right now."[19]

In Utah, the 221-member Goshute Tribe has signed a $1 million contract to accept nuclear material on its reservation. The state government, of course, has threatened to renew the Indian Wars before that happens. One Utah recycling company, Energy Solutions, has become so respectable it was allowed to sponsor the Utah Jazz's basketball arena—even though there were objections.[20] A group of Wyoming businessmen are suggesting a similar project at Owl Creek, where the entire repository would sit in an open-pit coal mine.[21] The proposal has been under review for more than a decade.

In fact, the whole idea of a central repository may be all wrong. What we should do is apply Amory Lovins' advice and practicing "distributed energy." Here is the way James Lovelock, originator of the "Gaia Hypothesis" and one of the world's most respected environmentalists, proposes handling the problem in *The Revenge of Gaia*:

> I have offered in public to accept all of the high-level waste produced in a year from a nuclear power station for deposit on my small plot of land; it would occupy a space about a cubic meter in size and fit safely in a concrete pit, and I would use the heat from its decaying radioactive elements to heat my home. It would be a waste not to use it.[22]

The only real "waste" at Yucca Mountain will be the heat wasted on the mountain walls. Disintegrating isotopes will raise the temperature of the storage caverns to 400°F, hot enough to boil water or run a district heating system. Why not put the material in cement casks and distribute them in basements around the country? Its energy could be used for heat, cooling and the even the co-generation of electricity. Such casks would be immune to fire or explosion and anyone trying to steal one would need a construction crane.

Of course there is always talk of shooting the last remaining material into space, but it was born of earth and here it should stay. If all else fails, we could return it to the interior of the planet, where it would disappear into the earth's molten core. We could drop it into volcanoes or rift valleys—anywhere that liquid rock breaks through the earth's solid crust. Japan already has plans to bury some of its waste in the Mariana Trench, where the Pacific and Philippines Plates are colliding seven miles beneath the surface, in the depths of the ocean abyss. There it would return to its place of origin—dust to dust, ashes to ashes, magma to magma.

Interlude

Yucca Mountain, Nevada

"As you can see up ahead, Yucca Mountain isn't really a mountain," says our guide as we near the end of an hour-long bus ride north from Las Vegas. "Those of you who know geology will recognize it's only a ridge.

"No one knows how it got the name 'Yucca' either," he continues. "There aren't many yucca plants around here. It's mostly mesquite bushes."

"How about Mesquite Ridge?" shouts one of the more high-spirited members of our party, a rake with blond highlights and a bright turquoise crewneck.

The Department of Energy gives monthly tours to Yucca Mountain these days, anxious to prove that—after almost 25 years—it still intends to open its Nuclear Waste Repository some day.

For the most part, the trip feels like an expedition into hostile Indian country. The whole state has been on the warpath over the project since the beginning. From the chatter on the bus, it seems surprising that we haven't already been surrounded by a posse of hostile Nevadans on horseback.

"See those buildings off on the left there," says our guide as we pass through the sagebrush. "They're brothels. As you may know, prostitution is legal in certain Nevada counties. The state has no trouble supplying them with water but for the first eight months after we arrived they wouldn't give us any. We used port-a-potties for a long, long time."

As it turns out, though, the brothels have their upside. Anticipating a surge in business, they are one of the locals that support the project.

The area north of Las Vegas has seen its share of nuclear activity. The repository sits right next to the former Nevada test site, an area is the size of Rhode Island that hosted 100 aboveground and 600 underground nuclear tests from 1951 to 1963. A larger Air Force bombing and gunnery range still surrounds the Test Site.

In 1982, after years of accumulation of spent fuel at reactor storage pools, the Nuclear Waste Policy Act authorized the Department of Energy to explore sites around the country for permanent storage of spent fuel rods. DOE looked at three—Yucca Mountain, Hanford, and Deaf Smith County, Texas—eventually choosing Yucca for its geology.

In 1994, after a decade of studies, a locomotive-like device with a 25-foot drill face began burrowing at 185 feet a day into the side of the mountain (ridge). One mile down it turned left, burrowed another three miles, then turned left again, re-emerging at a cliff face only five feet from its target. Inside the visitors' center, a video monitor shows the mole-like locomotive's whirling snout breaking through the rock wall like a diver returning to the surface as staff members in hardhats stand around and cheer.

That was 1997. Nothing much has happened since then. Now the mighty engine sits forlornly by the tunnel entrance, looking like a relic from the days of the golden spike. By the time the dozens of side tunnels that will honeycomb the mountain are ready to be bored, it will be long outmoded. "We tried to sell it but we didn't get any takers," says our guide.

With $10 billion sunk in the ground, Yucca is still waiting for an operating license, our guide informs us as we wait for a convoy of jeeps to take us to the top of the mountain. "If a license is ever issued, it will be at least seven years before the facility can open," he says. "Drilling the auxiliary storage tunnels will take about a decade." Not daunted, the Department of Energy has already proposed expanding the facility to hold 135,000 tons, at a cost of another $77 billion.

"We're not spending any government money out here," our guide reassures us as we start climbing aboard. "Every reactor pays a 1/10th of 1 percent tax per kilowatt-hour to pay for all this. They're just passing it through to their customers."

"So that's why my utility bill just doubled," says the rake. We hope he is just kidding.

The jeep convoy passes over some rough road before we get to the top. On the summit, we are greeted by a pony-tailed DOE staffer in a green shorts who starts giving a lecture the geology. "This entire region is solidified lava tuft from an extinct volcano," he begins, gesturing to the barren countryside that surrounds us. "The last earthquake was 90 miles away in 1872. The area receives almost no rainfall and the water table is 2,500 feet beneath the surface. The repository will extend down only 1,300 feet, leaving a 1,200-foot buffer below its floor. Groundwater in the region moves at a rate of only one foot per year. The containers are all earthquake proof and resistant to corrosion."

"So none of this material will end up in Las Vegas?" someone asks.

"It's not very likely."

"How do you know this thing won't collapse in a earthquake," someone asks.

"Experience with earthquakes around the world has shown that underground structures can withstand the ground motion. In actual tests at the Nevada Test Site, mine tunnels withstood underground nuclear explosions that are greater than any ground motion anticipated near Yucca Mountain. We've been studying this for fifteen years."

It all sounds good. Of course, all this drudgework can also be wiped away by a single phrase–"nation's nuclear garbage dump," or "another Chernobyl." In recent weeks, the newspapers have honed in on a memo someone wrote while trying to meet the D.C. Court's demand for predictions of geological and meteorological conditions at Yucca for the next million years. Somewhere, some one wrote something about "making up data" and now the press is screaming how "the government lied to us."

Far in the distance you can see the snow-covered 14,000-foot peak of Mt. Whitney, highest mountain in the United States until Mt. McKinley joined the Union in 1959. "My daughter climbed that in one day," says the woman standing next to me.

I raise my hand. "I want to read you something I cut out of the paper," I say. "It's from the *Tampa Tribune:*"

> The Energy Department faces the daunting, if not impossible, challenge of building a dump that can contain this nuclear nightmare for 1 million years.... Critics say one of the flaws is that the government bet all its time, energy and money on one site: Yucca Mountain, a volcanic ridge 100 miles north of Las Vegas that is a catastrophe waiting to happen.

The geologist has a pained look on his face. I realize he thinks I am being hostile. "I'm not saying I believe any of this," I quickly add. "I'm just curious as to how you respond to something like that."

"Well, I think what I've just said here..." he begins, but the group takes over.

"They never get it right." "They never know what they're talking about." "Actually, the Las Vegas papers are a lot worse than that," someone chimes in. The geologist lets them do the talking. I realize this whole crowd is very supportive—although I guess they're probably self-selected. People who are afraid of Yucca and unlikely to come out here.

Back down the mountain again, we don hardhats and enter the tunnel. A railroad track runs down the middle while electrical cables climb the walls. After about 500 yards we turn into a cathedral-like grotto. Another DOE scientist lectures us in the tones of a high school chemistry teacher. "When it arrives, the radioactive material will be vitrified—dissolved in molten glass and then hardened," he explains patiently. "Studies of glass artifacts in Egyptian tombs have shown they do not erode or leach over the millennia. The vitrified cylinders will then be encased in concrete casks and the casks fit into specially fitted holes drilled into walls like these."

"What are these big rooms for?" someone asks.

"We will be conducting scientific experiments to find further uses for the isotopes," he says. "There are many potential medical uses."

"My brother just had radiation treatment for his prostate," whispers an elderly woman standing next to me. "It scared him half to death."

Back at the visitor's center it is like a high school science fair, with eager young DOE officials at every table ready to show us how thoroughly they have done their work. Transporting spent fuel to these sites will be done by trucks and trains. The trucks carry dumbbell-shaped steel casks that can withstand almost any conceivable impact. Videotapes show them crashing into concrete walls at 60 mph and being blindsided by speeding freight trains. The casks do not rupture.

"Since 1960 2,700 shipments have carried spent fuel 1.6 million miles across the U.S.," says one green-clad blond young woman. "There have been eight accidents, none of which led to release of radioactive material. Europe has shipped 70,000 metric tons through densely populated areas and the U.S. Navy has transported 783 containers over a million miles since 1957 with a perfect safety record."

Yet all this was easily swept aside when Greenpeace toured the region with a van proclaiming the shipments would be a "Mobile Chernobyl."

At the souvenir table there are the usual t-shirts, mugs and baseball caps. I am looking for something that says, "I survived Yucca Mountain."

As we board the bus, the guide asks if anyone has ideas on how the DOE might market the project to Nevada residents.

"Why doesn't the state just put out a license plate saying `We solved global warming?'" I suggest. Our guide likes that but the Nevada residents in the party say opposition goes much deeper than that. "The press has people scared to death," says one young engineer.

On the trip back to Las Vegas, my seatmate turns out to be a retired science teacher who survived a Japanese prison camp in World War II and has written a book about it. He isn't having much luck with publishers and is delighted to find I am a fellow writer.

"I was the only GI in the camp who would talk to the guards," he recounts. "I didn't ask for any favors but I established personal relationship with them. We'd just talk about our families and stuff. All the other prisoners hated me and said I was a traitor. But it paid off. Near the end of the war the orders came down from the high command to kill us all, but the guards didn't do it. It was mainly because I had established a relationship with them. Afterwards, several of the other prisoners told me, 'You know, that was a very smart thing you did.' It's given me a lifelong interest in diplomacy.

"It doesn't work much on this issue, though," he lamented. "I went to a couple of those public hearings and it was impossible to talk rationally. Every time I tried to speak in favor of it, everyone shouted me down."

Chapter 24

France and the Future

Is there a land in which this vision of a world run on terrestrial energy has been fulfilled? Indeed there is. That country is France.

The French are a contrary people. They dropped out of NATO in 1966 under Charles de Gaulle, they didn't join the non-proliferation treaty until 1992, their army essentially refused to fight at the start of World War II, they think Americans are boorish and uncultured. They acknowledge that Americans have great power but believe we don't know what to do with it, while the French—unfortunately—have almost no power but know *exactly* what to do with it. In the case of nuclear energy, they have turned out to be right.

In May 2008 I took a weeklong tour of French nuclear facilities, from the reprocessing plant at La Hague to the Melox recycled fuel fabrication plant near Avignon. It was like a trip through Narnia. What only exists in theory on this side of the Atlantic is everyday reality over there.

Moreover, the people I met didn't express any particular smugness. "We're just putting into practice what you invented," they said over and over. They freely acknowledge the technological advances of mid-twentieth century America and—if anything—wish we would do more ourselves. In fact, Areva, the French nuclear company, is returning the favor by building a reprocessing plant at the old Barnwell site in South Carolina and proposing two new reactors at Calvert Cliffs, Maryland, in conjunction with Constellation Energy. The very day we sat talking with executives in Paris, Areva announced plans to build a new $3 billion uranium enrichment facility in Idaho Falls—another important link in the nuclear chain. The announcement caused a stir in France but didn't even make the Associated Press in this country. Stories in the Idaho papers were accompanied by the inevitable declaration from the group called the Snake River Coalition that no such a temple of idolatry would ever be built in their Garden of Eden.

Briefly, here's what nuclear power has done for France. It provides 80 percent of the country's electricity at the lowest rates in Europe. It gives France the second-lowest level of carbon emissions in Europe, behind only Sweden, a smaller country that is half hydro, half nuclear. It provides France's third largest import, behind only wine and agricultural products. It allows Germany and Denmark to posture that they are anti-nuclear while they have quietly abandoned the vow to shut down their own reactors and are importing nuclear electricity from France. Italy, much less skilled at this hypocrisy, actually closed down its three reactors, leaving the country with frequent blackouts. Italy now imports 70 percent of its electricity and horrified Europe in the spring of 2008 by announcing plans to build several new coal plants. When it switched plans to nuclear reactors two months later, Giuseppe Onufrio, director of Greenpeace Italy, called it a "declaration of war."[1]

In short, by marching to Charles de Gaulle's own drum, France has solved its energy and environmental problems and is about to show the rest of the world how to do it as well.

I have always found the first few hours of a trip to be the most vulnerable. When I went to Iraq a year ago, I arrived in Kuwait International Airport only to find a sign saying my baggage would not be arriving. I called British Airways and five minutes later realized my laptop was gone as well. In less than twenty minutes, I had lost 80 percent of my possessions. Fortunately I found my laptop sitting atop the phone booth where I had called BA. My baggage never arrived, however, which is why—thanks to small claims court—I am now flying to Paris for free.

Sure enough, this trip starts with a near-disaster as well. My tour guide for the trip is Jarrett Adams, a young American who has just joined the public relations staff at Areva's American headquarters in Baltimore. I am supposed to meet him at the Blue Train in Station St.-Lazare on the West side of Paris. From there we will catch the afternoon train to Avignon. I assume the Blue Train is a *train* and—through some misunderstanding of my bad French—my cabdriver ends up taking me to the Gare du Nord on the other side of the city. By the time I realize I am in the wrong place it is getting perilously close to departure. Our cell phones don't work yet but after a few frantic emails on hotel wireless cards, we finally rendezvous at the Blue Train—which is a *restaurant*—at Gare St. Lazare. It is amazing how quickly a lack of language can turn you into a lost child.

By late afternoon we are whizzing south on a "bullet train" that will make the 400-mile journey in a remarkable three hours. It is already getting toward the solstice and at this latitude there is not a trace of twilight as we rocket through the countryside. The most common crop is fields of a yellow flower that makes the whole landscape look like a well-tended garden. Our seatmate says it is a variety of hay fed to cattle. Further south, vineyards begin to prevail.

As the fields fly past, Adams fills me in on the French nuclear effort. "It started with Charles de Gaulle," he says. "You may recall, the French decided to build their own bomb in the 1950s and that meant they had to start enriching uranium. They were very smart, however, and saw the energy potential. France never had much coal or oil. They had a small foothold in the Middle East and North Africa but saw that slipping away as well. Long before America had its oil awakening, they were anticipating energy problems.

"After the Arab Boycott, the French government made an all-out commitment to nuclear. There was growing opposition worldwide but the government felt it didn't have any choice. They had a slogan. `We don't have any oil, but we have ideas.' That's the reason they're so far ahead of us. Now they're building for other countries. The Finns had a huge national debate from 2000 to 2002 when they realized they needed more energy. They decided on nuclear. Areva got the contract. It's the first European reactor built in twenty years. They're also selling to China, Africa and the Middle East."

As we approach Avignon, the landscape becomes mountainous. On one of the last ridges, we see our first windmill silhouetted against the sunset. French Premier Jean Sarkozy recently commented he thought France had done well to spare itself the sight of wind farms cluttering the landscape. At the same time, I have found the most pleasing vistas of nuclear cooling towers to be the ones taken against the background of French countryside.

Billing itself as the "Headquarters of the French Popes," Avignon is now a walled city besieged by tour buses. Our hotel claimed to be part of the French Pope's original complex, although it certainly had undergone a few renovations since the fourteenth century.

In the morning we head for the Melox plant in suburban

Marcoule. A very horizontal industrial establishment, it sits atop a small plateau overlooking the Rhone about 35 miles to the northwest. The building's glass-and-steel façade is striped with individual red and yellow lines that the brochure says are designed to make it blend with the surrounding landscape. Security is tight and the all-encompassing barbed-wire fences give it the feel of a low-grade prison. Belying this atmosphere, the walkways glisten with Mediterranean sunshine and knots of workers stroll casually in Areva's spanking white jumpsuit uniforms.

After threading our way through numerous identification procedures and radiation checks, we are ushered into an upstairs conference room where we meet Pierre Guelfe, chief engineer of the facility. Typically, Guelfe has a company Power Point ready to roll and—even more predictably—can't get the projector to work. I recall the story of Lou Gerstner's first tour through IBM, when every executive put up a slide show. Finally Gerstner walked up to one vice president, turned off the projector and said, "Just tell me about your business." Within twenty-four hours word had circulated among IBM offices all over the world—Gerstner had "turned off the projector." Taking my cue, I ask Guelfe to forget the Power Point and start peppering him with questions.

"What's the main operation you do here?" I ask.

"Our biggest task is recycling spent fuel from La Hague," he begins. "When the depleted fuel rods are removed from the reactors, they are shipped to La Hague for reprocessing. They let it cool down for a few years and then remove the uranium and plutonium. They ship the plutonium here. We take it and mix it with another stream of material, which is the scrap that is left over from uranium enrichment. The U-235 content of this is very low—as you know, U-235 is the fissionable isotope—"

"I know."

"But the plutonium is much more fissionable than the depleted uranium. So when we mix them together—you get a fuel that is very close to enriched uranium. It's called `Mixed Oxide Fuel'—`MOX.' We have twenty reactors here in France running on MOX and there are ten more in Germany and two in Switzerland. So we're putting plutonium and scrap uranium together. We use everything. We don't leave any waste."

"There's one thing I want to ask," I said. "I've read this several times but I want to make absolutely sure. The plutonium that comes out of a commercial reactor, that you separate from the fuel rod, that *cannot* be used to make a bomb, right?"

"That's right," he nodded. "You have four plutonium isotopes—Pu-239, Pu-240, Pu-241 and Pu-242. Of the four, only Pu-239 can sustain a chain reaction. The others are contaminants. The Pu-241 is *too* highly radioactive. It fissiles too fast so you can't control it to make a bomb. But you can use all of them to sustain fission in a MOX reactor."

I lean back for a second. "I don't know whether you know all this—I'm sure you do—but we completely ended reprocessing in the United States in the 1970s on the premise that if we extracted plutonium someone might use it to make a bomb. We were saving the world from nuclear proliferation. But in fact, from what you're saying, this is all wrong. You can't use plutonium from a commercial reactor to build a bomb?"

"You have to have a special kind of fast reactor that breeds only Pu-239," he said. "That's what the North Koreans did."

"So we've created this whole huge problem of `nuclear waste' on a false premise. And we're building this huge complex at Yucca Mountain basically for no purpose."

He gives a little Gallic shrug and smiles under his mustache. "That's right," he says. "The other thing we're doing here is recycling former Soviet bomb material," he continues, returning to his Power Point. "When the Soviets were making weapons they decided to stockpile all kinds of enriched uranium. Then the Soviet Union collapsed and all this bomb material was left sitting around. No one knew where half of it was and there were a lot people who didn't have any money and would be happy to sell it. So two American Senators—"

"Peter Domenici and Sam Nunn."

"Right. They negotiated a deal where we would `de-enrich' all this uranium down to reactor grade. And that is what we are doing now. They ship it to us here, we mix it with more scraps from uranium enrichment, and then we ship it to you. What is it they say? `One out of every ten light bulbs in the United States is being lit by a former Soviet weapon."

"You think people would be dancing in the streets."

"Swords into plowshares," he says with another little Gallic grin, "just like in the Bible."

We don our Areva jumpsuits and take a tour of the plant. As soon as we enter the first room we encounter a seven-foot cylinder painted yellow. "This is plutonium," says Marty Delphin, our guide. "It just arrived from La Hague."

I put my hand up against the container. Sure enough, it is warm. "So this is the dreaded plutonium?" I say. Delphin nods. "Plutonium is not a gamma emitter, right?"

"Just alpha," he says.

"Feels like energy!" I say.

"And it can't blow up!" adds Guelfe. Everyone laughs at that.

The working part of the plant is so narrow and cramped that it feels like the bowels of a submarine. At every turn there are glass compartments behind which some mechanical task is being performed in pelletizating of fuel. It is all somehow reminiscent of aquariums. Each unit has a glove box where workers can reach in and manipulate the process if something goes wrong.

"We have a different fuel for each country," explains Delphin. "Every reactor has a slightly different design so we have to mix the fuel to very tight specifications. These ones are headed for Japan," he says, indicating a sealed container.

Like the original uranium fuel, the MOX rods are only mildly radioactive. "Don't get too close," says Delphin as I start to do an inspection. "Just a precaution," he says apologetically.

I want to see the Russian material but we must rush back to catch the afternoon train. We're due in Paris for another round of appointments tomorrow morning.

The next day we meet at Areva's headquarters with Arthur De Montalembert, vice president for international and marketing and Jacques Besnainou, vice president for the back-end sector. Anne Lauvergeon, the president of Areva and rated one of the ten most powerful women in the world by *Time*, has a busy schedule and was unavailable for an interview.

De Montalembert is the kind of suave, debonair Frenchman you would expect to encounter anywhere in the upper reaches of the national bureaucracy. He talks in measured tones, the gray at his temples adding gravity to his words. Besnainou, on the other hand, is a pudgy, ebullient Sancho Panza with a hawk nose and a bubbling enthusiasm that counters de Montalembert's reserve.

They start showing us a Power Point and sure enough, it is the same one we saw at Avignon. "We already saw it," I say.

"Well, you've come at a very opportune moment," begins De Montalembert, quickly recovering his composure. "We're just announcing our new uranium enrichment plant in the United States. We've been considering five sites and we're now picking the winner."

"What were the five sites?"

"Columbus, Ohio, where you already have an enrichment plant, the Argonne Laboratory, near Chicago, Los Alamos, Hanford, and Idaho Falls, near the national laboratory," he says. I feel good. I've already visited three.

"So who's the winner?"

"I can't say until the announcement is made," De Montalembert smiles enigmatically.

"I promise I won't phone my rewrite desk." He smiles again but is not tempted. (It was Idaho Falls.)

"We are also building a reprocessing plant in South Carolina as part of the GNEP effort," he continues. "And we have plans to build two new reactors at the Calvert Cliffs site in Maryland."

"Do you think you'll run into opposition on any of this?" I ask.

"We know we will have the academy against us," he demures, choosing a nice French word to describe the universities. "I am just going down to debate Mr. Frank von Hippel of Princeton this afternoon. He says he is in favor of nuclear power but is against reprocessing."

"Basically, you've dropped the ball," chimes in Besnainou enthusiastically. "The U.S. was the leader in this technology for a long time but then you stopped. Glenn Seaborg invented the process for extracting plutonium in 1944. It was very difficult. He tried hundreds of solvents before he found one that worked."

"Sounds like Thomas Edison."

"Right. So the process hasn't changed at all since then," he says cheerfully. "You provided the theory. We've just done the engineering. We're keeping the torch alive."

"The younger generation will probably be different," adds de Montalembert. "Anne Lauvergeon spoke at Harvard last winter and there was a lot of opposition from the students. So she invited a group to come over and see what we are doing. We'll be showing them around next month."

"It's difficult to see why this is so hard to sell to America," says Besnainou with only a touch of regret. "You recycle household garbage. You're very good at that. Why not recycle spent fuel as well? We've cut our need for uranium 30 percent by reprocessing. There's so much energy left in there, we're calling spent fuel `the new uranium mines.'"

The next day we make our final visit to the reprocessing plant at La Hague. This time it's only 120 miles through the beautiful French countryside but takes two hours since the train doesn't travel as fast. As we approach the coast of the English

Channel, the landscape begins to undulate and I try to imagine American soldiers fighting along the hedgerows. As the train slows, I glance through the publicity brochure from the Paris headquarters:

> *Source of Life.* Since that gigantic nuclear explosion, origin of the universe, called the Big Bang, matter and energy have remained a united and faithful duo. Man himself is stardust.... Today the stars, the sun, and the burning core of the earth are ceaseless beds of nuclear reactions.... From distant stars to the earth's core, it continues its constructive work. Man has learnt to master one nuclear reaction, fission, taming it into a clean and inexpensive energy.

They understand theses things over here. Why can't we understand them as well?

The La Hague Reprocessing Center sits atop of a cliff just outside Cherbourg overlooking the English Channel. It's the same horizontal design, trying to look unobtrusive, with the prison-fence surrounding. Security here is even tighter, with an endless round of ID cards and personal codes that change every time we enter a new sector of the plant. Our guide is Christopher Naugnot, a mid-30s, brush-cut communications director whose English—like everyone else's—is very good.

Naugnot leads us into a conference room and wouldn't you know, there is the same old Power Point ready to roll. "We've got it memorized by now," I tell him. So he provides us with some brief details—the facility employs 5,000, 20 percent of the jobs in Cherbourg, it's been operating for twenty years, the locals love it. (We were to meet with the Mayor of Cherbourg but he had to cancel because of one of the ubiquitous French holidays.) "The U.S. has produced 50,000 tons of spent fuel and has designed Yucca Mountain to hold 70,000 tons," says Naugnot. "We've already recycled 24,000 tons at this facility."

"Why don't you just take all our fuel off our hands?" I ask.

"You're producing about 2,000 new tons a year. We can

only reprocess 1,700 tons a year at this facility so we'd be hard pressed. We'd much rather recycle in the United States."

"Are you going to do it?"

"That's what we're trying to do in South Carolina."

We don the familiar jumpsuits and start through the facility. Naugnot explains how the spent fuel casks are shipped by rail and offloaded into trucks, which bring them to the plant. "They enter through the basement and are then raised into a sealed room where they are cooled for a while before going into the storage pools. We'll see that next."

We climb a stairs and find ourselves standing outside a window looking through a thick yellowish glass into a room perhaps about 2,500 feet square, brilliantly lit from the inside and filled with about as much equipment as the average weight lifting room.

"Why is the glass yellow," I ask with one of those innocuous questions that usually lead someplace.

"It's treated with lead, for shielding the radiation."

Then suddenly, there it is before us. Like some sinister benthic organism being hauled out of the deep, a complete fuel assembly is slowing rising out of the floor, lifted by an overhead crane, until it reaches its full height of about thirty feet. With its steel frame and vertical black lines—the embedded fuel rods—it looks eerily like a miniaturized version of the World Trade Center. Yet its blank featureless face gives it the same soulless menace of the eye of a shark.

"What's the radiation coming out of that thing?"

Naugnot consults quickly with a nuclear engineer who speaks only French.

"Un million millirads," says the engineer. A million millirems. Quick calculation—that's 1000 rems. The highest exposure people got standing near ground zero at Hiroshima was only 500 rems. This is truly the most dangerous material on earth. Yet here we are, standing fifteen feet away, perfectly shielded by a foot of lead-laced glass. If we suffer the slightest exposure, our radiation badges will catch it when we leave.

"What happens if something goes wrong in there?" I ask.

"Right here," says Naugnot. On either side of the window there are two handles that attach to long mechanical arms extending twenty feet into the room. "You should see those guys work these things," he says. "We should have brought someone down to show you."

"How long has it been since someone was in that room?"

"Not since it was built. And they won't be in there again until years after it's decommissioned. If you walked in there now, you'd be killed instantly." But we have that concrete wall and leaded glass between us.

The next stop is the "swimming pool," a larger version of the storage pools that hold spent fuel at every American nuclear reactor. This one appears Olympic size. The blue Cerenkov glow is fainter, giving it the tint of one of those horrid children's Kool-Aid flavors. As we scan the perimeter I suddenly notice something wildly incongruous—life preservers lining the guardrail about every twenty yards.

"Anybody ever fall in?" I ask.

"Not yet," says Naugnot. "But if they did they wouldn't be harmed. The water protects you." I tell him the story about American workers occasionally taking a swim in the storage pools, much to the consternation of the NRC.

"We don't allow anything like that here," he says disapprovingly.

"How long do the rods stay in there?" I ask.

"Up to fifteen years."

"Fifteen years? Wow, I thought it would be a few months. So all this happens on a completely different time scale," I say.

"It takes a long time to get one of these operations started," he says. "But once you get going, it runs pretty smoothly. We don't have many lulls."

Next we see more behind-the-glass operations of the fuel assemblies being sliced into small sections and dissolved in nitric acid. Another solvent—Glenn Seaborg's discovery—then extracts the uranium and plutonium, which is shipped off to Avignon. Other actinides are drawn off by similar procedures.

What remains after that are the fission products—cesium, strontium and the others—all highly radioactive and compact. These are "vitrified"—dissolved into a molten glass that hardens the isotopes into solution. "The glass will last for thousands of years," says Naugnot. "There was some buried with the Egyptian Pharaohs that hadn't changed when the tombs were opened. The glass isn't protective—you need a lead-and-steel container for that." It is this vitrified material that is put in final storage.

And now at last we find ourselves standing in that "one room" in La Hague where the French keep all the nuclear waste from twenty-five years of producing 80 percent of their electricity. I have been talking about this room for months. Now I am standing in it.

It is a bit larger than I imagined. Somehow I had seen it as about the size of a small visitors' center. Instead it is more like a large basketball gymnasium. Still, it is one room. Lined up in the concrete floor are about forty manhole covers stenciled with Areva's triangular logo. All are so tightly sealed with no visible handles, that I wonder how they could be removed.

"They're magnetized," Naugnot explains, pointing to the ceiling. "See this large—how do you say it in English—"

"Gantry?"

"Yes, gantry. There's a magnetized crane that removes them. Inside the plug there's another cap with handles. The crane can grasp them as well. The canisters are very small. There's room for six in each ring. They're stacked six-deep beneath the floor. The total material stored here is ten grams for each French citizen—about the weight of a two-Euro coin."

And that is the sum total of what the French call "les dechets"—their nuclear waste. Even this storage is only temporary. The material can be retrieved any time the French Parliament decides it is economical to recycle more of the radioactive isotopes. The entire environmental footprint of twenty-five years of producing the France's electricity—the equivalent of all those sulfur sludge piles and billions of tons of carbon dioxide hurled into the atmosphere—is right here beneath our feet.

Nuclear power is so different from anything we have ever encountered before that it is still difficult for us to comprehend. The French have proved in practice what we can still only say in theory—there is no such thing as nuclear waste.

On September 25, 2007, NRG, of Princeton, N.J., filed the first license application since 1978 for a new nuclear power station in the United States. The plants would be built in Texas as merchant facilities—that is, without the guaranteed profits of regulated utilities. "For the first time in my lifetime, taking a step forward on energy will not mean taking a step back on the environment" said NRG President David Crane in making the announcement. (He's the one who says electric cars will be the best thing for the electric industry since air conditioning.)

> As we move forward, it is important to recognize that the nuclear industry of today is not the nuclear industry of the 1970s. Consider the improvements in design, construction and training and most important, the zero-tolerance safety culture of the American nuclear power industry today. Look at the spectacular improvements in operating performance and the self-policing that has achieved this improvement. Most important, consider the environmental alternative. In Texas this year we faced the prospect of 18 new coal plants, which would add as much carbon dioxide to the atmosphere as the entire country of Argentina. It is these three factors that convinced me that in aggressively pursing this project, we are pursuing the right course.[2]

Senator Pete Domenici, the Senator from New Mexico and a long-time supporter of nuclear, added, "It is my judgment that any effort to reduce carbon emissions in this country must include nuclear power. Nuclear power is clean, efficient, safe and across the globe other countries are taking advantage of it. This is not just a great day for nuclear power but for our environment as well."

Since that day, proposals for eight other reactors have come before the NRC, with twenty more in various stages of preparation. All this has flown beneath the radar but it will undoubtedly burst into the public consciousness very soon. Within the next few years we may be replaying the nuclear debates of the 1970s and 1980s—but this time with global warming at stake as well.

Can America embrace terrestrial energy, the greatest scientific discovery of the Twentieth Century, in order to solve our Twenty-first Century problems? It will all depend on public opinion.

America is not France or China, both of which are moving ahead rapidly with nuclear technology. In these countries, decisions are made by an elite with very little input from ordinary citizens. In America it will be completely different. There is not the slightest chance that the government or the "nuclear industry"—if such a thing exists—will be able to force the technology down the public's throat. It will happen only if Americans embrace nuclear power. If we are to do this, it will be because people have finally overcome their fears and are able to recognize terrestrial energy for what it is—a perfectly natural process that, although not without its risks and dangers, offers us the possibility of solving our energy problems without degrading the environment.

What will be the arguments against employing terrestrial energy? In recent years they have fallen into two general categories: (1) nuclear power is too dangerous, particular in an era of possible terrorist attacks; and (2) nuclear power is too expensive and will take too long to materialize.

The first argument plays on lingering fears of "another

Chernobyl" but adds the new element—that a reactor could be attacked with a hijacked airliner or some other weapon. Groups trying to shut down New York's Indian Point (which provides 20 percent of New York City's electricity) regularly argue, "If the September 11th terrorists had taken a turn and crashed their jetliner into Indian Point, the toll would have been much greater than at the World Trade Center."

Sadly, if the September 11th hijackers had tried to take out Indian Point it would have been a blessing. The results would have been a spectacular plane crash and nothing else. These thousand people might have lived.

In the 1990s, the Air Force attached a Phantom F4 jet to a railroad track and rammed it into a concrete barrier the thickness of a containment structure at 500 mph. The jet completely vaporized while the concrete wall was barely dented.[3] As Ted Rockwell, Admiral Hyman Rickover's biographer, puts it, "If you wanted to construct a battering ram, the last thing you'd use is a hollow metal tube." Critics argue that the plane was "only a jet fighter" and not a full-sized airliner, but what difference would it make? Airliners rarely travel 500 mph until they reach the upper atmosphere. A pilot trying to aim a jet liner at a target the size of a containment structure would barely be fly it at more than 200 mph so the impact would be far smaller. Nor would the explosion of several large fuel tanks make any difference. Jet fuel explosions come nowhere near the melting point of concrete and the heat would dissipate within a few seconds. Even then, the containment structure is only the *first* barrier. The solid steel reactor vessel would be equally impregnable.

"Well, then, they could aim it at the spent fuel pools," say the critics. True enough. Except that this would be like landing a jet liner in a backyard swimming pool. Even the best pilots would find it difficult. Most storage pools are inside a containment structure anyway. Even then, the fuel would only be exposed and would not burn. Chernobyl burned for four days only because its carbon moderator was so highly combustible.

There is nothing remotely similar at a storage pool. In any case, there will almost certainly never be another such terrorist attack. Remember, only three of the four September 11th hijackings were successful. On the fourth, United Flight 93, the passengers—knowing the other planes had crashed—revolted. The same would happen today.

Helen Caldicott, the venerable Australian opponent of nuclear power, has come up with the only really plausible scenario. In *Nuclear Power is Not the Answer*, her 2005 book arguing against reviving nuclear in order to cope with global warming, she writes:

> A 1000-kiloton bomb (1 megaton) landing on a standard 1000-megawatt power reactor and its cooling pools, which contain intensely radioactive spent nuclear fuel, would permanently contaminate an area the size of western German. [4]

True enough. But then a 1,000-kiloton bomb landing on *anything* would create a considerable amount of radioactive fallout. We'll just have to live with the risks.

The final argument against nuclear—now favored by opponents who want to appear practical and hardheaded—is that it costs too much. Often this is expressed with a faux sense of regret. Yes, nuclear power might play a role in preventing global warming but—too bad!—it's too expensive. This may be the first time in history that environmental groups have argued it is *not* worth spending money to protect the planet, but nevertheless. Most of this follows the Amory-Lovins-type vein, which says that we cannot pursue both nuclear *and* solar but must choose one or the other.

The utility industry is ready to go. At the 27th annual convention of Daniel Yergin's Cambridge Energy Associates in Houston in 2008, Alan Greenspan gave the convocation address. When he made an offhand remark that it was time to revive nuclear power, the entire room of more than a thousand utility executives and energy investors burst into thunderous applause.

Yet one question remains. Who will put up the money for the $5–7 billion needed to build a new reactor? So far Wall Street has shown very little interest. A few days after 2008 Republican Presidential candidate John McCain announced support for 45 new reactors, Admiral Frank "Skip" Bowman, executive director of the Nuclear Energy Institute, wrote an editorial stating that the federal government must ante up.

> These first projects need financing support—federal loan guarantees and assurances from state government that the company will be able to recover its investment.... Our nation will need something similar to the Clean Energy Bank concept being considered by some in Congress. This would be a government corporation providing loan guarantees and other forms of financial support to ensure capital for deploying clean-electricity technology.[5]

This isn't going to work. If nuclear can't stand on its own, it isn't worth doing. In fact, Wall Street is showing signs of interest in investing yet there is one thing holding it back. That is the nightmare scenario that as soon as one of these new plant proposals comes out of the starting gate, it will be immediately gang-tackled by environmental organizations and nuclear opponents who will tie it up in court and regulatory proceedings for the next fifteen years. Then we will be right back where we were in 1985. That is why the Bush Administration—which has done a lot for nuclear power—put through "regulatory insurance" in the 2005 Energy Policy Act.

The threat of endless regulatory and legal delay is no small matter. The Sierra Club, Natural Resources Defense Council and Environmental Defense have become major economic players. When KKR bought out TXU in 2007 and cancelled eight of eleven coal plants, it moved forward only by bringing aboard Environmental Defense as part of the coalition. The Sierra Club and NRDC constantly intervene in court and few projects can go ahead without their approval. All are lavishly funded by non-profit foundations and stocked with powerful legal staffs that can easily wage a war of attrition against any new energy proposals. Greenpeace International has a $150 million annual budget (the World Trade Organization only spends $100 million) and has influence all over the world.[6]

But that will not be decisive, either. The real battle of whether the greatest scientific discovery of the Twentieth Century can be used to resolve the problems of the Twenty-first Century will be fought in the court of public opinion. Will the public reflexively cringe thinking that a nuclear power plant is a bomb waiting to go off? Will it accept the scenario that one terrorist with a carving knife can set off a nuclear holocaust? Will it believe that every picocurie of radiation found in the vicinity of a nuclear reactor is the harbinger of a cancer epidemic? Will it recognize that because of its incredible energy density, the environmental footprint of terrestrial energy is infinitely smaller than the oceans of waste products produced by fossil fuels or the vast amounts of land that must be employed to gather solar energy?

The question will be for you to decide. Just as the Finns spent two years in national debate before deciding to go ahead with the Olkiluoto 3 Nuclear Plant at Eurajoki, so America will soon be engaged in an almost identical discussion over the future of terrestrial energy. I hope the material in this book has been helpful. The decision is up to you.

Epilogue

The Supernova

At later stages in their lives, stars also forged nitrogen, oxygen, and even iron. Then, literally exhausted, they expired and exploded, spewing their bounty of new elements into space. The largest and brightest stars bequeathed to the universe the heaviest of elements, including gold and uranium. Thus the stars carried on the work of Creation, hammering out a wide range of raw materials for future use.

—Dava Sobel, *The Planets*

I have spoken to this point about the power in the nucleus of the atom nuclear as "terrestrial energy," but that does not describe its true origins. The story goes back much further than the earth. It is time, at last, to trace terrestrial energy all the way back to its source—a star, in fact, not unlike our sun, but much bigger—a supernova. The energy stored in all the large atoms on earth is actually the residue of a star that exploded about 4.5 billion years ago.

You will recall that hydrogen, the simplest atom of all, is the most abundant element in the universe. In fact, for several billion years after the Big Bang, there was nothing except hydrogen and a little helium across the vast reaches of space. When enough of this hydrogen was drawn together by gravitational forces, however, the first stars started burning and nuclear fusion began its work of forging larger elements—carbon, nitrogen, and oxygen. As Hans Bethe, the great Nobel physicist, proved in 1937, the fusion of hydrogen into helium within our sun is actually a complex process that uses carbon and oxygen as catalysts.

As we survey the earth around us, however, we now find no free hydrogen and very little helium. Instead, we find the marvelous cornucopia of other elements from periodic table—sulfur, tin, phosphorous, titanium, bismuth, lead, mercury, aluminum—right up to the heaviest and most unstable of elements, thorium and uranium.

As physicists unraveled the mysteries of fusion and fission in the twentieth century, the question arose, where did all these larger elements come from? At first it was suggested they all formed in the Big Bang. The rate of uranium's radioactive decay indicated it was created 5 billion years ago and in 1950 this was believed to be the age of the universe.

As bigger telescopes revealed more and more distant galaxies, however, it became clear that the universe was much older. As theorists pinned down the details of the Big Bang, it also became clear that the temperatures and pressures needed to achieve the fusion of the largest atoms probably had not been achieved. Only hydrogen and helium were present at the creation.

The thermonuclear fusion in stars like ours explained how smaller atoms such as carbon, nitrogen and oxygen were formed. Still, that left most of the periodic table unexplained. The pressures and temperatures in a collapsing star could explain the fusion of elements up to iron. But all the elements above iron—the most stable nucleus in the universe—remained a mystery. How had they been created and then scattered across the universe so they could end up in such metal-heavy cinders as planet earth?

The only explanation could be that much of the earth—dashes of copper, nickel, zinc, silver, and gold—originally formed in the explosion of a "new star," the kind of "supernova" that has appeared a few times in human history but is now known to occur somewhere in the universe almost every second.

A supernova is star about three times the size of our sun that consumes its hydrogen very quickly and then runs out of fuel. Without the fusion energy pushing outward, the star collapses inward by force of its own gravity. Under these extreme pressures, nuclei are forged together into elements all the way up to iron, with twenty-six protons.

These heavier elements start to separate, with the heaviest tending toward the core. Then, as fusion energy again winds down and gravitational forces continue their contraction, the whole star collapses to a sphere only eighteen miles in diameter with three times the mass of our sun. As the elements are forced together, nucleus to nucleus, fusion energy reaches a flash point and the entire star bursts outward in a thermonuclear explosion that produces enough light to outshine an entire galaxy for more than a week.

In that single instant, all the elements heavier than iron are formed—sixty-six in all. Undoubtedly many transuranic actinides were formed as well—neptunium, plutonium, americium, curium, and so forth. Evidence of some of them has been detected in supernova explosions. Because of their relatively short half-lives, however, they have all long since disappeared. Thus, our solar system is a "second-generation" complex, formed out of the debris of another star that exploded 4.5 billion years ago.

And so at last we come to the origin of terrestrial energy. All the energy stored in the nuclei of the heavier elements on earth is the residue of this explosion. Terrestrial energy is really "solar energy," but from a different star. Yet the magnitude of that explosion was so vast that it has left us with enough stored energy to last for millennia. As James Lovelock, originator of the "Gaia" hypothesis, puts it:

> Perhaps the strangest thing about the Earth is that it formed from lumps of fall-out from a star-sized nuclear bomb. This is why even today there is still enough uranium left in the Earth's crust to reconstitute on a minute scale the original event.[1]

Our planet is a strange and unique place. While the universe is made up mostly of hydrogen and helium, on earth, oxygen, silicon, carbon and nitrogen are the dominant elements on the planet. Scattered among them, however, are traces of two tiny elements—uranium and thorium—that, as the glowing embers of the supernova, will be enough to provide humanity with enough energy for ages and ages to come.

Terrestrial energy. It is part of the plan.

Notes

Prologue

1. Alvin Weinberg, "Social Institutions and Nuclear Energy," *Science*, January 7, 1972:27.

Chapter 1

1. IPCC, "Executive Summary," Fourth Assessment Report, February 2007.
2. Bjorn Lomborg and Flemming Rose, "Will Al Gore Melt?" *Wall Street Journal*, January 18, 2007.
3. "Kyoto Protocol: France Trims Carbon Pollution in 2005," Agence France-Presse, January 11, 2007.
4. "The Kyoto Protocol: A Post-Mortem," *New Atlantis*, Winter 2004.
5. S.K. Solanki and M. Fligge, "Solar Irradiance since 1874 Revisited," *Geophysical Research Letters* 25, no. 3 (February 1, 1998): 341.
6. Nir J. Shaviv, "Carbon Dioxide or Solar Forcing?" http://www.sciencebits.com/CO2orSolar (emphasis in original).
7. David Deming, "Global Warming, the Politicization of Science, and Michael Crichton's *State of Fear*," *Journal of Scientific Exploration* 19, no. 2 (June 2005), www.scientificexploration.org/jse.php.

8. Michael E. Mann and others, "Global-scale Temperature Patterns and Climate Forcing Over the Past Six Centuries," *Nature* 392, no. 23 (April 1998).

9. Ross McKitrick, "What is the 'Hockey Stick' Debate About?" APEC Study Group, April 4, 2005:12.

10. S. McIntyre and R. McKitrick, "Corrections to the Mann et al. Proxy Data Base and Northern Hemisphere Average Temperature Series, 1998," *Energy & Environment* 14 (2003): 751.

11. J.R. Petit and others, "Climate and Atmospheric History of the Past 420,000 Years from the Vostok Ice Core in Antarctica," *Nature* June 3, 1999:429.

12. Lomborg, *Cool It*, 18.

13. Ibid., 38.

14. Ibid., 109–110.

15. National Research Council, *Abrupt Climate Change: Inevitable Surprises* (Washington, DC: National Academy Press, 2001).

Chapter 2

1. Teller, *Energy from Heaven and Earth*, 4ff.

2. Vaclav Smil, *Energy in World History* (New York: Westview Press, 1994).

3. William Manchester, *A World Lit Only By Fire* (Boston: Back Bay Books, 1993).

4. James F. Kasting, U.S. Global Change Research Information Offices, "The Carbon Cycle, Climate, and the Long-Term Effects of Fossil Fuel Burning," *Consequences* 4, no. 1 (1998).

5. Einstein, *Relativity*, 56.

Chapter 3

1. "Reactor Safety Study," Atomic Energy Commission, 1974.

2. David Suter, "All Things Being Equal," *Harper's*, June 1979.

3. See Bernard L. Cohen's excellent chapter, "The Fearsome Reactor Meltdown Accident," *The Nuclear Energy Option*, 73ff.

4. Svetlana Alexievich, *Voices of Chernobyl* (New York: Picador, 2006).

5. See for example, Zhores Medvedev, *The Legacy of Chernobyl* (New York: W.W. Norton, 1992).

6. D. Kinley III, ed., "Chernobyl's Legacy: Health, Environmental and Socio-Economic Impacts; The Chernobyl Forum 2003–2005," 2nd ed., International Atomic Energy Agency, 2005, 7.

7. Dr. David Santillo and others, eds., "The Chernobyl Catastrophe: Consequences on Human Health," Greenpeace, April 2006, 10.

Chapter 4

1. Barbara Tuchman, *A Distant Mirror: The Calamitous 14th Century* (New York: Alfred A. Knopf, 1978), 24.

2. Freese, *Coal: A Human History*, 33.

3. Ibid., 35.

4. Ibid., 37–38.

5. Ibid., 41–42.

6. Frederick Engels, *The Conditions of the Working Classes in England* (Moscow: Progress Publishers, 1980), 84.

7. Anthony Wohl, *Endangered Lives: Public Health in Victorian Britain* (Cambridge: Harvard University Press, 1983), 56.

8. Goodell, *Big Coal*, xx.

9. David McCullough, *The Great Bridge: The Epic Story of the Building of the Brooklyn Bridge* (New York: Touchstone Books, 1972), 351–352.

10. Suellen M. Hoy, "Municipal Housekeeping: The Role of Women in Improving Urban Sanitation Practices, 1880–1917," *Pollution and Reform in American Cities, 1870–1930*, Martin V. Melosi, ed. (Austin: University of Texas Press, 1980), 173–198.

11. David Stradling, *Smokestacks and Progressives: Environmentalists, Engineers and Air Quality in America, 1881–1951* (Baltimore: Johns Hopkins University Press, 1999), 17.

12. Bureau of Reclamation, "The History of Hydropower Development in the United States," United States Department of the Interior, www.usbr.gov/power/edu/history.html.

Chapter 5

1. www.umwa.org/job/blacklung.html.

2. Goodell, *Big Coal*, 158ff.

3. Esposito, *Vanishing Air*, 241ff.

4. Ibid.

5. "Cases of Contamination from Coal Combustion Waste," January 31, 2003, http://www.hecweb.org/Programs%20and%20Initatives/CCW/cases.htm.

6. See for example, "The Cap and Trade Success Story," Environmental Defense Fund, http://www.environmentaldefense.org.

7. Bill Hornaday, "Not enough coal on hand, utilities warn threat of rolling blackouts highlighted in industry letter," *Arkansas Democrat-Gazette*, May 6, 2006; Rebecca Smith, "Railroads Struggle to Deliver Coal to Utilities," AP Business Wire, March 15, 2006.

8. Robert Socolow, "Can We Bury Global Warming?" *Scientific American*, July 2005.

9. Robert Socolow, in discussion with the author, September 2006.

Chapter 6

1. David White, "The Petroleum Resources of the World.," *The Annals*, May 1920, 121.

2. Garret Hardin, "The Tragedy of the Commons," *Science* 162 (1968): 1243–1248.

3. "Horseless Carriage Days," Museum of American Heritage, Palo Alto, California, http://www.moah.org/exhibits/archives/horseless.html.

4. Anthony Sampson, *The Seven Sisters: The Great Oil Companies and the World They Shaped* (New York: Bantam, 1976).

5. Deffeyes, Hubbert's Peak, 4.

6. David Potter, *People of Plenty* (Chicago: Phoenix Books, 1954).

Chapter 7

1. Harold Barnett and Morse Chandler, *Scarcity and Growth: The Economics of Natural Resource Availability* (Baltimore: Johns Hopkins University Press, 1963).

2. Ibid., 11.

3. Julian Simon, *The Ultimate Resource* (Princeton: Princeton University Press, 1981); Paul Ehrlich, *The Population Bomb* (San Francisco: Sierra Club Books, 1969).

4. Freeman, ed., *A Time To Choose.*

5. Nicholas Wade, "Synfuels in Haste, Repent at Leisure," *Science* 205, no. 4402 (1979): 167–68.

6. Colin Campbell and Jean Laherrere, "The End of Cheap Oil," *Scientific American*, March 1998.

7. "Crude Oil—Uncertainty about Future Supply Makes It Important to Develop a Strategy for Addressing a Peak and Decline in Oil Production," GAO, February 2007.

8. "Congressional Peak Oil Caucus," (press release), March 29, 2007.

9. Simmons, *Twilight in the Desert*, 70.

10. Russell Gold, "As Prices Surge, Oil Giants Turn Sludge Into Gold," *Wall Street Journa*l, March 27, 2006.

Chapter 8

1. http://lsa.colorado.edu/essence/texts/naturalgas.html; http://www.pge.com/microsite/safety_esw_ngsw/ngsw/more/history.html.

2. Flora Tristan, *Promenades Dans Londres, 1840*, trans. D. Palmer and G. Pincetl, *Flora Tristan's London Journal: A Survey of London Life in the 1830s* (London: George Prior, Publishers, 1980).

3. Castaneda, *Regulated Enterprise.*

4. "The History of Regulation," NaturalGas.org, http://www.naturalgas.org/regulation/history.asp.

5. "Now, the Gas Crisis," *Newsweek*, February 7, 1977.

6. David Stockman, "The Wrong War: The Case Against National Energy Policy," *Public Interest*, Fall 1978.

7. Robert J. Michaels, "The New Age of Natural Gas: How the Regulators Brought Competition," *Regulation* 16, no. 1 (Winter 1993).

8. Biogems, "Save the Arctic Wildlife Refuge," Natural Resources Defense Council, http://www.savebiogems.org/arctic/.

9. "Balancing Natural Gas Policy: Fueling the Demands of a Growing Economy; Executive Summary," National Petroleum Council, 2003.

10. "Natural Gas Facts," American Petroleum Institute, http://www.naturalgasfacts.org/factsheets/cool_source.html.

11. Tyson Slocum, "Congress Sneaks in Language Undermining State and Local Control Over Hazardous, Liquefied Natural Gas Facilities," Public Citizen, November 24, 2004.

12. Stephen Pearlstein, "Enviros Need to Get with the Program," *Washington Post*, November 2, 2005.

13. Clifford Krauss, "Global Demand Squeezing Natural Gas Supply, *New York Times*, May 29, 2008.

14. Simon Romero, "Natural Gas: Big Worry This Winter," *New York Times*, November 15, 2005.

Interlude: Snowmass, Colorado

1. Lovins, et al., *Winning the Oil Endgame*.

Chapter 9

1. "A Responsible Energy Plan for America," Natural Resources Defense Council, April 2005, http://www.nrdc.org/air/energy/rep/rep.pdf.

2. Paul Hawken and others, *Natural Capitalism*, 10–11.

3. Arjun Makhijani, "Atomic Myths, Radioactive Realities: Why Nuclear Power is a Poor Way to Meet Energy Needs," *Journal of Land, Resources & Environmental Law* 24, no. 1 (2004).

4. "Households with Selected Appliances and Types of Main Heating Fuel," Annual Energy Review 2007. See Figure 2.6.

5. Vaclav Smil, *Energy as the Crossroads* (Cambridge: MIT Press 2003), 334.

6. Frank Grotton, "Energy Efficiency and the Rebound Effect: Does Increasing Efficiency Decrease Demand?" Congressional Research Service Report, July 2001.

7. Peter Huber and Mark Mills, *The Bottomless Well*, 120.

8. Clive Thompson, "The High-Performance Hybrids," *New York Times Magazine*, September 25, 2005.

9. Matthew Wald, "Automakers Use New Technology to Beef Up Muscle, Not Mileage," *New York Times*, March 30, 2006.

10. Andrew Ferguson, "A Nation of Dim Bulbs," *Weekly Standard*, December 31, 2007.

11. "Vision 2010," Alliance to Save Energy, http://www.ase.org/content/article/detail/2131.

12. "Energy Information Agency review of 'Vision 2010,'" Department of Energy. Made at the request of Congress.

13. "Assessment of Selected Energy Efficiency Policies," Energy Information Agency, May 2005, 13.

Chapter 10

1. Roger Hansen, "Water Wheels," http://waterhistory.org.

2. A. Trevor Hodge, "A Roman Factory," *Scientific American*, November 1990, 106–111.

3. Roger Hansen, "Water Wheels," http://www.waterhistory.org/histories/waterwheels.

4. http://www.americaslibrary.gov/cgi-bin/page.cgi/jb/gilded/hydro_1.

5. http://www.hetchhetchy.org; http://www.sierraclub.org/ca/hetchhetchy/.

6. Smil, *Energy at the Crossroads*, 32.

7. William U. Chandler, *The Myth of the TVA: Conservation and Development in the Tennessee Valley, 1933–1980* (Cambridge: Ballinger, 1984).

8. http://newdeal.feri.org/tva/.

9. http://users.owt.com/chubbard/gcdam/html/history.html; http://www.fwee.org/abhydro.html.

10. William Tucker, "Environmentalism and the Leisure Class," *Harper's*, November, 1977.

11. Reed McManus, "Down Come the Dams," *Sierra Club*, May/June 1998.

12. http://www.nwenergy.org/outreach/fact/dams_4ls.html.

13. "Wonders of the World: Edwards Dam," PBS Online, http://www.pbs.org/wgbh/buildingbig/wonder/structure/edwards.html.

14. "Dams and Development: A New Framework for Decision-Making," The Report of the World Commission on Dams, November 16, 2000.

15. Theo Stein, "Electricity Output Slashed as Powell Water Level Shrinks," *Denver Post*, July 5, 2004.

Chapter 11

1. *E Magazine*, January 2005.

2. Lester R. Brown, "Wind Power Set to Become World's Leading Energy Source," June 25, 2003, http://www.earth-policy.org/Updates/Update24.htm.

3. http://usinfo.state.gov/gi/Archive/2005/Apr/22-869267.html.

4. "A History of Wind Energy," Iowa Energy Center, November 17, 2006, http://www.energy.iastate.edu/renewable/wind/wem/wem-04_history.html; Gibbon, *The Rise and Fall of the Roman Empire*, Chapter LXI, Part III. See footnote 65.

5. "A History of Wind Energy," Iowa Energy Center, http://www.energy.iastate.edu/renewable/wind/wem/wem-04_history.html.

6. "Mr. Brush's Windmill Dynamo," *Scientific American* 63, no. 25 (December 20, 1890).

7. "Historic Wind Development in New England: Grandpa's Knob," New England Wind Forum, U.S. Department of Energy: Energy Efficiency and Renewable Energy.

8. "Wind Power Today," American Wind Energy Association, 2007.

9. James M. Taylor, "Kansas Audubon Society Opposes Wind Farm," *Environmental News*, January 1, 2005.

10. Richard Mertens, "In Wisconsin, Tilting at Windmills is a Serious Matter," *Christian Science Monitor*, April 25, 2005.

11. Phillip F. Schewe, *The Grid: A Journey Through the Heart of Our Electrified World* (Washington: Joseph Henry Press, 2007).

12. Peter Huber and Mark Mills, *Digital Power: Processors to Power Plants* (Great Barrington: Gilder Publishing Company).

13. Kevin Roderick and Eric Young, "Power Hungry in California," *Industry Standard*, January 15, 2001.

14. "Intermittent Wind Generation," California Energy Commission, June 1, 2004, 25.

15. Mel Olken, ed., "Wind Generation: Driving Technology, Policy, and Economics," *Power & Energy*, November–December 2007.

16. C. Johnson, "A Potentially Perfect Energy Source," November 22, 2006, http://mb-soft.com/public2/earthrot.html.

17. Eric Rosenbloom, "A Problem with Wind," September 2006, http://www.aweo.org.

18. Dr. V.C. Mason, "Wind Power in West Denmark. Lesson for the U.K.," October 2005, http://www.gardnermountain.org/docs/68.

19. Glenn R. Schleede, "Facing up to the True Costs and Benefits of Wind Energy," (address given at the annual meeting of the Associated Electric Cooperative, Inc., St. Louis, Missouri) July 24, 2004, 10.

20. "Final Report—2006 Minnesota Wind Integration Study," EnerNex Corporation, November 30, 2006.

Chapter 12

1. "Boeing Spectrolab Terrestrial Solar Cell Surpasses 40 Percent Efficiency," (press release), Boeing Company, December 6, 2006.

2. Amory Lovins, "More Profit With Less Carbon," *Scientific American*, September 2005.

3. Christina Page, "Not a Chip off the Old Block," RMI Solutions, Spring 2005.

4. Douglas Land and Laura Lind, "Green and Clean: The Energy, Environmental, and Economic Benefits of Fiber Glass and Mineral and Wool Insulation Products," Alliance to Save Energy, April 2001.

5. Donald L. Bartlett and James B. Steele, "The U.S. Is Running Out of Energy," *Time*, July 21, 2003.

6. Howard C. Hayden, *The Solar Fraud* (Pueblo: Vales Lake Publishing, 2002), 179–180.

7. "Explosions Rock Solar Energy Plant," UPI, January 10, 1990; "Authorities Evacuate Area After Blast at Solar Plant," Associated Press, February 27, 1999.

8. Marianne Lavelle, "Power Revolution," *U.S. News & World Report*, November 5, 2007.

9. Gregory Dicum, "Plugging into the Sun," *New York Times*, January 4, 2007.

10. Matt Scanlon, "Not Your Mother's Solar Power Anymore," *Mother Earth News*, December–January 2001.

11. Lawrence Scanlan, "Power Switch," *Canadian Geographic*, May 1, 2001.

12. Lance Gay, "Boom Time for Solar Energy—Or Is It?" Scripps Howard News Service, December 16, 2005.

13. Kerry Cavanaugh, "Low Output Clouds Solar Power in L.A.," *Daily News of Los Angeles*, January 2, 2004.

14. "Average Retail Price of Electricity by State, 2007," Energy Information Administration, http://www.eia.doe.gov/cneaf/electricity/epm/tables5_6_a.html.

15. Jeff Donn, "Can Solar Power Work? Pioneering Community Answers Yes and No," Associated Press, June 5, 2004.

16. Lovins, "More Profit With Less Carbon," *Scientific American*, September 2005.

17. "Time of Day Pilot Program," Colorado Spring Utilities, http://www.csu.org/customer/rates/rate_residential/page7450.html.

18. Ken Zweibel and others, "A Solar Grand Plan," *Scientific American*, January 2008.

Chapter 13

1. http://www.worldbank.org/html/fpd/energy/geothermal/technology.htm; Adler Planetarium and Astronomy Museum, Chicago, Illinois.

2. Ibid, http://www.worldbank.org.

3. Ibid.

4. Tim Flannery, *The Weather Makers* (New York: Atlantic Monthly Press, 2005), 276.

5. John Sass and Sue Priest, "Geothermal California," GRC Bulletin, September–October 2002.

6. "The Untimely Death of Salter's Duck," *Green Left Weekly*, no. 64 (July 29, 1992).

7. http://www.earthsci.org/mineral/energy/wavpwr/wavepwr.html.

8. Magnus Linklater, "Power from the Waves," *Scotland on Sunday*, July 10, 2005; Michael Kanellos, "Energy on the High Seas," CNET, http://business2-cnet.com.com/FAQ-Energy-on-the-high-seas/2100-11392_3-6202562.html.

9. Jill Davis, "Alexander's Marvelous Machine," *OnEarth Magazine*, Spring 2005.

10. Lovins, *Soft Energy Paths*, 44.

11. James Jordan and James Powell, "The False Hope of Biofuels," *Washington Post*, July 2, 2006.

12. Joseph DiPardo, "Outlook for Biomass Ethanol Production and Demand," U.S. Department of Energy, http://www.eia.doe.gov/oiaf/analysispaper/biomass.html.

13. Matthew Wald, "Is Ethanol for the Long Haul?" *Scientific American*, January 2007.

14. "Biofuels from Switchgrass: Greener Energy Pastures," Oak Ridge National Laboratory, 2005.

15. Nathanael Greene, "Growing Energy: How Biofuels Can Help End America's Oil Dependence," Natural Resources Defense Council, December 2004.

16. Lovins, *Winning the Oil Endgame*.

17. Rich Calder, "S.I. Pol's Bill Greasing Way for 'New' Fuel," *New York Post*, May 1, 2007.

18. "Turkey in the Tank: High Price of Gasoline is a Boon for Biofuels," *Wall Street Journal*, October 28, 2005.

19. Searchinger and others, "Use of U.S. Croplands for Biofuels Increases Greenhouse Gases Through Land Use Change," *Science* 319, no. 5867 (February 29, 2008); Fargione and others, "Land Clearing and Biofuels Debt," *Science* 319, no. 5867 (February 29, 2008).

20. Laura Blue, "Can Airplanes Fly on Biofuels?" *Time*, February 25, 2008; Michael Grunwald, "The Clean Energy Scam," *Time*, April 7, 2008.

21. Patrick Barta and Jane Spencer, "As Alternative Energy Heats Up, Environmental Concern Grows," *Wall Street Journal*, December 5, 2006.

22. Friends of the Earth and others "The Oil for Ape Scandal: How palm oil is threatening orangutan survival," September 2005.

23. James Kanter, "Amid Doubts, Europe May Ban Some Biofuels," *New York Times*, January 15, 2008.

24. Simon Johnson, "Straight Talk: The (Food) Price of Success," *Finance and Development Magazine*, December 2007.

25. Edith M. Lederer, "UN Expert Calls Biofuel "Crime Against Humanity," Associated Press, October 27, 2007.

26. Lester Brown, "Supermarkets and Service Stations Now Compete for Grain," Earth Policy Institute, July 13, 2006.

14. Ibid.

15. Ted Trainer, *The Final Energy Crisis,* ed. Andrew McKillip with Sheila Newman, 280–28.

16. Darley, *High Noon for Natural Gas,* 204.

17. Lovins, *Soft Energy Paths,* 151.

Chapter 17

1. Hans Christian von Baeyer, *The Fermi Solution,* 3ff.

2. Fermi, *Atoms in the Family,* 89.

3. Richard Feynman, "There's Plenty of Room at the Bottom: An Invitation to Enter a New Field of Physics," *Engineering and Science,* February 1960, http://www.zyvex.com/nanotech/feynman.html.

Chapter 18

1. Bernard Cohen, *The Nuclear Energy Option,* 55.

2. http://www.accessexcellence.org/AE/AEC/CC/historical_background.html.

3. Ibid.

4. Teller, *Energy from Heaven and Earth,* 137; "Edward Teller, Father of the Hydrogen Bomb," Academy of Achievement, http://www.achievement.org/autodoc/page/tel0int-4.

5. Dietz, *Atomic Energy,* 25.

6. Jungk, *Brighter Than 1,000 Suns,* 69.

7. Enrico Fermi, "Artificial Radioactivity by Neutron Bombardment," Nobel Prize speech, 1938.

8. Teller, *Energy from Heaven and Earth,* 142.

9. Ibid., 144; Jungk, *Brighter Than 1,000 Suns.*

10. "Einstein's Letter to Roosevelt," The Atomic Heritage Foundation, http://www.atomicheritage.org/index.php?option=com_content&task=view&id=172&Itemid=116.

11. Jungk, *Brighter Than 1,000 Suns,* 115–116.

12. Richard Feynman, *Surely You're Joking, Mr. Feynman* (New York: Bantam, 1986).

13. Teller, *Energy from Heaven and Earth,* 146.

14. Paul W. Frame, "Why Do They Call It That? The Origin of Selected Radiological and Nuclear Terms," Oak Ridge Associated Universities, http://www.orau.org/PTP/articlesstories/names.htm.

15. Enrico Fermi, "Fermi's Own Story," amended to Corbin Allardice and Edward R. Trapness, "The First Pile," Department of Energy, http://www.hep.uchicago.edu/cp1.html.

16. Robert Serber, *The Los Alamos Primer: The First Lectures on How To Build*